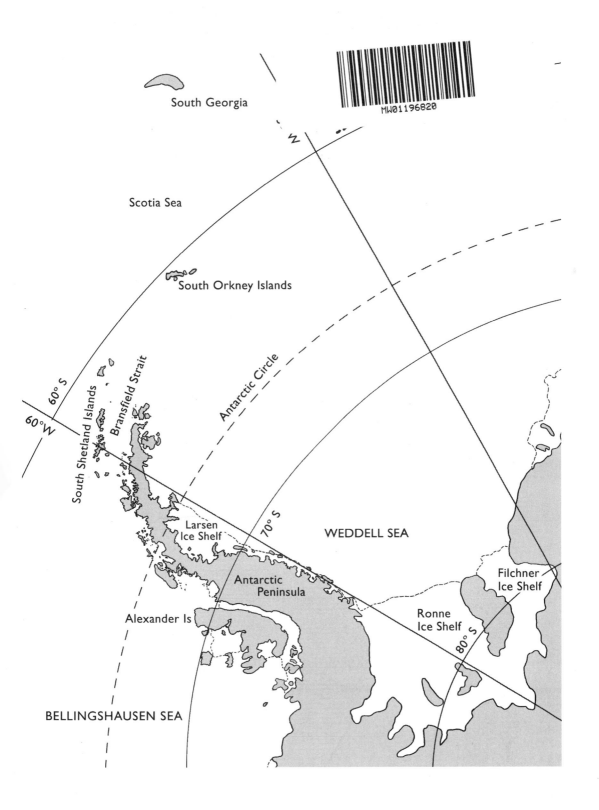

THE ANTARCTIC PENINSULA REGION

Cover photo: The *Belgica* beset in the ice of the Bellingshausen Sea.
(photo from Lecointe, *Au Pays des Manchots*, 1904)

THE STORIED ICE

Exploration, Discovery, and Adventure in Antarctica's Peninsula Region

Joan N. Boothe

REGENT PRESS
Berkeley, California

Copyright © 2011 by Joan N. Boothe

PAPERBACK
ISBN 13: 978-1-58790-218-5
ISBN 10: 1-58790-218-4

HARDBACK
ISBN 13: 978-1-58790-224-6
ISBN 10: 1-58790-224-9

E-BOOK
ISBN 13: 978-1-58790-181-2
ISBN 10: 1-58790-181-1

Library of Congress Control Number: 2011923350

All rights reserved under International and Pan-American Copyright Conventions. No part of this book may be used or reproduced in any manner whatsoever without the written permission of the publisher, except in the case of brief quotations embodied in critical articles and reviews.

The author and publisher gratefully acknowledge the permissions granted to reproduce the copyrighted material in this book. Every effort has been made to trace copyright holders and to obtain their permission for the use of copyrighted material. The publisher apologizes for any errors or omissions and would be grateful if notified of any corrections that should be incorporated in future reprints or editions of this book.

First Edition

1 2 3 4 5 6 7 8 9 10

Manufactured in the U.S.A.
REGENT PRESS
Berkeley, CA 94705
www.regentpress.net
regentpress@mindspring.com
www.joannboothe.com
joannboothe@joannboothe.com

Table of Contents

Preface vii
Introduction 1

Ch. 1 – In Search of a Southern Continent: Efforts Prior to 1819 9
Ch. 2 – The Continent Found: 1819–1821 27
Ch. 3 – The Sealers' Age of Discovery: 1821–1839 41
Ch. 4 – Three Great National Expeditions: 1837–1843 57
Ch. 5 – Quiet Decades in the South; the New Hunters: 1844–1896 72
Ch. 6 – De Gerlache and the First Antarctic Night: 1897–1899 84
Ch. 7 – Nordenskjöld's Saga of Survival: 1901–1903 99
Ch. 8 – Bruce and the *Scotia*, Bagpipes in the South: 1902–1904 116
Ch. 9 – Charcot and the *Français* Explore the Antarctic Peninsula: 1903–1905 127
Ch. 10 – Whalers and Politics: 1904–1918 139
Ch. 11 – Charcot's Return with the *Pourquoi-Pas?*: 1908–1910 151
Ch. 12 – Filchner's Battles in the Weddell Sea: 1911–1912 164
Ch. 13 – *Endurance*, Shackleton's Triumphant Failure: 1914–1916 177
Ch. 14 – The Decade Following World War I: 1919–1927 198
Ch. 15 – The First Aviators Arrive: 1928–1936 213
Ch. 16 – The Wintering Explorers Return: 1934–1941 231
Ch. 17 – World War II, New Bases, and Political Conflict: 1940–1955 249
Ch. 18 – The International Geophysical Year and the Antarctic Treaty: 1955–1959 268
Ch. 19 – The Antarctic Treaty Era: Antarctica After 1959 278

Appendix A: Antarctic Timeline 305
Appendix B: Antarctic Firsts 315
Appendix C: Glossary of Terms, Abbreviations, and Acronyms 327
Sources and Notes 331
Literature Cited 349
Index 361

Preface

Several years ago, I found myself talking to a new acquaintance who had recently been to the Antarctic Peninsula. He told me he had had a wonderful time there because the scenery and wildlife were so spectacular. His only disappointment had been that the area was so lacking in historical interest. I was stunned. The human story of where he had been is extraordinarily rich. The names on the map of the region fairly shout it out—the Drake Passage; the Bransfield and Gerlache Straits; Paulet, Wiencke, and Dundee islands; the Danco Coast; the Larsen, Ronne, and Filchner Ice Shelves; the Scotia and Weddell Seas; and all the others. I remembered my own first trip to Antarctica, a trip just like that taken by my disappointed acquaintance and so many others. I too had gone to the South Shetland Islands and the Antarctic Peninsula. When I stood on the deck of my ship in the evening light and saw the South Shetlands for the first time, I found myself overwhelmed with emotion, thinking about all that had happened there through the centuries. My new friend had been cheated! But the seed for this book was planted. I would write a history to tell the story of Antarctica's Peninsula Region, that part of the Antarctic that the vast majority of visitors, particularly tourists, travel to, and where so many explorers and expeditions had achieved so many Antarctic "firsts."

So rich is this history that a book of a reasonable length can cover only the highlights of its 500 years. I have thus had to simplify, as well as make hard choices about what to include or leave out. Two appendixes provide limited additional detail, as well as context for what was happening elsewhere in Antarctica during all these years. The first is an Antarctic Timeline, separated into happenings covered by this book and those that took place elsewhere in the South Polar Region. The second lists Antarctic Firsts. For the reader who wants still more, the Sources and Notes section provides suggestions for further reading.

This book has been years in the research and writing, and during that time, I have received help and encouragement from many people. This includes friends and relatives who knew little or nothing at all about the subject before reading the manuscript as well as a number of people who are acknowledged experts in the field. I owe much to all of them.

Readers of the very rough and excessively long first draft were brave indeed and their comments were especially valuable. Claire Brindis, Marion Blumberg, Dot

Cato, Jim Kempenich, and Paul Hook all made their way through to the end and provided me with much appreciated feedback. Sandy Briggs returned her copy of the draft with dozens of pages of careful comments and questions, all insightful and to the point. Rick Dehmel, who has deep knowledge of the literature of Antarctic history, read carefully, then spent hours talking with me about the book as he contributed another set of questions and suggestions, all of them valuable. Two expert readers—Louise Crossley and Robert Headland—also read portions of the first draft and provided critical commentary. David Rumsey showed me the Brue map, a wonderful discovery. To all of these people, many of them with very busy schedules, I am deeply grateful.

Other expert and inexpert friends read and commented on later drafts. Among those to whom the subject was at least partially new, I much appreciated feedback and encouragement from Karen Ireland, Walter Minnick, Jacquie Samuels, and Bob Whitby. And Carole Ann Peskin, my cabinmate on an Antarctic trip to the Ross Sea, was amazing. She spent untold hours carefully reading and filling my margins with thoughtful commentary.

Several experts in Antarctic history were extremely generous with their time, making many important contributions in their reading of later drafts. Erica Wikander, at the time Executive Vice President of Quark Expeditions, offered strong encouragement after reading an early draft. Kristin Larsen, who worked for years in the Antarctic with the United States Antarctic program, made a crucial suggestion about naming from which I am most grateful. Art Ford, John Splettstoesser, and the late Colin Bull—major figures all in the world of Antarctic science and history—read carefully, corrected errors, suggested additions to enrich the text, and generally provided a wealth of valuable comments. Damien Sanders, a former member of the British Antarctic Survey and now a history lecturer on Antarctic cruise ships, supplied information that I had not found in my reading. R.H.T. Dodson, a member of Finn Ronne's 1947–48 expedition, read Chapter 17 and provided valuable input based on personal knowledge. I am particularly grateful to all of these people for their encouragement, especially that from Art Ford, who became a strong support, almost a mentor, as the project went on. Two expert reviewers—John Behrendt and another who chose to remain anonymous—provided feedback that was much appreciated, asking questions, pointing out errors, and suggesting changes and additions that would strengthen the book.

I must also thank those involved in the finishing and publication of this book. Bill Carver, my wonderful editor, was marvelous to work with. His corrections to syntax, questions about content, clarification of language, polishing of my writing, and so much else, were invaluable. My agent, April Eberhardt, worked hard for me and supplied useful comments for which I am very grateful. I owe a great debt to Paul Veres for his hard work on the maps. I was a demanding taskmaster on them

PREFACE

and he came through beautifully. Barbara Templeton's careful reading of the final copy was invaluable in catching copy errors and inconsistencies. And my publisher, Mark Weiman of Regent Press, was a delight to work with, always there for me.

Finally, I must thank my husband, Barry Boothe. He read and commented on at least four different drafts and has been enthusiastic and encouraging throughout the years that I have been working on this book. Without him, it would never have happened.

Despite the many who have read this manuscript, it is almost inevitable that errors remain. If there are such, I am fully responsible. I would, however, appreciate hearing from readers who spot errors so that I can make corrections for a possible later edition.

JOAN N. BOOTHE
San Francisco, California
July, 2011

Introduction

In mid-July 1898, seventeen men stared with hungry eyes across a frozen sea as "a fiery cloud separated, disclosing a bit of the upper rim of the sun." It was their first sight of the sun in over two months. One later wrote, "For several minutes my companions did not speak. Indeed, we could not . . . [find] words with which to express the buoyant feeling of relief, and the emotion of new life. . . ."[1]* These were the men of Adrien de Gerlache's icebound *Belgica* expedition, the first human beings to live through the black months of an Antarctic night and witness an Antarctic sunrise. They had opened a new chapter in a story that had begun hundreds of years earlier, when men first began searching for a speculative continent of riches in the Southern Hemisphere.

More than 2,000 years before the *Belgica*'s men saw the sun peep over the Antarctic icescape, Greek philosophers theorized that a large landmass must fill much of the southern part of the globe, to balance the known landmasses in the Northern Hemisphere. And one of those philosophers, Parmenides, extrapolating from the fact that voyagers to the far north had found ice-covered seas and land, concluded that the far south would be similarly frigid and uninhabitable. Someone among the Greeks, possibly Aristotle, named this South Polar Region Antarktos, in opposition to the North Polar Region, which had been christened Arktos after a prominent constellation in the far northern sky. The Greek theory of a Southern Continent survived the fall of Rome, but the original concept of a forbidding, ice-covered land did not. Instead, medieval men conjured a more enticing belief in a vast, rich, even inhabited, landmass in the Earth's high southern latitudes. This myth would inspire numerous exploratory voyages in the centuries to come.

The early Antarctic story is largely one of false discovery. Beginning early in the sixteenth century, sailor after sailor—explorers, naval officers, merchants, hunters, even pirates—discovered austral land and declared it the northern shores of the sought-for continent.

Then someone else would sail south of the putative discovery, proving that it too was just another island, and the possible shores of the theoretical Southern Continent would be pushed farther south.

These early southern voyagers challenged the unknown in tiny, cramped

---

* All numbered notes are in "Sources and Notes," following the Appendixes.

sailing vessels, wholly dependent on the currents and winds to power their progress. Robert Falcon Scott, an early twentieth-century British explorer of Antarctica's Ross Sea Region, wrote of them, "In the smallest and craziest ships they plunged boldly into stormy ice-strewn seas; again and again they narrowly missed disaster; their vessels were wracked and strained and leaked badly, their crews were worn out with unceasing toil and decimated by scurvy. Yet in spite of inconceivable discomforts they struggled on...."[2]

Little by little, that struggle replaced myth with fact: not only is Antarctica, a landmass bigger than Australia, the most inaccessible of the seven continents, it is also the coldest, driest, windiest, and highest in average elevation—the most inhospitable place on Earth for humankind. Before the explorers arrived, no human being had even set foot on its land, let alone lived there. But visitors to the Antarctic also learned other truths about the white continent. The waters and shores of the Southern Ocean encircling the Antarctic regions teemed with wildlife that people could exploit, and the entire Antarctic offered unique opportunities for scientific study. Gradually, people also came to appreciate Antarctica's mystical allure, came to see that the Southern Continent is indeed a place of riches: fascinating and unique wildlife, magical light, unimaginable colors, haunting sounds, spectacular landscapes, and seemingly limitless horizons.

The Storied Ice tells of how humankind learned all this about Antarctica's Peninsula Region, the focus of much early exploration and the most visited part of the far south for much of its history. From north to south, the area that falls within our story includes South Georgia Island, the South Sandwich Islands, the South Orkney and South Shetland island groups, the Antarctic Peninsula, and the Weddell Sea and its ice-choked coasts.

All these locations lie south of the Antarctic Convergence (or Antarctic Polar Front, the modern scientific term), the ecological boundary in the Southern Ocean where cold surface waters flowing north from the Antarctic Continent meet and drop below the warmer, saltier, less-dense waters from the subtropics. The region of the Convergence, which follows a track rather like that of a drunken snake around the circumference of the South Polar Regions between roughly 48° and 62° S latitude, is typically about 20 to 30 miles wide, a narrow band over which there are abrupt changes in water and air temperatures.* The climate of far southern lands depends hugely on whether they are south or north of this band. For example, Cape Horn, the island at the tip of South America at about 56° S, is north of the Convergence. Although the weather there is chilly and wet, it is far from polar. East of Cape Horn, the Convergence takes a jog north. The consequence is that South

---

*All distance measurements cited in this text are in standard English measures. Thus miles, for example, are ordinary statute, rather than geographic or nautical miles.

INTRODUCTION

Georgia Island, although two degrees north of Cape Horn at 54° S, is south of the Convergence and has a near-polar climate.

Well south of the Convergence lies the Antarctic Circle, that imaginary line on the Earth at approximately 66° 33' S where, each year, the sun does not rise for one day in mid-winter or set for one day at mid-summer. A significant portion of the Peninsula Region lies north of this latitude, and these places typically have a milder climate than that found elsewhere in the Antarctic. Here, summer temperatures are often above freezing, and the sun rises, if only for a very few hours, in the middle of winter and correspondingly sets briefly even during the height of summer. Compared to most of the world, however, these locales are cold and desolate. They are definitely part of the Antarctic. The entire Peninsula Region is a place of snow, ice, and marvelous southern fauna, and of lands and seas that have more in common with the rest of the South Polar Region than with any other part of the globe. No native peoples have ever lived there; vegetation is scarce, and what little exists is limited primarily to mosses, lichens, and grasses; the only land animals are minute insects; and ice blankets almost all the land. At sea, floating ice is a constant companion—pack ice from the winter freeze, icebergs from tidewater glaciers, and those signature ice forms of the Southern Ocean, massive tabular icebergs sometimes many miles long that calve off ice shelves and drift inexorably wherever the currents and winds take them. The Peninsula Region shares one more thing with the rest of Antarctica—a fascinating and colorful history.

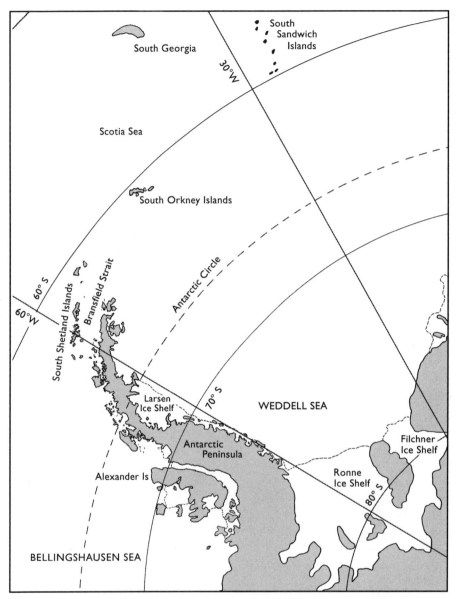

The Antarctic Peninsula Region — *The subject region of* The Storied Ice, *Antarctica's Peninsula Region was the focus of much early exploration, and the northern portion of the Antarctic Peninsula has been the most visited portion of the far south for much of Antarctica's human history. From north to south,* The Storied Ice's *purview includes South Georgia Island, the South Sandwich, South Orkney, and South Shetland island groups, the Antarctic Peninsula and adjacent islands, and the Weddell Sea and its ice shelves and ice-covered coasts.*

Argentina, Chile, and Great Britain have made competing, overlapping claims to nearly all this region (discussed in Chapters 17 and 19).

Unlike the northern Antarctic Peninsula, the Weddell Sea is among the least visited areas in the Antarctic. The water is often clogged with pack ice, and in many years much of this sea is virtually impenetrable, sometimes even with the most powerful icebreakers. The southern portion of the Weddell Sea is covered with massive ice shelves that extend several hundred miles north from the continental coast.

INTRODUCTION

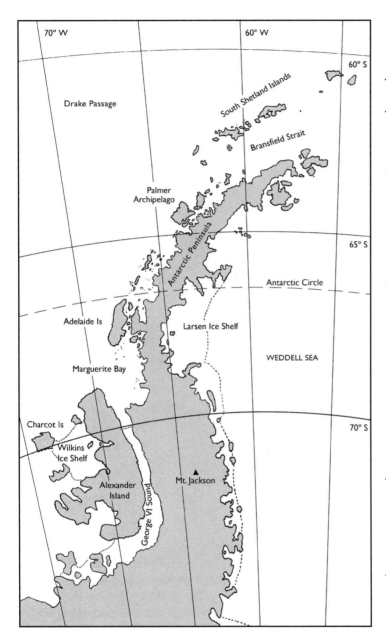

The Antarctic Peninsula — *a 1,000-mile-long finger of land stretching northeast from the main body of the Antarctic Continent. Flanked by ice shelves, particularly on the east, the Peninsula is almost entirely covered in ice. Some of that ice, however, is now retreating. The northern parts of the Larsen Ice Shelf, on the Peninsula's east coast, have collapsed in recent years, and are thus not shown on this map. (Other maps in this book, dealing with times before the present, show this ice shelf as extending much farther north.)*

In addition to being ice-covered, the Peninsula is very mountainous. Near the southern junction with West Antarctica, the Peninsula rises to as much as 10,000 feet in elevation. The most lofty mountain on the Peninsula is Mt. Jackson, nearly 10,500 feet high.

Several significant island groups and islands adjoin the Peninsula on the west side. From north to south, these include the Palmer Archipelago, a group of several large and many small islands separated from the Peninsula mainland by the Gerlache Strait; Adelaide Island, which forms the northern edge of Marguerite Bay; and Alexander Island, separated from the mainland by George VI Sound.

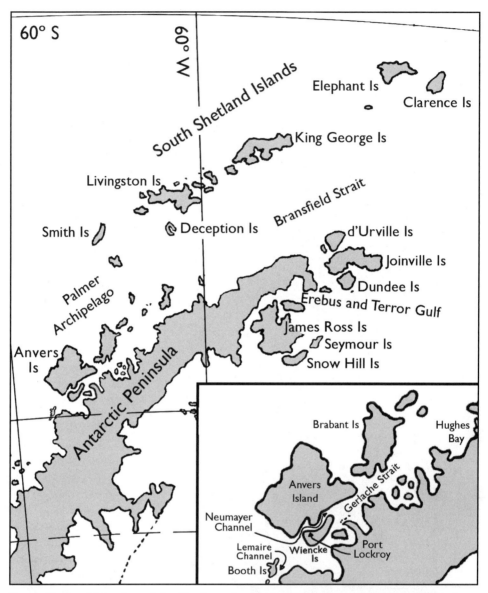

South Shetland Islands and the Northern Antarctic Peninsula — *The South Shetlands are a 300-mile-plus chain of eleven large and many smaller islands lying just north and east of the Antarctic Peninsula. Smith Island is the farthest west of the group while Clarence Island anchors the east end. The islands are more than 90% snow and ice covered, and several are quite mountainous. The tallest mountain in the South Shetlands is Smith Island's 6,500-foot-high Mount Foster, which remained unclimbed until 1996. Several of the islands are volcanically active, including Deception Island, where there were major eruptions in 1967, 1969, and 1970 (see Chapter 19). King George Island, the largest in the group at 44 miles long, is home to more research stations than any other single place in the Antarctic regions. As of winter 2010, there were eight year-round government bases on the island and several summer-only operations. King George Island was also the site of the first Antarctic marathon, run in 1995.*

The small inset map highlights a portion of the Peninsula Region where much of the activity described in The Storied Ice *took place.*

INTRODUCTION

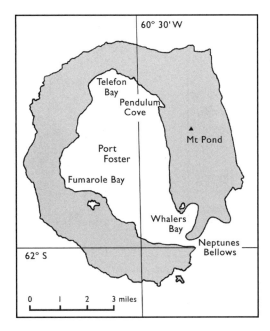

Deception Island — *One of the smallest major islands in the South Shetland archipelago, this place's significance in the Peninsula Region story far exceeds its diminutive size. An active volcano, which last erupted in 1970, Deception Island's land is shaped roughly like a doughnut, with a narrow bite, or entrance, into the drowned volcanic crater that sits inside the island walls.*

Deception Island has been a center of human activity since its discovery in 1820, used by sealers and then later by scientists, explorers, whalers, and as a home for government research bases. Today it is one of the most visited tourist stops.

The island, approximately 8 miles in diameter, is more than half covered in glaciers. The highest point is Mt. Pond, which rises to approximately 1,750 feet.

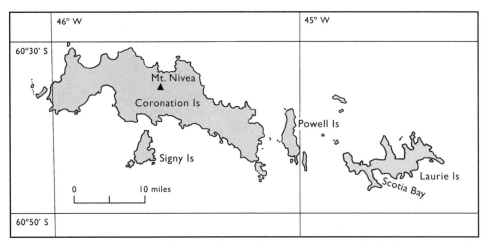

The South Orkneys — *A small archipelago of four mountainous islands plus many smaller islets. The group, which is 90% covered in year-round ice, is centered at 60° 40' S, 45° 15' W, about 250 miles east of the South Shetlands. Coronation Island accounts for more than three-quarters of the approximately 230 square miles of land in the group and is home to Mt. Nivea, the group's tallest mountain at approximately 4,150 feet.*

The South Orkneys are home to the longest existing permanent Antarctic region scientific base, Argentina's Orcadas Station, which is at the head of Scotia Bay on Laurie Island.

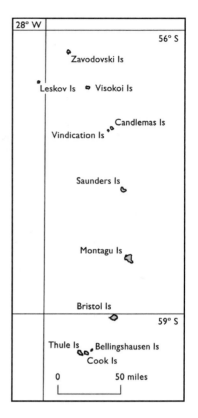

South Sandwich Islands — *A 200-mile-plus north-south chain of eleven major islands, ranging from Zavodovski Island at 56° 20' S at the northern end to the three-island Southern Thule group at 59° 30' S in the south. The total land area of just over 100 square miles is about 85% ice-covered. All the islands are volcanic, and many show evidence of current or recent volcanic activity. The tallest point in the group is Mt. Belinda, on Montagu Island, nearly 4,300 feet high.*

Zavodovski Island is home to the world's largest chinstrap penguin rookery, but these birds nest on nearly all the islands in the group in large numbers, totaling an estimated 1.5 million in a recent survey.

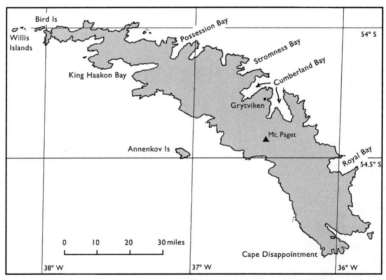

South Georgia — *At 54° 17' S, 36° 30' W, South Georgia is approximately 960 miles from the nearest point on the Antarctic Continent. Mountains fill much of the interior of this 110-mile-long island, which is roughly half permanently covered in ice. The highest point is 9,625-foot-high Mount Paget, and twelve other mountains exceed 6,500 feet. The island is home to huge breeding colonies of fur and elephant seals, and king, gentoo, and macaroni penguins. It is also one of the world's most important breeding places for wandering albatross.*

CHAPTER 1

In Search of a Southern Continent: Efforts Prior to 1819

Humankind's story in the Antarctic Peninsula Region opens well to the north, near the southern tip of South America. It was here that men first claimed to have seen a coast of the Southern Continent, and here where they found sea routes that would eventually lead to the discovery of land south of the Convergence.

It is possible that a Chinese fleet visited the area around 1421 or 1422, finding and sailing through what we now call the Strait of Magellan. The fleet may also have visited the Falkland Islands, sailed south to the Antarctic Peninsula, continued northeast to South Georgia, and then sailed across the South Atlantic in high latitudes. The chief evidence for all this is speculative inference from the Piri Reis map, which shows land in the far south near the actual location of the Antarctic Peninsula.[1] Published in 1513, this work includes portions copied from a now-lost early-fifteenth-century world map.

The first documented southerly voyage, however, took place in 1502. A letter attributed to Amerigo Vespucci, the pilot, claimed that his Portuguese ship had reached 50° S along the east coast of South America—well toward its southern extremity—and had sighted land stretching farther south. Although even contemporary evidence cast doubts on the high latitude reached, there was now no question that there was a large landmass on the west side of the South Atlantic Ocean, an addition to the land that Columbus and others had already found much farther north. Was this part of the great Southern Continent? Or perhaps evidence that it might exist yet farther south?

Only a year or two later, another Portuguese voyage gave the world its first actual claim of sighting the speculative southern land. The ship's captain said he had seen it across a channel of water beyond the Americas, at about 40° S.[2] The claim that water separated the Americas and the Southern Continent was equally important. If valid, it suggested a westward sea route to the East Indies, an alternative to the Cape of Good Hope passage around Africa that the Portuguese had established a decade earlier.

FERDINAND MAGELLAN DISCOVERS A STRAIT

In 1517, Ferdinand Magellan (Fernão de Magalhães), a 37-year-old Portuguese mariner, traveled to Spain after his own king had refused to finance an attempt to reach the East Indies by sailing west, around the Americas, rather than by the Cape of Good Hope route that Magellan himself had already used. The Spanish crown agreed to fund the voyage, and so, in September 1519, Magellan set off from Spain with a small fleet. His five ships, the smallest of about 135 tons and perhaps 75 feet long, the largest of approximately 200 tons, carried about 250 men. Magellan reached South America off today's Brazil at the end of November. He then sailed south, carefully examining every promising river, estuary, or any other apparent westward opening. In late March 1520, having found nothing but dead ends, a discouraged Magellan camped for the winter on the desolate coast of what would come to be called Patagonia, after the name Magellan's men gave the local natives.

Magellan resumed his voyage in October 1520 with only four ships, having lost one during winter on a reconnaissance trip. Only three days after sailing, he reached 52° S and spotted a headland (today called Cape Virgins) at the north of another possible westward opening. A storm soon swallowed the two ships sent to examine this new possibility. Days passed. Then, just as Magellan was beginning to think his ships lost, they returned. Their excited captains reported that they had sailed west for three days, and open water still lay ahead.

Full of hope, Magellan sailed his four ships into the opening, past two narrows into a wide basin where more openings beckoned him on. He split his fleet to try both. He, himself, took two ships southwest to examine what appeared to be the main passage. He ordered the other two, including the vessel with most of his supplies, to explore two other options to the south. During the first night, Estavo Gomez, the pilot of the ship with the provisions, led a mutiny—a renewal of unrest that had plagued the expedition throughout the winter. Without Magellan there to quell the mutiny, Gomez took command of the ship and sailed for home.

Magellan, in the meantime, pressed on, for over 100 miles. The way ahead looked difficult after he rounded Cape Froward, now known to be the southernmost point of the South American landmass, and Magellan sent a longboat to scout ahead. Her men returned jubilantly three days later. They had, they said, reached the end of the channel. Only open sea lay beyond. An excited Magellan retraced his route to deliver the triumphant news to the other two ships. But to his dismay, he found only one. After days spent vainly searching for the missing vessel, he gave up and returned west to continue with his depleted fleet.

Magellan reached the western opening to the strait at the end of November. The way to the Indies lay before him, somewhere across what he now named the Pacific, or peaceful, Ocean—a name that those who followed Magellan here would often have cause to question.[3] It had taken him 38 days and 310 miles to sail

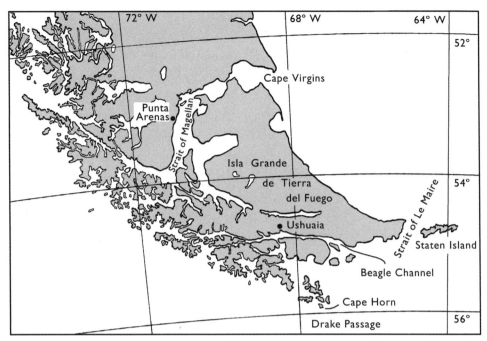

Strait of Magellan, Tierra del Fuego, and Cape Horn Region — *The Strait of Magellan separates the South American continent from the Tierra del Fuego archipelago. The archipelago itself consists of one dominant island, Isla Grande de Tierra del Fuego, several other large islands, and numerous smaller ones, all separated from one another by a bewildering maze of channels. The island known as Cape Horn is the farthest south landmass in the archipelago. To the south of Cape Horn lies the infamous Drake Passage. The Argentine city of Ushuaia, from which most Antarctic tour ships depart, is on the south coast of Isla Grande.*

through the strait, both time and distance extended by the delays, false turns, and exploration necessitated by being the first European to navigate this passage. Magellan called his strait Todos los Santos (All Saints' Strait), a title soon forgotten when others renamed it the Strait of Magellan. A form of the name he gave the land south of the strait did persist—Tierra del Fuego (Land of Fire), a name suggested by the native-set fires he saw glowing in the night.

Then came the sail across the Pacific, a long and desperately hungry voyage because not only had Gomez taken the bulk of the expedition's provisions, but the distance was also far greater than anyone had imagined. Many of Magellan's men died from starvation before the ships finally reached the Philippines in March 1521. There the survivors spent a month recovering before beginning the voyage to Europe via the Cape of Good Hope. Magellan, himself, never reached home. He died in the Philippines, killed by natives there along with several others in his party. Only one ship from the fleet, the 85-ton *Victoria* with a bare eighteen men, completed history's first world circumnavigation. She reached Spain in September 1522.

Magellan's voyage set the stage for accurate, not speculative, knowledge of the Antarctic, because it proved conclusively that the southernmost part of the globe could only be reached by sea. Whether men would find land there or simply more ocean remained an open question. The voyage, however, reinforced the belief in an extensive southern land, because Magellan's surviving men declared the northern coast of Tierra del Fuego to be part of the theoretical continent.

The discovery of a sea route between the Atlantic and Pacific oceans excited the courts and counting houses of Europe, but it rapidly became clear that Magellan had been lucky in his relatively problem-free passage. Every one of the next six attempts to follow Magellan encountered great difficulties, and after the Spanish established a regular land route across the Isthmus of Panama, mariners abandoned Magellan's great discovery for decades.

This changed in 1577, when England entered the picture. Up to this point, the English had been largely absent from Southern Hemisphere exploration. Now, Queen Elizabeth decided to join the countries exploring the Pacific. Because Spain controlled the overland route connecting the Atlantic and Pacific, Francis Drake, the first Englishman to embark for the South Pacific, had to use the only alternative the English knew—the Strait of Magellan.

Francis Drake Finds an Open Sea to the South

Drake left England in December 1577. After spending the winter of 1578 camped on the Patagonian coast just where Magellan had, he resumed his voyage in mid-August. He sailed his three-ship fleet into the Strait of Magellan a few days later. Four days after entering the strait, Drake's men landed on an island. There they found huge numbers of

> strange birds, which could not fly at all . . . : in body they are less than a goose, and bigger than a mallard, short and thick set together, having no feathers, but instead . . . a certain hard and matted down; . . . their feeding and provision to live on, is in the sea, where they swim in such sort, as nature may seem to have granted them no small prerogative in swiftness. . . .[4]

This was the first English-language description of penguins.

It took Drake only sixteen days to reach the Pacific, the shortest passage anyone had made through Magellan's strait. Then his luck ran out. A storm that hit the night he exited the strait continued virtually unabated for a month. At the end of September, massive waves swamped one of the ships. Those aboard the other two vessels, including Drake on his flagship, the *Golden Hind*, were helpless in the wild winds to offer aid. Impotent, they could only watch and listen in horror as their companions drowned.

When the winds at last eased off in early October, Drake's two surviving vessels approached land just north of the western end of the strait. Then a new squall hit, forcing both ships out to sea and separating them. The next morning, with his consort nowhere in sight, Drake concluded that she was lost. In fact, she was perfectly safe. Her captain, John Winter, had taken her a few miles into the Strait of Magellan to ride out the storm. There he burned fires on shore as a signal, in hopes that Drake was safe and would find him. Winter waited for weeks, then gave up and sailed for home, back eastward through the strait, thus becoming the first to travel Magellan's passage in both directions.

In the meantime, Drake sailed on alone in the *Golden Hind.* Again, storms took control. This time they blew him southeastward. On October 18, a lull in the weather allowed Drake to anchor his battered ship in the south of the Tierra del Fuego archipelago. Here he took on wood and water and encountered a group of native peoples with whom he traded, human beings who were the most southerly that Europeans had ever met. The natives were friendly, but the weather was not, and the results were fatal for several of Drake's men. Renewed gales drove the *Golden Hind* out to sea, leaving eight sailors marooned on shore. These men eventually made their way to southern Patagonia, on the east coast of South America, but within two months, all but one were dead, victims of encounters with hostile natives. The single survivor, Peter Carder, at last reached England more than nine years after parting from Drake. Once home, his fortunes changed entirely: Queen Elizabeth received him at court to hear his story and awarded him a handsome sum of money.[5]

As for Drake, he continued battling the violent winds. On October 24, 1578, he at last located a small island with a safe anchorage. The voyage account related, ". . . at length we fell with the uttermost part of land towards the South Pole. . . . The uttermost cape or headland of all these lands, stands near in 56 deg. without which there is no main, nor island to be seen to the Southwards: but that the Atlantic Ocean, and the [Pacific], meet in a most large and free scope."[6] The winds had driven Drake to a revolutionary discovery—the open sea below Tierra del Fuego, that infamously difficult stretch of ocean now known as the Drake Passage. Historians have long debated just where Drake had been. Perhaps it was Cape Horn he saw, perhaps some other island in southern Tierra del Fuego. Unfortunately, it is impossible to know, given the uncertainty of his longitudes.

Drake left the high southern latitudes behind when the stormy weather finally relented. As he sailed north along South America's west coast, he plundered Spanish settlements and cargo ships, and then continued on to North America and the coast of today's California. Drake sailed home across the Pacific and reached England in September 1580, completing the first English world circumnavigation.

Another Englishman, John Davis, made a different discovery a few years after Drake's voyage. Davis was attempting to enter the Strait of Magellan in August 1592 when storms drove him far to the east, to an uncharted island group in the South Atlantic at about 51–52° S, several hundred miles to the east of today's Argentina. It was the small archipelago now known as the Falkland Islands (or Islas Malvinas if one is from a country, such as Argentina, linked historically with Spain). Others may have seen them earlier, but Davis's sighting is the first reasonably well documented one. Although the Falklands lie north of the Antarctic Convergence, they are significant for the Peninsula Region story in several ways. Southern fur sealing began in these islands in the late eighteenth century. A few years later, the Falklands would become an important staging area for ships sailing much farther south. And it was from the Falklands that the British would administer their claim to nearly the entire Antarctic Peninsula Region.

Seven years after Davis chanced on the Falklands, another voyage led contemporary writers to report that land had been discovered much farther south. In June 1598, a large Dutch fleet left Europe to sail into the Pacific via the Strait of Magellan. The fleet reached the Pacific in early September 1599, but shortly after, a storm scattered the ships. The gales reportedly drove one, the *Blijde Boodschap* (Good News), to 64° S. There her captain, Dirk Gherritz, was said to have seen mountainous snow-covered land. Historical accounts that take this report seriously credit it as a sighting of the South Shetlands or, less frequently, the islands of the Palmer Archipelago to the west of the Antarctic Peninsula. But had Gherritz in fact seen anything, or even reached so far south? Even some contemporary evidence raises doubts. True or not, the account of this voyage was the first to claim that a ship had sailed south of 60° and seen land.

In the meantime, people in Europe were beginning to think about Drake's report of open water beyond Tierra del Fuego. Perhaps that meant there was an alternative sea route between the Atlantic and the Pacific that could replace Magellan's ferociously dangerous strait.

WILLEM SCHOUTEN AND JACOB LE MAIRE ROUND CAPE HORN

At the outset of the seventeenth century, Dutch politics introduced another motivation to seek a new way past South America. It was then that the States General of the Netherlands granted the Dutch East India Company (VOC) a monopoly on trade with the East Indies, via either the Cape of Good Hope or the Strait of Magellan. Isaac Le Maire, a VOC director, urged the company to look for additional routes. When his fellow directors showed no interest, Le Maire formed his own company and mounted an expedition to make such a search.

Le Maire chose Willem Schouten, a man who knew the East Indies well, for his captain. Le Maire himself remained in Holland, but he sent his son Jacob along to look out for his interests. The expedition's two vessels left the town of Hoorn in June 1615 with 80 men and wintered on the coast of Patagonia before beginning their search. While they were cleaning the ships during winter, a fire destroyed the smaller of the two vessels, the *Hoorn*.

Schouten and Jacob Le Maire resumed the voyage in mid-January with the surviving ship, the *Eendracht*. A few days later, they intentionally bypassed the Strait of Magellan and continued south along the coast of Tierra del Fuego. On January 25, they scored their first discovery, a landmass to the east that the two Dutchmen named Staten Land to honor the States General. (Today, we know it as Staten Island, or Isla de los Estados in Argentina.) By evening, the *Eendracht* reached the southern end of the land. Schouten and Le Maire had found a way from the Atlantic to the open ocean below the Americas. They christened the passage between Tierra del Fuego and Staten Island that they had sailed through the Strait of Le Maire.

Now the Dutch headed into virgin waters. Everything behind them to the north was known, and soon they would be in the Pacific, already crossed by several other expeditions. For where they were, however, there were no charts or records, not even rumors. Not only that, they were sailing west out of sight of land through freezing sleet and snow. Schouten kept a fire blazing below deck to warm the men and served hot drinks whenever possible. And he furnished the crew with special clothing—woolen vests, leather boots, jackets, and oiled caps—to fight the cold. The captain's efforts helped, but nothing could fully compensate for the men's frigid watches and their struggles with icy riggings and frozen sails.

Schouten saw land on January 29, 1616. It was, he wrote, "covered over with snow, ending with a sharp point, which we called Cape Hoorn."[7] This name, for what is in fact a small island, presumably honored the Dutch town from which the expedition had sailed, or perhaps it was for the lost ship of the same name, or maybe for both. Schouten did not say.

When the Dutch reached the west coast of Tierra del Fuego nearly two weeks later, they had completed the first rounding of Cape Horn. Like Magellan, they had had a much easier time in their pioneer voyage than many who would follow them. It had taken only two and a half weeks, and despite miserable weather, they had been in control of their ship the entire time. The Cape Horn route, a passage

The route of the *Eendracht* through the Strait of Le Maire and around Cape Horn — *as drawn by the Dutch navigators who first made this voyage.* (*Originally published in Schouten,* The Relation of a Wonderfull Voiage, *1617. Illustration from reproduction in facsimile edition, 1966*)

that would become legendary for its dangers—especially when sailed east to west against the prevailing winds and currents as its discoverers had—would seldom be so kind. For some who followed the Dutch, it would take months to complete the passage as crews battled storms that could drive a ship hundreds of miles off course, or worse, to the bottom of the ocean.

Once in the Pacific, Schouten headed west and eventually reached Batavia (present-day Jakarta) in the East Indies. The voyage effectively ended there, because the local VOC authorities, refusing to believe that the expedition had found an alternative to the proscribed Strait of Magellan, seized the *Eendracht*. They then sent Schouten and Le Maire home as prisoners aboard a VOC ship. Schouten reached Holland early in July 1617, but young Le Maire had died en route. His distraught father was not even left with the compensation of enjoying a monopoly on his voyage's discovery. Despite his efforts to keep the route secret, the news of a new passage spread quickly in Dutch ports, and then, of course, beyond.

Schouten and Le Maire's voyage proved that Tierra del Fuego was only a group of islands, opened a new way into the Pacific, and reduced the importance of the Strait of Magellan. Most important for Antarctic discovery, the Cape Horn route meant that ships would be sailing much farther south than before. But as for the Great South Land of legend, people continued to believe. Favorite theories die hard, and Staten Land gave the mapmakers something new to cling to. They quickly drew it in as another coast of the Southern Continent, somehow supposed to curve around so far south of Cape Horn that its true character had been too distant for Schouten and Le Maire to see. It would be 27 years before someone would pass to the east and south of Staten Island and thus cut it off from supposed land farther south. At that point, the Strait of Le Maire lost its importance, because navigators realized they could sail east of Staten Island to reach the Cape Horn route.

The Antarctic Convergence Is Crossed

Cape Horn and the passage around it were great discoveries, but with the possible exception of Gherritz, who may or may not have seen something, no one had yet reported land south of the Antarctic Convergence. In 1675, someone at last made an unquestionable landfall there. Anthony de la Roché, a British merchant captain, was rounding Cape Horn eastward in April when adverse winds and currents swept him far off course to the east, to uncharted land between 54° and 55° S. Historians generally credit this as a sighting of South Georgia, which lies at 54° S, 36° W. De la Roché, who did no exploring there despite remaining anchored for fourteen days, identified his unnamed discovery as yet another tip of the Southern Continent.

South Georgia's discovery, the first definite landfall below the Convergence, was a major event in Antarctic history. No one else would see South Georgia, however, for over 80 years. Then, in June 1756, nature repeated itself as winds drove the

León, a Spanish ship captained by Gregorio Jerez, there after rounding Cape Horn. It was a passenger on Jerez's ship who gave the world its first description of this land, desolate and ice-covered, "filled with steep mountains of a frightful aspect, and of so extraordinary a height, that we could hardly see the summits, although at a distance of more than six leagues."[8]

In fact, ice permanently blankets more than half of this 110-mile-long, banana-shaped island, and dozens of glaciers pour down from the mountains that form its central spine.

Land was not the only Antarctic feature that those sailing around Cape Horn were finding in the 1600s. At the close of 1687, a few years after de la Roché's discovery of South Georgia, a storm drove the English pirate Edward Davis to nearly 63° S in the Drake Passage. There, the ship's surgeon, Lionel Wafer, wrote, "We met several islands of ice; which at first seemed to be real land."[9] These were almost certainly tabular icebergs, and this is one of the earliest reports of them.

Another visitor soon provided a more complete description. He was Edmond Halley, the English astronomer best remembered today for his identification of the comet named after him. In 1699–1700, he led the first true research voyage to the Southern Ocean, or at least its fringes. In addition to directing him to pursue scientific work, Halley's orders called for him to sail south until he reached the coast of the supposed Southern Continent.

Halley departed England in September 1699 in command of the naval vessel *Paramore* and first headed south across the Atlantic to South America. Near the end of the year, he set out for the Southern Ocean. On January 28, 1700, the air and water temperatures dropped sharply, indications that the *Paramore* had probably crossed the Antarctic Convergence. Halley spent the next few days creeping along in thick fog. On February 1, he reached 52° 24' S a number of miles to the east of South Georgia and sighted "three islands as they then appeared; being all flat on top, and covered with snow. Milk white, with perpendicular cliffs all round them. . . . The great height of them made us conclude them land. . . ." Clearer weather the next day allowed Halley to approach this supposed land. Now, he realized, it was "nothing but . . . ice of an incredible height" with a cliff at least 200 feet high.[10] The largest ice island, he wrote, was 5 miles long. Halley turned north after ten days because the *Paramore* was not designed for contact with ice. He reached England in September 1700, never having found any sign of the Southern Continent—indeed, never having searched for it.

Even though ships were rounding Cape Horn regularly by the late 1760s, many of them reaching well south of 60° S, men had made only two landfalls below the Antarctic Convergence—South Georgia in the Peninsula Region and Bouvet Island, discovered by Frenchman Jean-Baptiste-Charles Bouvet in 1739, well to the east, in the South Atlantic at 54° 25' S, 3° 31' E. What really existed in the far south remained an open question. More than half of the Pacific remained unexplored. Was there a large continent somewhere in its high southern latitudes? Or did the waters farther south just repeat the island clusters already found farther north? And what about the South Atlantic and Indian Oceans? So much was still unknown.

Advances in maritime techniques after the middle of the eighteenth century made possible the long exploratory voyages necessary to answer these questions. Two developments were crucial: the introduction of foods that helped crews ward off scurvy, and improved navigation methods and equipment.

Scurvy, a vitamin-deficiency disease, had begun to haunt sailors in the fifteenth century, when exploratory and other voyages began to spend many weeks at sea, dependent on salted meat and other dried and preserved foods. This horrific disease, which results in agonizingly swollen limbs and joints, bleeding gums, and sometimes severe mental illness, results when people fail to ingest vitamin C for more than a few weeks. Unlike most mammals, humans cannot metabolize this vitamin. Thus, it must come from a diet containing it, usually fresh fruit or vegetables, but also raw or very lightly cooked meat, in particular livers and kidneys.[11] Although scurvy is always fatal unless the sufferer receives sufficient vitamin C in time, it is easily cured with no lasting effect by gaining the vitamin. It was only in the early twentieth century that all this was finally understood. In the mid-eighteenth century, however, Britain's James Lind, a naval surgeon, published *A Treatise of the Scurvy,* a work reporting that he had found that the disease could be effectively treated with citrus fruit, onions, and vegetables—precisely what was in fact needed to prevent scurvy.

Better navigation technique was equally critical. For centuries, those who had dared to sail far from land had real problems knowing just where they were. The compass told mariners their direction, but determining their location in a featureless ocean was much more difficult. Men had long known how to find their north-south position, or latitude, by combining the date with the height of the sun at midday. Development of the sextant in the mid-eighteenth century greatly improved sun sight accuracy. That left the question of east-west position, or longitude. Early sailors who passed beyond sight of land relied on estimates of their speed and direction to determine longitude, but were often wildly off. Although the 1767 publication of the *Nautical Almanac* by the British Astronomer Royal, Nevil Maskelyne, made it possible to determine longitude using astronomical sights, the necessary calculations were cumbersome and difficult. A more practical method became available at almost the same time when another Englishman, John Harrison, invented the marine chronometer, a precise timepiece that could withstand the rigors of a sea voyage. Harrison's chronometer made it possible for sailors to determine the difference between their own local noon and the time at a known reference location—Greenwich, England, in the case of late-eighteenth-century British sailors and virtually everyone today. The navigator then translated the time difference into miles.

A British naval captain, James Cook, was a pioneer in using these advances in provisioning and navigation. In 1772, he set out to make the first serious attempt to learn the truth about the Great South Land.

JAMES COOK'S HIGH-LATITUDE CIRCUMNAVIGATION OF THE ANTARCTIC

Cook was a remarkably able man, a determined officer who had risen from the ranks to captain His Majesty's ships. From 1768 to 1771, the Royal Navy had sent him to the South Pacific to observe a transit of Venus. He had also carried secret orders to search for the Southern Continent. His search, in fact, was quite limited, concentrated in the South Pacific and about New Zealand, but Cook did correctly surmise that no Southern Continent extended into the temperate regions. When he reached home in 1771, he persuaded the British government to send him on a new expedition to investigate just what did exist in the south.

Cook prepared for his second voyage with great care. He selected his officers from the many eager to sail with him. He also took an artist, who would produce a marvelous visual record of the voyage, trained naturalists, and two astronomers who could use the heavens to determine latitude and longitude. And, the first exploring captain to do so, he carried several of the newly introduced marine chronometers. Cook took equal care with his provisioning, choosing his stores with an eye to combating scurvy and making a point of picking up fresh food along the way whenever possible. Finally, at his recommendation, the Admiralty outfitted the expedition with ships from the merchant fleet, vessels that could safely sail close to strange coasts, that were strong enough to handle violent weather and drifting ice, and offered ample cargo capacity for a long voyage.

The expedition left England in July 1772 with about 200 men aboard two ships—the *Resolution* under Cook's personal command and the *Adventure* under Capt. Tobias Furneaux—and sailed first to the Cape of Good Hope at the tip of Africa. There, Cook began his southern exploration. During the Southern Hemisphere summers of 1772–73 and 1773–74, he sailed clockwise around roughly two-thirds of the Antarctic regions, all of this route entirely outside the Peninsula Region. Cook pushed his ships to higher latitudes than anyone else had reached, turning back or retreating only when ice stopped him, and on January 17, 1773, he became the first to cross the Antarctic Circle. Cook spent both winters exploring the South Pacific and refitting in New Zealand, but his ships became separated after the first winter. Before they could be reunited, the *Adventure* lost ten men to Maori cannibals, and Furneaux immediately took his ship home to England. Meanwhile, Cook, unaware of his consort's fate, completed the voyage alone in the *Resolution*. During the second summer, he reached a new farthest south of 71° 10' S, at 106° 54' W, a

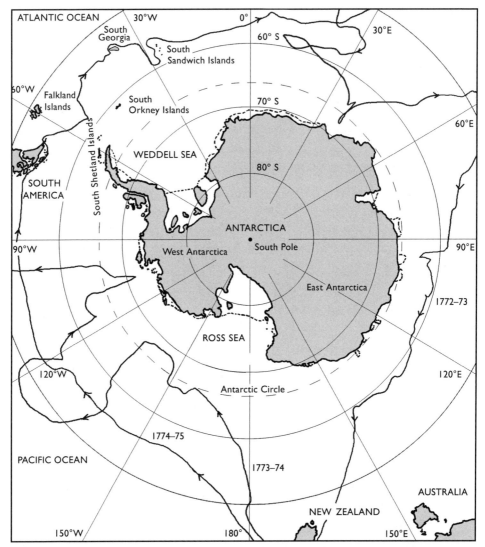

Captain James Cook's track on his 1772–75 high-latitude Antarctic circumnavigation — *Cook started from the Cape of Good Hope and headed east, making several deep dips southward. He spent both the 1773 and 1774 winters in New Zealand, then sailed across the Pacific to Cape Horn in 1774–75, on to the first landing on South Georgia and discovery of the South Sandwich Islands. He crossed his original track south of the Cape of Good Hope in late February 1775 to complete the circumnavigation.*

southern record that would stand for nearly 50 years.

Cook spent the most important parts of his voyage's third high-latitude season in the far north of the Peninsula Region. He departed New Zealand in November 1774, sailing east across the Pacific directly toward Cape Horn along latitudes 54 to 55° S. After two weeks spent surveying and re-provisioning his ship at Tierra del Fuego, he sailed from Staten Island on January 3, 1775, in search of the speculative southern land shown on his charts. He was also looking for the land reported by de la Roché in 1675 and the *León* in 1756. Three days later, Cook reached 58° 9' S, 52–53° W, a place where his chart showed land. With nothing in sight, he turned northeast in search of de la Roché's and Jerez's landfall. It was an ironic decision. During the two previous summers, Cook had pushed south to the ice edge in places where the continent lay far to the south. Now he was turning back when he was, in fact, due north of the tip of the Antarctic Peninsula, at the very point where the Antarctic Continent is most easily reached. Perhaps if his exploration had begun from Cape Horn rather than from Africa, Cook might have probed farther south here, as he had in his first two summers of exploration when he was still fresh and eager. Instead, he missed his best chance to discover what he was looking for. Cook's third summer, however, did yield more in the way of discovery than either of his first two had.

On January 14, midshipman Thomas Willis spotted something that was either a very large iceberg or land. Cook's men started laying bets, ten-to-one for ice, five-to-one for land. By afternoon, Cook wrote, "It was now no longer doubted but that it was land and not ice which we had in sight: it was however in a manner wholly covered with snow."[12] It was his first sight of an Antarctic landscape in nearly three years of searching. Cook named this landfall, at the northern extremity of the South Georgia island group, Willis Island, after the first man on board to have seen it. (There are actually several islands there, now called the Willis Islands.)

Cook reached South Georgia itself two days later, and began sailing down the northeast coast. On January 17, precisely three years after achieving history's first crossing of the Antarctic Circle, Cook made the first landing on South Georgia. It was on a beach in the place he named Possession Bay, because it was here that he claimed the land for Great Britain.

After landing, Cook continued down South Georgia's northeast coast, charting

Possession Bay, South Georgia — *the bay where Cook landed and claimed South Georgia for Great Britain, as painted by William Hodges, Cook's expedition artist aboard the* Resolution. *(From Cook,* A Voyage Towards the South Pole . . . , *1784)*

and naming geographic features. He reached land's end three days later at a place he christened Cape Disappointment. This land, he now knew, was just another island. He named it the Isle of Georgia in honor of King George III. Despite the name he gave the southernmost cape, Cook wrote, "I must confess the disappointment . . . did not affect me much, for to judge of the bulk of the sample it would not be worth the discovery."[13] Rather than the rich and fertile Southern Continent of legend, it was a place where "The inner parts of the country [were] . . . savage and horrible: the wild rocks raised their lofty summits till they were lost in the clouds and the valleys laid buried in everlasting snow. Not a tree or shrub was to be seen, no not even big enough to make a tooth-pick."[14]

Cook, however, thought that there was probably more land to the south, because he believed, as did the scientific world of the day, that ice formed only on land. There was, he said, thousands of times too much ice in the ocean for South Georgia—the only southern land he had seen in three years—to have produced it all. And so, for the first time that summer, Cook sailed south in search of whatever he might find.

On January 31, Cook discovered land just north of 60° S. It was the southern end of the South Sandwich Islands, a 220-mile-long chain of eleven volcanic islands. Fog made it impossible to see this landfall in detail, but thinking that it might be more than simply islands, Cook named it Sandwich Land for Lord Sandwich, then First Lord of the Admiralty. This desolate place offered only ice and rocky cliffs, with virtually no harbors, a prospect that Cook described as "the most horrible coast in the world."[15] He later wrote jointly of this discovery and of South Georgia that they were "lands doomed by nature to everlasting frigidness and never once to feel the warmth of the Suns [sic] rays, whose horrible and savage aspect I have no words to describe. . . ."[16]

Cook sailed about the area for a few more days and discovered several more islands, but the fog was unrelenting, and ice held him well off the coasts. On February 6, he ended his explorations and headed north. The South Sandwich Islands were the last land Cook saw until reaching Cape Town seven weeks later.

The world's concept of the far south had to change after Cook's voyage, the first ever high-latitude circumnavigation of the Antarctic regions. Although he had failed to find a Southern Continent, he had set its possible limits. If it did exist, it had to be south of his track—often south of 60° S. In these latitudes, Cook told the world, all was ice-covered and hostile. There was, he said, nothing of value there. Further exploration would be worthless. His own reports, however, contradicted this assessment. Cook and his men had seen unanticipated riches in the south—the abundant seals and whales of the Southern Ocean—and they described them for all to hear. Cook had slain the myth, but his voyage generated a new, and valid, motive for men to venture south.

The First Sealers Arrive in the South

Efforts to exploit the Southern Ocean's wildlife began nearly simultaneously with Cook's return home. Sealing in particular would have major consequences for Antarctic discovery in the nineteenth century. Hunting, however, began with New England whalers, who arrived in the Falkland Islands, on the northern fringe of the Peninsula Region, in the early 1770s. Although there primarily for whales, they quickly realized that the southern elephant seal, especially the blubber-rich adult males that could be as large as three to four tons, were a valuable target as well. Fur sealing then evolved naturally out of elephant sealing, and both forms of sealing soon led hunters south of the Antarctic Convergence, to South Georgia. (The whalers, however, were not among them initially. For them, with whale stocks in the South Pacific and in the Arctic adequate to meet industry demands, there was no need to brave the ice. It would only be near the end of the nineteenth century that whalers would make the first serious attempt to expand their hunt farther south, into the Antarctic.)

The Americans and the British dominated the sealing industry. Although some accounts give the impression that the Americans were the more important of the two through 1818, this largely reflects the fact that more information is available about their efforts. Not only have more American logbooks survived, but several Americans also published narratives about their own voyages. Accurate records for sealers from any country, however, are scarce. This is not entirely accidental. Because the fur sealers typically killed all the animals they found, they were constantly looking for new grounds and tried to keep their discoveries secret, at least until they had exploited each new find.

The first documented pure sealing voyage to the south set sail in 1784. That year, the *United States,* an American ship out of Boston, went to the Falklands. The captain's original objective was elephant seal oil, but when he reached the islands and saw the great masses of fur seals, he decided to focus on them. His employers, disappointed when the ship arrived in Boston in 1786 with 13,000 fur seal skins rather than the anticipated barrels of elephant seal oil, sold the cargo for $.50 a pelt. The buyer then sent the skins to Calcutta, whence the next purchaser shipped them on to Canton, China, in 1789. There they fetched $5 each. The many American ships in Canton that year carried home the sensational news of the sale. Fur sealing in the Southern Hemisphere took off, and the fur seals of the Falklands, and then farther south, were doomed to near extinction on beach after beach.

John Leard captained a British voyage to the Falklands and the Cape Horn region at roughly the same time as the voyage of the *United States.* After he returned to London in July 1788 with 6,000 fur seal skins, Leard proposed to the government that the British establish a sealing operation off the southern coast of Patagonia—an industry, he said, that should be controlled, in the interests of conservation. In particular, he wrote of how vital it was "not to kill the Females when

The sealers' targets. Left: a bull elephant seal (at South Georgia, 2009). Right: a bull fur seal (at South Georgia, 1995) — *The males of both these species are far larger than the females they gather around themselves in harems. Sealers, however, took both males and females. (Author photos)*

with Young—nor should the Young [be taken] until about one Year old. . . ."[17] Although Leard was more concerned with creating a viable long-term industry than he was with saving seals, his ideas were enlightened for the times. The official response, if any, no longer exists. Whatever it was, nothing came of Leard's proposal, either to establish an official British sealing industry off Patagonia or to decree a conservation regime.

In the Falklands, however, private British ships soon joined the Americans taking fur seals. Although the work was relatively simple once a man knew what he was doing, some limited experience was necessary to do the job effectively. Profitable sealing meant knowing where to find the seals, how to kill and skin them, and ways to prepare, preserve, and stow the cargo. Early on, when no one had the requisite knowledge, the sealers had to learn by trial and error.

A splendid example of what could happen when an inexperienced crew set to work was recounted by Edmund Fanning, a 23-year-old Stonington, Connecticut, man who sailed as first mate on the *Betsey*, a ship that made a fur sealing voyage to the Falklands in 1792. Not only had no one aboard ever been sealing, no one even knew what a fur seal looked like. The first animals these neophyte sealers saw upon landing were sea lions, 300 of them. Thinking they were fur seals, they set right to work. Fanning wrote of what happened next: " 'Sir!' exclaimed Mike [the boatswain], as we were wondering whether our small vessel . . . could carry many thousand such mammoths, 'do you think these overgrown monsters are seals?' Surely they are, he was quickly answered. . . ." Matters grew worse as the men stalked the supposed fur seals. The animals "set forth a roar that appeared to shake the very rocks on which we stood, and in turn advancing upon us in double quick time, without any regard to our persons, knocked every man of us down with as much ease as if we had been pipe stems, and passing over our fallen bodies, marched with the utmost contempt to the water." By then, the men had figured out "to our entire

American sealers in the Falklands — *A near-contemporary illustration of American sealers relaxing from their work of taking fur and elephant seals in the Falklands. On the distant shore, one can just make out five groups of female seals clustered in harems with bulls guarding them. In left center, a sealer feeds a bird, perhaps a penguin, into a try-pot. (From Fanning,* Voyages and Discoveries in the South Seas . . . , *1833)*

satisfaction, that these were not fur seals."[18] This was Fanning's first sealing voyage, but not his last. He would return many times and become a major figure in the Peninsula Region story.

The hauls by the *United States* and by Leard's crew were experiments, but successful ones. The inept voyage Fanning experienced was one of the first dedicated American fur sealing ventures. From these and other voyages, American and British shipowners alike recognized the potential for fabulous profits, and began sending out more and more ships explicitly for fur sealing. Their captains soon had to look beyond the Falklands, however, because, as Leard had forecast, uncontrolled slaughter quickly devastated the local fur seal population. One of the first places they moved on to was South Georgia. Before long, dozens of ships were heading there, carrying gangs of sealers who ruthlessly searched the island beaches for seals. A number of these men remained ashore for months at a time, some even through the icy winters, typically living in makeshift shelters created from upturned whaleboats.

South Georgia fur sealing peaked at the beginning of the nineteenth century. The 1801–02 summer saw about 30 ships there, most from New England because the early days of Britain's conflict with France reduced the number of British ships in the sealing trade. But war was not the only factor shrinking the sealing fleet at

South Georgia in the years that followed. The sealers had done their work so well, taking more than 1,200,000 fur seals from the island by one contemporary estimate, that few of the animals remained.[19] After the summer of 1801–02, most of the remaining sealers in the Western Hemisphere moved north of the Convergence to the Cape Horn area and to islands off the west coast of South America. There they repeated the slaughter that had already laid waste to the fur seal populations at the Falklands and South Georgia. By 1808, the replacement beaches were themselves nearly bare of seals. The British, swept up in the Napoleonic Wars, had already left the field. Now nearly all the Americans quit as well, and in 1812, the outbreak of a new conflict, this one between Britain and the United States, pulled the few remaining American sealers home.

Chapter 2

The Continent Found: 1819–1821

In 1815, when the war between the United States and Great Britain ended, a number of sealers, now largely in search of elephant seals, returned to South Georgia. But in February 1819, a momentous discovery hundreds of miles to the south shifted the sealers' focus far deeper into the Peninsula Region.

WILLIAM SMITH DISCOVERS THE SOUTH SHETLANDS

William Smith, a 28-year-old British merchant captain, sailed from Buenos Aires in January 1819 to round Cape Horn en route to Valparaíso, Chile. Smith, an experienced captain, had made the same trip several times and was familiar with ice from sailing the waters about Greenland. So when he encountered strong headwinds as he began rounding the Horn, he deliberately headed his ship, the *Williams,* south in search of better weather. It was a fateful decision.

On February 19, 1819, Smith sighted what he thought was land, seen through a snowstorm. The weather improved the next day, and he headed in for a closer look. It was indeed land, at what he estimated to be 62° 17' S, 60° 12' W, and fur seals crowded the visible beaches. Smith had discovered the South Shetland Islands. His marine insurance policy, however, contained a clause denying coverage for anything other than standard merchant enterprises. He thus resumed his voyage without further exploration.

Smith reached Valparaíso on March 11 and immediately reported his discovery to Captain William Shirreff, the senior British naval officer on South America's Pacific coast. Shirreff was unimpressed. Not only was he skeptical of the discovery, but he had more pressing concerns. The Spanish colonies were rebelling, and he had his hands full protecting the interests of local British merchants. Others in the Valparaíso British community also had doubts. John Miers, a local British mining engineer who wrote a contemporary account of Smith's find, reported that "all ridiculed the poor man for his fanciful credulity and his deceptive vision. . . ."[1] Despite this scornful reception, the news of a possibly significant discovery began to circulate.

It took Smith until May to assemble another cargo. He then left Valparaíso and returned around the Horn, this time bound for Montevideo. Smith, his pride apparently overcoming his insurance concerns, was determined to verify his discovery en route, even though that meant sailing south in winter. When he reached about 62° S, however, there was no land in sight and ice was forming on the water. Wisely, he bowed to the season and turned north.

When Smith reached Montevideo in July, he again reported his discovery. Although the official British reaction was as cool as it had been in Valparaíso, local commercial captains responded quite differently. Excited reports of Smith's sighting of fur seals had preceded him across the continent, and an eager group of American merchants and sealers offered to pay him for their location. Smith, who wanted to land and claim his find for Great Britain first, turned the Americans down.

Smith left Montevideo for Valparaíso at the end of September, heading around Cape Horn for the third time in a year. Once again, he detoured to investigate his discovery. This time the effort paid off. On October 15, Smith sighted land in roughly the same place he had seen it eight months earlier. He and a few of his men went ashore the next day, on the landmass now known as King George Island. There they raised the British Union Jack and took possession of the land for Great Britain.

In contrast to his hurried departure from the islands in February, this time Smith took several days to explore his discovery, and he soon realized that his landing site was on one of a chain of islands. (There are, in fact, eleven large and many smaller landmasses in the South Shetlands. Together, they make up a mountainous, ice-covered island arc extending for over 300 miles in a west-southwest to east-northeast direction just north of the Antarctic Peninsula.) By the time Smith sailed north to report that his February sighting had been no mistake, he had roughly mapped 150 miles of island coasts. He had also satisfied himself that tens of thousands of fur and elephant seals inhabited the beaches.

Smith reached Valparaíso in late November and repeated his story. This time the British, both official and otherwise, expressed serious interest in the discovery that Smith had named New South Shetland. It was the report of seals that most excited them, but John Miers's embroidered account of Smith's observations—dispatched to England in early 1820—made everything about the islands sound promising. He even wrote that Smith thought he could see pine trees there.

THE FIRST SEALERS IN THE SOUTH SHETLANDS

Two ships had already sailed to exploit the new seal beaches before word of the discovery reached Europe or the United States. When Smith departed Montevideo in September 1819, Joseph Herring, first mate on the *Williams*, remained behind and persuaded British merchants in Buenos Aires to send a sealing expedition to Smith's islands. The ship, the *Espírito Santo*, with Herring aboard to point the way, sailed first to the Falklands and then south. On December 25, 1819, several men landed in the South Shetlands, possibly on what we now call Rugged Island. Under the impression that they were the first to set foot in the islands, they, like Smith, raised the Union Jack and claimed the land for Great Britain.

The *Espírito Santo* soon had company. It was a ship sent by Edmund Fanning, the same man who had mistaken sea lions for fur seals in 1792 and had later become one of the most successful American sealing captains. By 1819, although retired from the sea himself, he was still involved in the industry as a shipowner. Fanning, eager to find new sealing grounds, believed that there might be unexploited beaches at the Aurora Islands, a discovery reported by several Spanish ships in the late eighteenth century. He also thought Gherritz's supposed far south landfalls worth investigating. Fanning thus dispatched James Sheffield, captain of the *Hersilia*, with orders to search for the Aurora Islands. If Sheffield found no seals there, he was to sail south to 63° S along the longitude of Cape Horn and then turn east to look for Gherritz's land. These instructions could have led to Sheffield's discovering the South Shetlands had Smith not found them first.

The *Hersilia* sailed from Stonington, Connecticut, in late July 1819, reaching the Falklands in October. Here the story becomes confused. What is certain is that Sheffield dropped off Nathaniel Palmer, his 20-year-old second mate, and a crew member to gather fresh provisions. He then sailed in search of the Aurora Islands. Fanning's version is that Sheffield found the Auroras (which do not in fact exist) but no seals; returned to the Falklands to pick up the two men; and then proceeded to 63° S as instructed.

Palmer, however, claimed some credit for Sheffield's route from the Falklands. By his account, the *Espírito Santo* arrived while he was in the Falklands and he realized she was a sealing vessel. He then made a point of establishing good relations with her crew, persuading them to tell him a little about where they were going. When Sheffield returned, Palmer passed along what he had learned. Sheffield then left the Falklands in pursuit of the *Espírito Santo*.

No matter how he found them, Sheffield arrived in the South Shetlands in mid-January 1820. There, he soon met up with the *Espírito Santo*, whose crew welcomed the Americans, assuring them that there were enough seals for both ships. (This friendly reception would not be repeated the next summer, when there were far more ships in the region.) The *Hersilia*'s men took 9,000 sealskins in less than sixteen days and could have taken many more had they not run out of curing salt. The *Espírito Santo* did even better.

Sheffield sailed the *Hersilia* into Stonington on May 21, 1820. Although the news of the new sealing grounds had already reached the United States, the actual arrival of a ship with a rich cargo fueled the excitement. New England shipowners at once started organizing fleets to sail south for the 1820–21 season. So, too, did the British.

The news had reached Britain near the end of April, and not only the sealers took immediate note. French cartographer Adrien Brue was preparing new maps for an upcoming edition of his world atlas when the news arrived. In June 1820, he carefully placed a substitution patch on his map of the southern portion of the Western Hemisphere, over what had been a blank space far to the south of Cape Horn. Now, instead of nothing, the map included a broad tip of land labeled Nou^{au} Shetland Austral (New South Shetland). That one little patch was historic, the first representation on a published map of far south Antarctic region land based on actual discovery.[2]

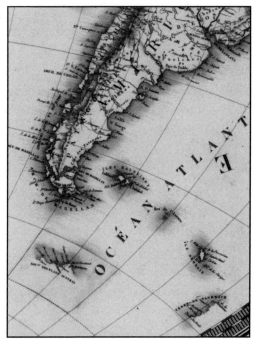

Brue's June 1820 Map — *This extract from Brue's June 1820 map of the southern portion of the Western Hemisphere shows not only South Georgia and the portion of the South Sandwich Islands discovered by Cook in 1775, but it also shows the South Shetlands, quite accurately located. As can be seen by the slightly darker lines outlining the two sections on which the South Shetlands lie, this portion of the map was added after the rest of the map had been completed. Brue has been careful here to show that it is not yet known whether the new discovery is insular or something much more, the tip of a larger landmass to the south. (Image from DavidRumsey.com, David Rumsey Historical Map Collection)*

EDWARD BRANSFIELD SIGHTS THE ANTARCTIC PENINSULA

Another ship had also been in the South Shetlands in the summer of 1819–20. Captain Shirreff, having rethought his position after Smith left Valparaíso in May, had decided to send a naval expedition to investigate the reported discovery. When Smith returned in November, Shirreff chartered the *Williams* and engaged Smith as pilot. The rest of the ship's complement was drawn from the Royal Navy. Shirreff named Edward Bransfield commander and provided him with three midshipmen to help in surveying. A naval surgeon, Adam Young, was the ship's doctor and naturalist.

Shirreff ordered Bransfield

> . . . to explore every harbour that you may discover, making correct charts there of. . . . And you will ascertain the truth of the account brought here of an uncommon abundance of the Sperm whales, otters, seals & c. . . . Ascertain the natural resources of the land for supporting a colony. . . . and if [the land] should be inhabited you will minutely observe the character, habits, dress and customs of the inhabitants. . . .[3]

Bransfield was also to do scientific work and to determine if the South Shetlands were islands or part of a continent. Finally, he was to claim all the land he saw. These orders were obviously far more work than Bransfield could hope to carry out in one short voyage, not to mention evidence of continued wishful thinking about southern lands.

Bransfield reached the South Shetlands in mid-January 1820 and began sailing along the northern shores of the islands. When he could see land through the intermittent fog, it was everywhere the same—forbidding and ice-covered. Bransfield passed Smith's landing site on January 22 and sailed into a large bay on the southern coast of the same island. Here he landed, raised the Union Jack, and, as Smith had, claimed the entire region for Great Britain. Dr. Young, for his part, collected a few natural history specimens and noted that the only vegetation consisted of bits of stunted grass, a few mosses, and some lichens, a far cry from the glowing description that John Miers had sent home.

Five days later, Bransfield left what he named King George Bay on King George Island and resumed his survey. Bad weather persuaded him to head south, away from the South Shetlands, on the morning of January 30, 1820. When the haze lifted that afternoon, the men aboard the *Williams* unexpectedly spotted land to the southwest. It was the tip of the Antarctic Peninsula, the roughly 1,000-mile long finger of mountainous land reaching north from the Behrendt Mountains on the main body of the Antarctic Continent. Two days later, Bransfield once more sighted mountains to the south. He inked them in on his chart, connecting these peaks by a dotted line to the land he had seen on the 30th. It was a lucky guess: his line closely coincides with the actual coastline of the Antarctic Peninsula.

Bransfield then sailed about the South Shetlands for a few more weeks, discovering and naming a number of locations. He also attempted a brief probe south, to the east of the Antarctic Peninsula, before pack ice blocked him at 64° 50' S, 52° 30' W. The *Williams* left the south on March 18. After Bransfield reached Valparaíso nearly a month later, he and the other naval men returned to their own ships. As for Smith, he took the *Williams* home to England to fit her out for a fifth voyage south, this time for sealing in the summer of 1820–21.

Unfortunately, the full significance of Bransfield's voyage went unrecognized for nearly a century by even the British because the naval authorities in Valparaíso lost the *Williams*' logbook soon after the voyage. Only Bransfield's chart reached the British Admiralty, and even that soon became buried in the archives. In July 1917, however, William Speirs Bruce, an early-twentieth-century Peninsula Region explorer, would rediscover the chart while delving through old Admiralty records. Excitedly, he would write to a friend, "To-day I have definitely proved that Edward Bransfield, R. N., . . . discovered the Antarctic Continent."[4]

But had Bransfield made such a claim in 1820, it could only have been a guess.

For centuries, people had erroneously proclaimed land sightings in the south to be coasts of the theoretical Southern Continent. When Bransfield did in fact see the continent, he had no more basis than his predecessors for declaring the land continental. He could just as easily have been seeing the coast of a large island. At least one of Bransfield's men realized this. Summing up what they had seen, the ship's doctor, Young, wrote, ". . . from the . . . almost constant fogs in which we were enveloped, we could not ascertain whether [the land] formed part of a continent, or was only a group of islands."[5] As for Bransfield, he had made no effort to investigate the matter. Even if he had, it would have been nearly impossible for him to reach an accurate conclusion with the time and ship available to him. The coastline of Antarctica is so elusive, so often difficult to approach or even see, that it was not until the twentieth century that observational evidence would definitively link together the disparate pieces that men had seen, at last firmly establishing that Antarctica is indeed a continent.

One contemporary reporter did publish the kind of claim that others had made for centuries before Bransfield's genuine sighting. While Bransfield was in the south, John Miers, the British mining engineer in Valparaíso, sent home his glowing account of Smith's discoveries, a fantasy Bransfield had already demolished by the time Miers's article appeared in the *Edinburgh Philosophical Journal* in October 1820. The article opened, ". . . a large Southern Continent is about to be discovered."[6] Miers went even further, suggesting that the new discoveries were somehow linked with Cook's Sandwich Land. Another explorer, however, had already ruled that out months earlier.

Fabian von Bellingshausen Circumnavigates Antarctica

That explorer was a Russian naval captain, Fabian Gottlieb von Bellingshausen, who had set out on a great Antarctic voyage shortly before Bransfield reached the South Shetlands.[7] His expedition came about because of Russia's need for a supply route to its vast far eastern territories. The two options were to the south and to the north, and in March 1819, Czar Alexander I announced that he was sending fleets to explore both polar regions. He gave Bellingshausen command of the southern voyage, an effort explicitly designed to complement James Cook's 1772–75 Antarctic circumnavigation. Where Cook had gone far to the south, Bellingshausen would take a somewhat more northerly route, and where Cook had taken a more northerly path, Bellingshausen would go farther south.

Born in 1778 in Estonia, Bellingshausen was a highly experienced naval officer who had already circumnavigated the Earth, from 1803 to 1806 as a junior lieutenant in a Russian fleet. Like Cook, he was a superb seaman and navigator as well as a thoughtful and observant man. One of his great regrets regarding his Antarctic voyage would be that the two civilian scientists who were to have joined him changed their minds too late to be replaced. Despite this disappointment, Bellingshausen would return with significant scientific results, a tribute to his own ability and determination.

Bellingshausen left Russia in July 1819 with about 190 men aboard two ships, his flagship, the *Vostok* (East), and the *Mirnyy* (Peaceful). After several northern stops en route, he headed for the Southern Ocean to begin his Antarctic voyage approximately where Cook had left off, at South Georgia and the South Sandwich Islands.

The expedition reached South Georgia on December 28.[8] Bellingshausen then began his exploration by sailing down the island's southwest coast, because Cook had already surveyed the northeast. Although he christened a number of places, Bellingshausen did not land, because "the land was inhabited only by penguins, sea elephants and seals; there were but few of the latter, since they are killed by the whalers [sic]. . . . We saw not a single shrub nor any vegetation; everything was covered with snow and ice."[9] Bellingshausen apparently liked the look of South Georgia no more than Cook had.

Fabian von Bellingshausen — *Portrait painted around the time of the voyage. (From reproduction in Bellingshausen, ed. Debenham,* The Voyage of Captain Bellingshausen, *1945)*

On December 31, he headed for Cook's Sandwich Land. Here, he would survey from the east to complement Cook's efforts along the western side of the group. It was a course choice that would lead him to discoveries of his own.

New Year's Day 1820 began with snow, but by late morning the weather briefly cleared enough for Bellingshausen to glimpse a small island absent from Cook's chart. He christened it Leskov Island in honor of the *Vostok*'s third lieutenant. Three days later, the thick weather eased off again and the Russians spotted two more uncharted islands. One was an imposing conical mountain that Bellingshausen named Visokoi (High), the other a volcanic mass belching thick, stinking smoke. He called the latter Zavodovski in honor of the *Vostok*'s captain.

Bellingshausen crept closer to Zavodovski Island the next day, and Zavodovski himself led a small landing party. After forcing their way ashore through tens of thousands of chinstrap penguins thronging the beach, the men walked about halfway around the island before returning to the ship. It was the horrible smell of penguin guano, Zavodovski said, that forced him to cut the visit short. Perhaps the penguins were responsible for the island's stench, perhaps not. Stinking gases spew

33

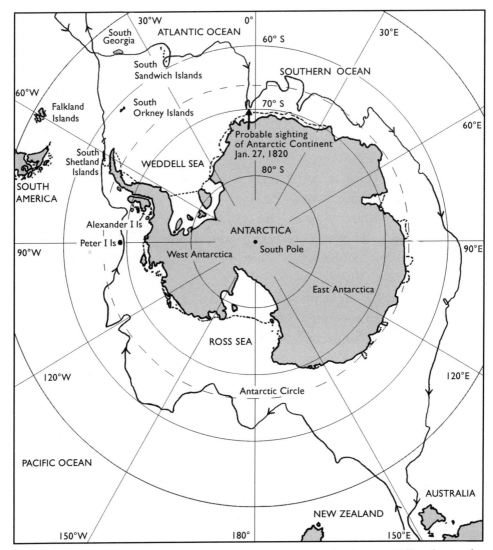

Bellingshausen's route on his 1819–21 Antarctic circumnavigation — *Bellingshausen began his circumnavigation near South Georgia, which was his first landfall. From there, he sailed to the South Sandwich Islands, then continued east, dipping far south toward the coast of east Antarctica, where he almost certainly saw land on January 27, 1820. He spent the 1820 winter in Australia, resuming his voyage the next summer. He arrived in the South Shetlands at the beginning of February 1821 only to find dozens of sealing vessels already there. He completed his circumnavigation when he crossed his late-1819 track in February 1821.*

from vents along Zavodovski Island's shores, a fact reflected in names later visitors bestowed on island features: Mount Asphyxia, Acrid Point, Stench Point, Fume Point, Noxious Bluff. . . .

The Russians reached the southern end of the South Sandwich Islands chain on January 8. For the next few days, Bellingshausen carefully surveyed every bit of land he saw—hazardous work because of nearly constant fog. Shortly before leaving the area, he could at last see well enough to realize just how dangerous that

THE CONTINENT FOUND: 1819–1821

work had been. His two ships had been sailing "among ice whose proximity we had only discovered by the noise, [and] we were amazed at the great number of icebergs and at our luck in having escaped disaster."[10]

When Bellingshausen finished in the South Sandwich Islands on January 17, he had definitely determined that Cook's discovery was a group of islands, unconnected to any other land. The survey work he had done was so accurate, and the profile sketches by his artists so clear, that the British would use both in their first, 1930, edition of *The Antarctic Pilot*.[11]

The Russians sailed eastward after leaving the South Sandwich Islands. On January 27, 1820, three days before Bransfield saw the Antarctic Peninsula, Bellingshausen almost certainly sighted the coast of East Antarctica, at about 69° S, 2° W. He never claimed to have seen land here, but his description of what he saw corresponds so closely to what the land looks like in that area that modern scholars agree that he must in fact have done so. Bellingshausen spent the winter in Port Jackson (now Sydney), Australia, and while there, received sketchy reports of Smith's South Shetlands discovery. He resumed his Antarctic voyage the following summer, continuing eastward around the unseen Antarctic Continent to his south. On January 20, 1821, Bellingshausen discovered a small landmass that he named Peter I Island. It lay several hundred miles to the west of the Antarctic Peninsula, in what is now called the Bellingshausen Sea. At about 68° 50' S, it was the first land anyone had reported seeing south of the Antarctic Circle.

Bellingshausen made his next great discovery a week later. On January 27, shortly after he neared the southern reach of the Antarctic Peninsula's west coast, he saw an ice-covered landmass clearly against a cloudless cobalt sky. He named

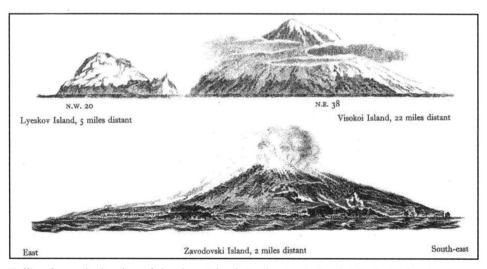

Bellingshausen's sketches of the three islands in the South Sandwich chain that he discovered. *(Illustration from reproduction in Bellingshausen, ed. Debenham,* The Voyage of Captain Bellingshausen, *1945)*

this land—also well south of the Antarctic Circle—Alexander I Land, a "land," he wrote, because "its southern extent disappeared beyond the range of our vision."[12] This discovery remained on the maps as a possible part of the Antarctic Continent for more than 100 years. That it is actually a massive island (today called Alexander Island) was only conclusively determined in 1940.

After this discovery, Bellingshausen turned north to escape the ice. When he crossed the Antarctic Circle on January 31, he concluded two full weeks spent south of the Circle, during which time he had sailed across 28° of longitude. It was an unprecedented achievement. Bellingshausen then shifted to a northeasterly course and headed for the South Shetlands. He arrived three days later, joining dozens of other vessels already there, sealing ships that had been in the area for months. He would soon find himself in conversation with the captain of one of these vessels.

The Great Seal Rush in the South Shetlands

The hundreds of men Bellingshausen encountered in the South Shetlands in the summer of 1820–21 had rushed there from Great Britain and the United States when word of Smith's discovery reached their homelands. Most found the climate and ice-covered landscape an unpleasant surprise. These men had come because the seals were here, not because they had any burning desire to explore or experience the Antarctic, and the enchantment that so many would later feel escaped them. Icebergs were a hazard, not a marvel to wonder at, and the glowing golden edges of ice floes were inadequate compensation for the grave danger they posed to frail wooden ships. One British sealer wrote home describing the South Shetlands as a "detestable place . . . , *detestable* [sealer's emphases], I say, because I am certain it was the last place that ever God Almighty made . . . as the snow never quits this place, even now in midsummer. . . ."[13] Seals, however, covered the beaches, and the men were soon knee-deep in blood and blubber.

The first ships to arrive found so many animals that Robert Fildes, a British captain, reported that "it was impossible to haul up a boat without first killing your way . . . and it was useless to try to walk through them if you had not a club in your hand to clear the way, and then twas better to go two or three together to avoid being run over by them. . . ."[14] Soon, however, there were more ships and sealing gangs than there were seal beaches. On several occasions, the resultant intense competition for seals led to violent clashes between armed work gangs.

The fierce competition also led to shipwrecks, because the sealers could not afford to remain offshore in bad weather or limit their efforts to the easily accessible rookeries. At least seven ships, a large proportion of the several dozen there, were wrecked during the 1820–21 season. But despite the rivalries that contributed to these accidents, everyone appreciated the need for mutual support. Fellow sealers came to the rescue on each occasion.

THE CONTINENT FOUND: 1819–1821

Shipwrecks or storms sometimes stranded men on shore for days. Though most were soon rescued, one group—the chief officer and ten men from the British vessel *Lord Melville*—was not. These men found themselves marooned on King George Island when their ship, driven off in a storm, failed to return. Since no one else realized they were there, they were stranded when the fleet sailed away at the end of summer. Dependent on whatever they could glean from the land, their food and fuel came from penguins and seals, and their winter shelter probably incorporated an overturned boat, local stones, and whatever materials they had landed with. All survived until the next summer, but it cannot have been pleasant, this first wintering below 60° S. James Weddell, a fellow sealer, commented, "notwithstanding every precaution they could take . . . they suffered severely. . . ."[15]

But it was the fur seals who really suffered. The hunters slaughtered them indiscriminately in their rush to beat the competition. They killed more than 300,000 adults in the 1820–21 and 1821–22 seasons. And at least 100,000 pups died along with their mothers.[16] When Bellingshausen arrived in the South Shetlands and heard reports of the carnage, he commented perceptively, ". . . there could be no doubt that round the South Shetland Islands just as at South Georgia . . . the number of these sea animals will rapidly decrease."[17] His judgment was painfully accurate. Indeed, the decline was already under way. In a letter to the editor of the *New Haven Journal* dated February 18, 1821, Daniel W. Clark, first mate of the *Hersilia*, wrote, "We are now loaded with fur skins, having taken upwards of 18,000. . . . As for getting another cargo in these islands, it is utterly impossible—for there is scarcely a seal left alive."[18]

NATHANIEL PALMER SIGHTS THE ANTARCTIC PENINSULA AND MEETS BELLINGSHAUSEN

The rapid devastation of the South Shetlands seal population and the competition among the sealers soon led to searches for new hunting grounds in the islands. Nathaniel Palmer, the *Hersilia*'s second mate the previous summer, was now the 21-year-old captain of the *Hero*, a 47-foot-long, 40-ton scout ship for Edmund Fanning's Stonington fleet. When Fanning's five ships reached the South Shetlands in early November, their overall commander, Benjamin Pendleton, quickly realized that he faced serious competition. And so, on November 14, he sent Palmer on an exploratory cruise in the *Hero* to locate sealing beaches the Stonington fleet could call its own.[19]

Palmer spent his first full day riding out a blizzard at Deception Island, near the west end of the South Shetland chain. Then, on November 16, 1820, he headed south across the Bransfield Strait, the channel separating the South Shetlands from the Antarctic Peninsula. The day was clear, and Palmer could see distant mountains, mountains that were almost certainly on the Antarctic Peninsula mainland.

The young sealer, however, was unaware that the land he saw was continental, nor does he seem to have been particularly interested in the question. To Palmer, discoveries were important only if they were home to seals. Others, however, ignorant of Bransfield's and Bellingshausen's sightings nearly ten months earlier, would cite the events of this day to claim that America's Nathaniel Palmer had been the first to see the Antarctic Continent.

John Davis, another American sealer, definitely saw the Antarctic Peninsula a few weeks later. Davis, too, was searching for unexploited seal beaches when he sailed southward from the South Shetlands in late January 1821. He ventured a few miles farther than Palmer had, into waters that no one else had reported exploring. Here Davis saw new land. Unlike Palmer, he commented specifically on it, noting in his logbook that he thought what he saw was part of a continent. And he did more. On February 7, 1821, Davis landed a gang of men on a beach at what today is known as Hughes Bay. This is the earliest clearly documented landing on the Antarctic Continent. Davis's men, however, did not linger. There were no seals in sight.

At nearly the same time that Davis made his historic landing, Palmer was back at Deception Island. Shortly after midnight on a misty February 4, 1821, he routinely rang the ship's bell. To his surprise, another bell answered. It happened again an hour later. Then, in the morning, the *Hero*'s mate heard voices. The mystery was solved when the fog lifted and Palmer's men saw two large warships. They were Bellingshausen's vessels, just arrived in the South Shetlands. After both Bellingshausen and Palmer showed their colors, the Russian sent out a boat and invited Palmer aboard the *Vostok*. There the two men talked, with Mikhail Lazarev, the *Mirnyy*'s captain, acting as translator. The Russian and American accounts agree to this point, but their versions of the two captains' conversation differ substantially.

In 1833, Edmund Fanning wrote of what he claimed Palmer had told him about his conversation with Bellingshausen. Fanning's version ran,

> To the commodore's interrogatory if he had any knowledge of those islands then in sight . . . [Palmer] replied . . . that they were the South Shetlands, at the same time making a tender of his services to pilot the ships into a good harbor at Deception Island. . . . The commodore thanked him kindly, 'but previous to our being enveloped in the fog,' said he, 'we had sight of those islands, and concluded we had made a discovery, but behold, when the fog lifts, to my great surprise, here is an American vessel . . . ; we must surrender the palm to you Americans. . . .' His astonishment was yet more increased, when Captain Palmer informed him of the existence of an immense extent of land to the south. . . . [The] commodore was so forcibly struck . . . that he named the coast then to the south, Palmer's Land. . . .[20]

Bellingshausen's version is far less dramatic and correspondingly more plausible:

> Mr. Palmer . . . informed us that he had been here for four [sic] months' sealing

in partnership with three [sic] American ships. They were engaged in killing and skinning seals, whose numbers were perceptibly diminishing. There were as many as eighteen vessels about at various points, and not infrequently differences arose amongst the sealers. . . . Mr. Palmer soon returned to his ship, and we proceeded along the shore.[21]

From the Russian's perspective, it was simply a meeting with a sealer who volunteered information about the South Shetlands, a discovery he already knew about.

Whatever the truth of Palmer's meeting with Bellingshausen, William Woodbridge of Hartford, Connecticut, published an atlas in September 1821 that included a map showing land designated Palmer's Land to the south of the known islands. And decades later, when the Antarctic Peninsula first appeared on maps, American charts would label it the Palmer Peninsula.

Bellingshausen began his own exploration of the South Shetlands after meeting with Palmer. On February 6, a small group of Russians went ashore on one of the islands. The men returned with geological specimens, bits of vegetation, three seals, and several penguins. Unfortunately, the seals fought each other once aboard ship, and the men had to kill two to protect their skins for specimens. They spared the third for the artist to sketch, as well as in hopes of taking it home alive. The sketch made it to Russia; the seal did not. Bellingshausen had no better luck transplanting the penguins. He commented wryly, "Apparently a bread and meat diet does not suit these birds."[22] Or perhaps it was the fact that a warmer climate suited the penguins no better than the cold of the south had his now-dead kangaroo, picked up in Australia, which he had allowed to hop about loose on the deck of the *Vostok*. On February 9, concerned for the health of his crew and his now-leaking ships, Bellingshausen set his course for home.

Bellingshausen's voyage, begun and ended in the Peninsula Region, was a worthy successor to Cook's. Like his British predecessor, the Russian had sailed the Antarctic with great skill and care, always carefully noting what he saw. During the course of his voyage, history's second high-latitude Antarctic circumnavigation, he had discovered new land in the South Sandwich Island group; almost certainly seen the coast of East Antarctica; spent far more time than Cook had south of the Antarctic Circle as well as discovering the first lands seen in those latitudes; and returned with valuable observations on high-latitude ice and weather conditions. Even so, his massive voyage account, which included a magnificent atlas containing the results of his detailed surveys, did not reach the non-Russian-speaking world until an abridged version appeared in German in 1902. The British Admiralty did note the discoveries and eventually included limited extracts from Bellingshausen in its publications, but the full narrative was not available in English until 1945. The unfortunate result of these decades of delay was that Bellingshausen's hugely successful voyage had only a limited impact on the evolution of Antarctic exploration.

Within the space of less than a year, from the end of January 1820 to November of the same year, those aboard ships led by three men—Bellingshausen, Bransfield, and Palmer—had almost certainly sighted the Antarctic Continent. (Others would definitely see it in early 1821.) Although neither Bransfield nor Palmer claimed to have seen a continent, and Bellingshausen did not even claim to have seen land in January 1820, others would proclaim continental sightings on behalf of all three men, assertions leading to squabbles, persisting for years, over who was first. The Americans, Fanning in particular, began beating the drums for Palmer almost immediately, and for decades he was the only claimant. The British finally began to champion Bransfield after Bruce rediscovered his chart in 1917. More than 30 years later, in 1949, the Russians at last entered the debate, declaring Bellingshausen the hero after concluding that his primacy might be a useful basis for an Antarctic claim. Today, all of this has settled down, with modern Antarctic scholars generally giving Bellingshausen the credit for his sighting three days before Bransfield and nearly ten months before Palmer.

There is, however, one other man to consider—William Smith, the British merchant captain who had discovered the South Shetlands in 1819. As Otto Nordenskjöld, the Swedish leader of a 1901–03 exploring expedition to the Antarctic Peninsula, wrote in 1904,

> Even if we allow that some groups of ocean-islands of a purely Antarctic nature were already known; even if it be possible that both Sheffield and Bellingshausen would shortly, and independently of Smith, have discovered the same district . . . and if some day it should be proved that earlier navigators had already seen these tracts, or that an American sealer had visited the place at an earlier date without making the fact known . . . still, it is undeniable that [Smith] was the first who, in a most indisputable manner, made acquaintance with a part of the Antarctic Continent. It is, of course, true that the South Shetlands are merely a group of islands, but it is a group so intimately connected with the neighbouring continent, whose tops are visible in certain places in clear weather . . . that the first sealer who devoted a few weeks to fishing around these islands must, of a necessity, have discovered the mainland too. Without at all desiring to depreciate the value of the observations made in these regions during the years which followed, I wish to express the opinion that none of them can, in the slightest degree, be compared in importance with William Smith's discovery.[23]

Chapter 3

The Sealers' Age of Discovery: 1821–1839

Nonsealers—Smith, Bransfield, and Bellingshausen—made most of the important Antarctic discoveries from February 1819 through the summer of 1820–21. That changed the following summer. From 1821–22 until the late 1830s, sealers, searching for new hunting grounds, were responsible for every significant addition to the Antarctic map.

George Powell and Nathaniel Palmer Discover the South Orkneys

The success of the 1820–21 season in the South Shetlands excited shipowners and their captains alike, and an even more frenzied seal rush ensued. Roughly a hundred ships, far more than had been seen the previous summer, reached the islands in the summer of 1821–22. These second-season fortune hunters, however, found the seal population already severely depleted. Despite this, a few of the more enterprising captains managed to make a modest profit from the season. One was George Powell, a 25-year-old Englishman who had returned for a second summer in the islands.

Shortly after Powell reached the South Shetlands in early November 1821, he used his scout ship, the 59-ton *Dove,* to land a sealing gang on Elephant Island. Powell came back at the end of the month to find that his men had taken only 150 fur seals. Nathaniel Palmer, also back for another season, was there at the same time. This year he was captain of the *James Monroe,* a new scout ship for the Stonington fleet. His shore gang had had no better success than Powell's had. The Englishman, who had decided to look for new seal beaches farther east, suggested to Palmer that they search together, because it would be safer with two ships. Palmer agreed.

On December 6, 1821, the effort paid off, at least in terms of discovery. That dawn, Powell's lookout sighted land more than 200 miles to the east of the South Shetlands. It was the South Orkneys, a group of four small mountainous, ice-covered islands centered at 60° 40' S, 45° 15' W. As Powell and Palmer sailed toward them, their ships had to thread their way through a crowded maze of icebergs. This would be a common experience in these islands in later years. Many would find them difficult to reach, guarded as they so often are by ice. The first men to see the South Orkneys, however, soon found their way to the shore of the largest island.

There Powell, a man who appreciated discovery for its own sake, landed and claimed the entire archipelago for Great Britain. He named his landing place Coronation Island in honor of Britain's newly crowned King George IV. There is no evidence that the American Palmer objected to the name or to Powell's claim for America's rival, nor did he send anyone from his own ship ashore. The beach held no seals, and that was all the youthful Palmer cared about.

After the brief ceremony ashore, both captains resumed their search for cargo. Their efforts led to the discovery of several more islands, but very few seals. Five days later, they gave up and returned to the South Shetlands. There Powell and Palmer rejoined their separate fleets and returned to sealing. Powell, however, still had more on his mind than simple hunting. In addition to collecting sealskins, he assembled information for what would become the first reasonably accurate chart of the area. When he left the south at the end of February with 4,440 skins, Powell had concluded the most commercially successful British voyage for the season, but his most important results were the discovery of the South Orkneys, his narrative of the voyage, and the chart. By November 1822, R. H. Laurie, a London publisher, was selling Powell's chart. It showed the South Orkneys as part of the South Shetlands with the name Powell's Group. That name, possibly bestowed by Laurie, did not last long.

The South Orkneys had been ripe for discovery in the summer of 1821–22. Only six days after Powell and Palmer made landfall there, Michael McLeod, a

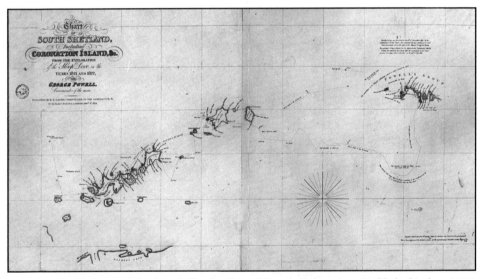

George Powell's chart of the South Shetlands and South Orkneys — *Published in late 1822. Note the name "Powell's Group" for the South Orkneys (on the far right) and the indication of land to the south of the South Shetlands, named here as "Palmer Land." The large islands in the center, separate from the main body of the South Shetlands, are from left to right, Elephant and Clarence Islands. (From reproduction in Hobbs,* The Discoveries of Antarctica . . . , *1939. Originally published by R. H. Laurie, 1822)*

sealing captain in command of James Weddell's scout ship, sighted the islands from about 60 miles away. McLeod believed he had made an exciting new discovery and reported it as such to Weddell when he returned to the South Shetlands. Weddell himself visited the South Orkneys in February 1822 and gave them the name they have today.

JAMES WEDDELL BESTS COOK'S FARTHEST SOUTH RECORD

Weddell, like Powell, was one of the few sealing captains who made money in the 1821–22 season, and his small profit was just enough to convince him and his two partners in London to try again the next summer. Since most other shipowners had given up, Weddell's voyage was one of a very few to the Antarctic regions in 1822–23.

Weddell, an adopted Scotsman, was an expert navigator who had greatly impressed his superiors during his years rising from simple seaman to the rank of master in Britain's Royal Navy. In 1818, he had voluntarily resigned to become a sealer. Thirty-five years old by late 1822, he was one of the most experienced captains in the South Shetlands. He was also better educated than most sealers and deeply interested in exploration and discovery—an interest reflected in his nautical instruments. In addition to the usual compasses and sextant, he carried barometers, thermometers, and three chronometers. The chronometers alone had cost £240, a major investment for the day.

In mid-September 1822, Weddell left England with the same ships he had commanded the previous season. These were the 160-ton *Jane,* with a complement of 23 men under Weddell himself, and the 65-ton *Beaufoy,* with 14 men under a new captain, Mathew Brisbane. Time lost en route repairing a leak in the *Jane* delayed Weddell so severely that it was December 30 by the time he was ready to leave South America. This was far too late, he reasoned, to try his luck in the South Shetlands. Seals there were scarce, and earlier arrivals would have claimed the few remaining good beaches. Instead, Weddell decided, his best hope was to begin his season with a search for new sealing grounds. He would start in the South Orkneys.

Weddell reached the South Orkneys on January 12, 1823. Three days later, he spotted six of what the sealers called "sea leopards" on shore and sent his men to take them. Their fur was nearly worthless, but Weddell, curious about these animals, saved a few skins and skulls to take home for trained scientists to examine. (Several years later, one of these scientists would name this seal species the Weddell seal.) The search for fur seals was more disappointing: it produced only a few scattered animals. Still, there were those few. Perhaps, Weddell hoped, they were stragglers from some nearby, more populated beach. He thus decided to investigate when some of his men reported that they had spotted a range of mountains to the southeast.

On January 23, Weddell sailed away from the South Orkneys. The supposed

James Weddell's "drawing from nature" of a Weddell seal — *The animal that the sealers, including Weddell, called a "sea leopard" because of its mottled, leopard-like spots. This name sometimes led to confusion in the early literature with leopard seals, an entirely different species of seal. (From Weddell,* A Voyage Towards the South Pole . . . , *1825)*

mountains proved to be only a line of huge icebergs, but Weddell chose to continue looking, and promised a reward of £10 to the first man who sighted genuine land. This substantial offer, Weddell wrote, was to be

> the cause of many a sore disappointment; for many of the seamen, of lively and sanguine imaginations, were never at a loss for an island. In short, fog banks out of number were reported for land; and many, in fact, had so much that appearance, that nothing short of standing towards them till they vanished could satisfy us as to their real nature.[1]

Weddell spent nearly two weeks zigzagging south, north, and south again, searching for land in areas near previous vessels' tracks. Then, on February 4, he changed tactics and headed due south into unexplored waters, into the heart of what we now call the Weddell Sea. This huge embayment in the Antarctic Continent, to the east of the Antarctic Peninsula and south of the Atlantic Ocean, mirrors the Ross Sea on the opposite side of the continent. Unlike the Antarctic Peninsula and the South Shetlands, which are the most visited places in Antarctica, the adjacent Weddell Sea is one of the least traveled to, because it is so often clogged with dense pack ice. James Weddell, however, knew none of this, and he was about to enjoy a rare year in which the sea that would be named for him would open its doors.

Long days of thick fog, biting wind, and freezing snow tormented the crews as they stood watch on the icy decks of the *Jane* and the *Beaufoy*. Weddell had the cooking stoves moved below to warm the men's quarters and dry their frozen clothes. That helped, but he could do nothing about the fact that every excited cry of "land ho" ended in disappointment—a disappointment the men felt personally because their wages were a proportion of the value of the catch. New land meant the possibility of new seal beaches; no land meant no pay. Not only were the conditions

THE SEALERS' AGE OF DISCOVERY: 1821–1839

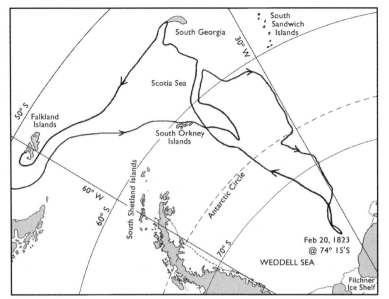

James Weddell's track on his 1822–23 sealing voyage, including his route deep into the Weddell Sea to a new record south.

brutal, but pushing south through the ice of the Weddell Sea was also dangerous. On one occasion, the ships narrowly escaped a collision with an iceberg in the fog, and glancing blows from ice floes were routine. On February 10, at about 66° S, 32° 30' W, an iceberg studded with rocks and dirt generated another excited report of land. But this too was a false alarm, the bitterest disappointment so far.

A week later, icebergs nearly blocked the ships' way south. Then conditions changed completely. When Weddell crossed the 70th parallel, the sea was almost ice-free. He reached 71° 34' S on February 17. He had passed James Cook's southernmost latitude and now held the record.

The next evening, Weddell wrote, "was lovely and serene, and had it not been for the reflection that probably we should have obstacles to contend with in our passage northward, through the ice, our situation might have been envied." Only the ice to the north worried him now because, amazingly, where they were "*not a particle of ice of any description was to be seen* [Weddell's emphasis]."[2]

February 20 was Weddell's last day sailing south. His noon sight put the ships at 74° 15' S (at 34° 16' W), a new farthest south by over 200 miles. It was another magnificent clear day, this time with no ice in sight except for three tabular icebergs. Achieving a new record south was hugely exciting for the explorer in Weddell, and he was sorely tempted to go on. But Weddell was also a sealer, on a voyage on which he hoped to make a profit. Because he, like Cook, incorrectly believed that the ice in the sea formed only on land, he took the ice-free water about him as evidence that there was no significant land to the south. And that, in turn, meant no possibility of seal beaches. Moreover, he was running short of fuel and provisions. It was time to turn back.

45

The *Jane* and the *Beaufoy* "*in latitude 68° South, passing to the Southwards through a chain of Ice Islands. Feb^r 1823. . . . From a Sketch by Cap. Weddell.*" (Weddell, A Voyage Towards the South Pole . . . , 1825, caption and text on plate facing p. 34)

But before he began the return sail, Weddell raised a flag and fired a salute to celebrate the new southern record and his discovery of the body of water that he named the George IV Sea. (Seventy-five years later, the name would be changed to honor Weddell, at the suggestion of Karl Fricker, a German geographer.) Weddell wrote, "These indulgences, with an allowance of grog, dispelled [my men's] gloom, and infused a hope that fortune might yet be favourable."[3] Although the grog was probably more important than the waving flag for most of the men, perhaps a few appreciated the day's historic importance.

It was not long before Weddell met ice on the northward sail, and his splendid weather soon deserted him as well. By the time he reached South Georgia on March 12, tense weeks of sailing through treacherous ice-strewn waters devoid of land had given Weddell a perspective quite different from that of Cook or Bellingshausen. He wrote, "Notwithstanding the forbidding appearance of this land, every one, I believe, in the two vessels, feasted his eyes upon [the island]. . . ."[4]

Weddell spent more than a month anchored at South Georgia's Undine Harbor refreshing his ships and men. There he found fresh water, greens, and birds for the pot, all most welcome after so long at sea. Weddell also took time for scientific observations of the local fauna. The king penguins, in particular, enchanted him. He wrote of them, "In pride, these birds are not surpassed even by the peacock, to which in beauty of plumage they are indeed little inferior. . . . [Their] frequently looking down their front and sides in order to contemplate the perfection of their exterior brilliancy, and to remove any speck which might sully it, is truly amusing to the observer."[5] Weddell's marvelous wildlife descriptions are so good that the ornithologist Robert Cushman Murphy, who visited South Georgia in 1912–13, wrote, ". . . nothing of my own observations would lead me to change a line of Weddell's [description]."[6]

After wintering in the Falklands, Weddell returned south in 1823–24, this time to the South Shetlands, in hopes of salvaging some profit for the voyage. The *Jane* and the *Beaufoy* were two of the very few ships there that summer, a season plagued with unusually heavy ice. Weddell spent a largely futile month fighting the ice and weather before admitting defeat and taking his battered ships and discouraged men home.

The two ships reached England in July 1824. Weddell, anticipating considerable skepticism over his report that he, a sealer, had reached three degrees farther south than the legendary Cook, asked his chief officer and two seamen to swear under oath that his logbooks were true. They readily did, and that, along with his carefully detailed records, convinced the public. At the time, however, no one fully understood how amazing Weddell's achievement was. It was more than a matter of going beyond Cook. He had set his record in the central Weddell Sea. No one would match his latitude anywhere in this body of water for 80 years. To achieve what he had, Weddell had to have encountered an extraordinarily light ice year.

Weddell's friends persuaded him to write a book about the voyage, to add credence to the farthest south record. *A Voyage Towards the South Pole . . .* appeared in 1825. It was an important book, not only as a voyage narrative, but also because Weddell included insightful analyses of Antarctic discovery throughout the Peninsula Region. He also published his charts and sketches, in effect sailing directions for where he had been, and used the pulpit his book offered to argue for a conservation approach to exploitation of new sealing grounds. Weddell, a highly unusual sealer in so many ways, closed the second (1827) edition of his book with these words: ". . . if I have contributed, by my private adventure, to the advancement of hydrography, I conceive that I have only done that which every man would endeavour to accomplish, who, in the pursuit of wealth, is at the same time zealous enough in the cause of science to lose no opportunity of collecting information for the benefit of mankind."[7] He was probably optimistic about "every man," and especially about his fellow sealers.

Weddell may have had company in the Weddell Sea in early 1823. Benjamin Morrell, an American sealer, also claimed to have reached a high latitude there. Morrell and his ship, the *Wasp,* came to the Weddell Sea after what the sealer described as a lengthy single-season voyage. In order, he said, he had visited South Georgia, Bouvet Island, and the Iles Kerguelen in the far south of the Indian Ocean, sailed for several thousand miles at a very high latitude along the unseen coast of East Antarctica, and spent a few days in the South Sandwich Islands. Morrell wrote that he left the latter on March 6 and headed south into the Weddell Sea, reaching

70° 14' S, 40° 3' W, before he turned back. He also said that he had sighted land at about 67° 41' S, 47° W, and sent the *Wasp*'s boats after seals he saw on shore. He had reached, he said, the north cape of this landmass—which he called New South Greenland—at 62° 41' S, 47° W. In fact, no land exists anywhere near his stated positions.

History has accepted Weddell's account. Morrell's is another matter, because some of his claims for the voyage are simply impossible. As a result, a number of his contemporaries and many later historians have dismissed him entirely. Several went so far as to call him an out-and-out liar. That is probably unfair. Some elements of Morrell's voyage are quite possible, and he was in fact a highly successful sealer and a skilled ship's captain. True or not, Morrell's ghostwritten 1832 book, *Narrative of Four Voyages . . .* , sold well.[8]

Antarctic fur sealing was already in serious decline when Weddell and Morrell made their 1822–23 voyages. The two previous summers had seen a human invasion around the South Shetlands—several thousand men and perhaps as many as 150 ships. But this was an ephemeral invasion, one by men who were pillagers, not settlers. They wiped out entire seal populations on island after island, and when there was nothing left to take, all but a few hopeful stragglers departed. Most of the years from 1822 through roughly 1900 would see a maximum of four or five expeditions, sealing or otherwise, in the Peninsula Region, and many years would see none at all. By the late 1820s, the fur seal destruction in the South Shetlands was so complete that when H.M.S. *Chanticleer* was in the islands at the end of the decade, her men saw not a single fur seal.

THE *CHANTICLEER*: A SCIENTIFIC EXPEDITION AT DECEPTION ISLAND

The *Chanticleer*, a British naval vessel, spent two months in early 1829 at Deception Island as part of a wide-ranging scientific expedition. The voyage's principal objective was to complete investigations begun in 1822 to determine the shape and size of the Earth using a series of pendulum experiments. Thirty-two-year-old Henry Foster, a fellow of Britain's Royal Society and one of the leading scientific officers of the Royal Navy, commanded the effort. He was the first true scientist to visit the Antarctic Peninsula and South Shetlands. William H. B. Webster, the ship's surgeon, doubled as expedition naturalist.

Foster sailed south across the Drake Passage at the end of December 1828 and reached the South Shetlands in early January. After several days spent exploring the far northern coast of the Antarctic Peninsula, he sighted Deception Island on January 9. He knew that the sealers had used its sheltered harbor, and it seemed well located for his observations. Did it also offer a safe anchorage where his ship could remain for several months while the men worked on shore? He decided to investigate. The *Chanticleer* wove her way toward the island through a maze of icebergs.

By noon, Foster was close enough for his men to row a small boat into the island's central harbor. After passing through the dramatic, narrow cleft in the island walls we now call Neptunes Bellows, Foster and his companions found themselves in a remarkable place. It was the sea-filled crater of a volcano. The way they had entered, through a small bite out of one side of the doughnut-shaped island, affords the only access to its splendidly sheltered interior.

Outside Deception Island, it was a stunningly beautiful day, and those left behind on the *Chanticleer* spent the afternoon fascinated by the drifting icebergs. Their reaction, that of a scientific party there to study and learn, was in marked contrast to the way so many sealers had responded to the ice. Even a small iceberg that brushed the ship was enchanting. It was, William Webster wrote, "of a beautiful cerulean colour, perfectly translucent, with veins of an elegant verditer. In fact the whole was splendid and magnificent, and its variegated colours afforded us a treat which it was worth coming even to South Shetland to witness."[9]

Foster was equally pleased with what he saw inside Deception Island's walls. He returned to the ship in the evening and reported that he had found an ideal anchorage. Two days later, with the *Chanticleer* safely moored in what he eventually named Pendulum Cove, Foster landed his instruments and began his pendulum observations.

The men of the *Chanticleer* were there for science, but they posed just as much a danger to the local residents as the hunters had. All seals, not just the entirely missing fur seal, were scarce, but the men found a huge population of penguins. They killed thousands, eating some immediately and salting down the rest for the future.

Webster and the crew also spent time exploring the island. One result was the first study of Deception Island's geology. This was a place, Webster wrote, that "consists of one mass of black volcanic ashes and sand. . . . The whole island . . . betrays proof of its having been ejected from beneath the surface of the globe; and the various materials, as might be expected from such a convulsion of nature, are scattered about in all parts of the island in the utmost confusion. . . . huge masses of cinders and ashes . . . which imagination converts into the refuse of Vulcan's forge." Although no one saw any sign of an active crater, Deception Island was still a place where "volumes of smoke and steam are rushing from the peaks of snow-clad hills, while prodigious masses of ice and snow are standing on the verge of boiling springs." Webster correctly concluded that "the subterranean fire is merely abated and not extinguished."[10] Deception Island was a dormant volcano, not an extinct one.

Foster deposited minimum and maximum thermometers at Pendulum Cove before he left Deception Island in early March. He attached a note requesting that any finder record the results and send them and the instruments to the British Admiralty. Then it was on north for the rest of the voyage, all of it to be pursued outside the polar regions.

The *Chanticleer* anchored at Pendulum Cove — *as drawn by Lt. Kendall, expedition artist and survey officer. (From Webster,* Narrative of a Voyage . . . , *1834)*

Although the *Chanticleer* expedition's time in the Peninsula Region was short, it was significant, because Foster and his men had carried out the first extended scientific land study south of the Antarctic Convergence. The effort is perhaps best remembered today, however, for the names it left on the map. Not only does the title Pendulum Cove survive, but Deception Island's enclosed harbor is now known as Port Foster.

AN AMERICAN SEALING / EXPLORING / SCIENTIFIC EXPEDITION

The *Chanticleer* was the only ship in the South Shetlands in early 1829. A year later, another expedition was alone in the area. This 1829–30 voyage of three American ships, the *Seraph, Annawan,* and *Penguin,* was a combined sealing and scientific/exploration venture, one that brought two key men from the heyday of South Shetlands sealing back south. Benjamin Pendleton, who had been in charge of Edmund Fanning's Stonington fleets in 1820–21 and 1821–22, was captain of the *Seraph* as well as overall expedition leader. Nathaniel Palmer commanded the *Annawan.* His brother Alexander was captain of the *Penguin.*

Pendleton's small fleet sailed on the strength of politics, commercial hopes, and scientific objectives. The political impetus had come from American whalers, sealers, and merchants who had been petitioning the government for years to send naval ships to the Pacific to gather navigational information. Sealers had provided the commercial motive; they wanted a government expedition sent south to discover new sealing grounds. The scientific stimulus had its genesis in what modern knowledge would judge a crackpot theory. In the early 1820s, John Cleve Symmes began promoting an expedition to the polar regions to investigate his theory of a hollow Earth with holes in the poles. Pressure on the government increased significantly when Jeremiah Reynolds, a very persuasive speaker and expert publicist, agreed to help Symmes. Reynolds, who in fact had reservations about the hollow Earth theory from the beginning, broke with Symmes in 1827. By then, however, he was firmly committed to promoting an American Antarctic exploratory and scientific effort.

During much of 1828, it looked as if Reynolds would get his expedition. Planning and organizing were moving along, appointments were made, a ship was selected, and a bill was introduced in Congress. Then things fell apart. Andrew Jackson was elected president in November, and he, unlike his predecessor, John Quincy Adams, felt that the government should concentrate its resources on exploration of the United States itself. Jackson vetoed the expedition funding as soon as he was inaugurated president in March 1829.

Although there would be no government expedition, the idea for an exploring effort survived, because the work done to lobby the government had generated so much public interest. Edmund Fanning took the lead in organizing a private replacement. It was to be an ambitious two-ship program using the *Seraph* and *Annawan,* vessels owned respectively by Pendleton and Nathaniel Palmer. Alexander Palmer's *Penguin,* technically on a separate sealing voyage, would work with the two expedition ships. The plan was that Pendleton and Nathaniel Palmer would spend a brief time sealing in the South Shetlands to cover expenses. They would then do the real work of the expedition—exploring and science.

The New York Lyceum for Natural History sponsored the scientific program and provided a $500 research grant to 31-year-old James Eights, a scientifically inclined physician. Private individuals lent the expedition books, charts, and instruments. Other than Eights, the scientific team consisted of the untrained Reynolds and three equally amateur assistants.

The *Seraph* and *Annawan,* the latter carrying the five-man scientific group, left New York separately in August 1829. Pendleton and Nathaniel Palmer planned to rendezvous at Staten Island (off Tierra del Fuego)—where Alexander Palmer was already at work sealing—before heading for the South Shetlands together. Unfortunately, the *Seraph* was late. The *Annawan* and *Penguin,* under the two brothers Palmer, waited for her for over a week, well past the agreed date, and then went on.

When the Palmers reached the South Shetlands on January 20, their crews immediately began hunting seals. A few men, including Reynolds, landed on Elephant Island on the 22nd. Temporarily marooned there until the 26th because of bad weather and drifting ice, they spent the nights, as sealers often did, under overturned boats, using seal blubber for fuel. They would not be the last to live on the island this way. Nearly 90 years later, a much larger group would shelter on Elephant Island under their boats for more than four months.

While the crews hunted, Eights commenced his scientific work. He found the South Shetlands both fascinating and forbidding, writing, "Although many of the scenes about these islands are highly exciting, the effect produced on the mind, by their general aspect, is cold and cheerless . . . , for on their lonely shores the voice of man is seldom heard; the only indication of his ever having trod the soil, is the solitary grave of some poor seaman near the beach. . . ."[11] He did, however, find other sorts of life to investigate. The most significant were the first known specimen of a flowering plant from south of 60° S and a ten-legged sea spider of a type never seen before. And to his surprised delight, Eights also discovered bits of fossil wood.

At one point while sailing about the islands, Eights noticed large rocks embedded in icebergs. This, he thought, must be the explanation for the granite boulders, so different from the local rocks, that he had seen on the South Shetland beaches. The boulders, he concluded, had been "brought . . . by the icebergs from their parent hills on some far more southern land. . . ."[12] This was an insightful statement. Not only did Eights see these rocks as evidence of additional land to the south, but he was also proposing an explanation for the existence of erratics, rocks that bear no geological relationship to local formations. At the time, world geologists were still puzzled over the existence of similar boulders in the Northern Hemisphere.

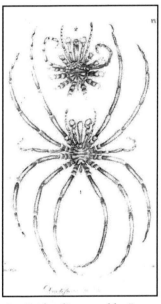

Sea Spider discovered by James Eights in the South Shetlands — *Eights published an account of this discovery in 1837, illustrated with this drawing, but contemporary scientists, especially those in Europe, questioned his finding. Eights's report was not fully credited until William Speirs Bruce, of the 1902–04 Scotia expedition (see Chapter 8), reported finding a similar spider in the South Orkneys in 1903. (From reproduction in Quam,* Research in the Antarctic, *1971. Originally published, Eights, "Description of a New Animal . . . ," 1837)*

The sealing in the South Shetlands was far less successful than James Eights's efforts, and although the expedition's organizers felt that science and exploration were the venture's most important objectives, the crew had quite a different perspective. One of the legacies of President Jackson's veto was that they were civilians rather than naval personnel. That, in turn, meant they had signed on under the usual sealer terms, their pay to come from a share in the voyage's profits. They were, to put it mildly, unhappy. Nathaniel and Alexander Palmer, both sealers at heart, sympathized. Thus the two Captains Palmer left the South Shetlands on February 22 to cruise west in search of new land with seals. The weather was nasty, the sea rough, icebergs frequent, and they found no land. On March

23, they abandoned the effort and headed north.

As for the *Seraph,* Pendleton had arrived late at Staten Island and found the other ships gone. He continued on to the South Shetlands alone, but because he and Nathaniel Palmer had failed to establish a specific rendezvous point there, locating the *Annawan* and *Penguin* would be a matter of luck. Pendleton spent a month in the islands and never saw another sail. Then, his sealing as fruitless as the Palmers' had been, he too headed west in search of new land. His effort there proved to be equally unproductive, and he quickly gave up. When he reached the coast of Chile in early May, Pendleton at last found the *Annawan* and *Penguin.*

Nearly everyone judged the venture a failure: the expedition had discovered no new land, had found few seals, and the crew had nearly mutinied. The only real accomplishments belonged to James Eights, but his work alone was sufficient to make this a highly significant expedition. He had brought home thirteen cases of specimens full of fossil wood, rocks, lichens, and marine animals, and later reported his results in five published scientific papers. Sadly, his achievements were virtually ignored for many years, but in the twentieth century, Eights at last began to receive the recognition he deserved, both in scientific circles and on the Antarctic map.[13] Today the Eights Coast, a section of the West Antarctic coastline between the Peninsula Region and the Ross Sea, honors him. And a U.S. scientific base that operated on the far southernmost portion of the Antarctic Peninsula plateau for several years in the early 1960s was named Eights Station.

The expedition was important in another, negative, way. It clearly demonstrated the folly of attempting to finance a scientific/exploration venture to the Antarctic through concurrent sealing. Fanning and Reynolds thus renewed their campaign for a government-financed expedition. Ultimately, their efforts would lead to the American national exploring expedition described in the next chapter.

JOHN BISCOE CIRCUMNAVIGATES ANTARCTICA

Although most sealers quit the far south after the mid-1820s, a few shipowners remained optimistic. The South Shetlands had replaced South Georgia. Perhaps there was another South Shetlands to be found somewhere else in the Antarctic. In 1830, Enderby Brothers, a London whaling firm, dispatched a speculative expedition to search for it. The Enderbys were unusual shipowners, as interested in exploration as they were in voyage profits. Thus they explicitly ordered their captain, John Biscoe, to explore as well as hunt in high latitudes.

Biscoe left England at the end of July 1830 with a total complement of 29 men aboard two small ships. He personally commanded the 150-ton, 74-foot-long *Tula.* George Avery was captain of his consort, the 49-ton, 52-foot-long *Lively.* Late in November 1830, Biscoe sailed south from the Falklands to the South Sandwich Islands and began the Antarctic portion of what would be a two-year voyage.

It was December 10 when Biscoe reached the position where his chart showed Leskov, Visokoi, and Zavodovski islands, discovered by Bellingshausen in 1820. The chart proved to be wrong, and it took Biscoe eleven days to locate the islands. He had confirmed Bellingshausen's discovery, but that was the only return for his time spent searching: he had seen no seals of any sort. Biscoe continued battling gales, snow squalls, fog, and the masses of sea ice in the region until near the end of the month, but when he at last reached the main body of the South Sandwich group on December 29, there were again no seals. Biscoe gave up and sailed on to the east and south.

Nearly two months later, in late February 1831, he sighted land along the coast of East Antarctica. This was the first *reported* sighting of the Antarctic Continent outside the Peninsula Region, and Biscoe, who fully appreciated its historic importance, named his find Enderby Land in honor of his employers. His next several months proved horrific. On the return north, a storm separated the two ships. Then most of Biscoe's crew began dropping from scurvy. As one after another died, Biscoe limped north to Tasmania with barely enough men to sail the *Tula*. Avery had an equally desperate voyage in the *Lively* before rejoining Biscoe in the north. The survivors spent the 1831 winter in Hobart recovering. Biscoe recruited local men to replace his lost crew members, and then resumed hunting with his two ships in mid-October 1831, first for seals to the south of New Zealand. After three months with little success, he turned east to begin his voyage home, via Cape Horn with a detour to the South Shetlands en route.

Biscoe intentionally took a far southerly route because he hoped to find land to the west-southwest of the South Shetlands. His gamble paid off when he reached the Peninsula Region. On February 15, 1832, he spotted land at 67° 15' S, 69° 29' W. Biscoe named it Adelaide Island in honor of Britain's queen. More than 200 miles south of the South Shetlands, it was an exciting find, though not quite as exciting as Biscoe supposed. Unaware of Bellingshausen's Peter I Island and Alexander I Land discoveries, he mistakenly thought his Adelaide Island to be "the farthest known land to the southward."[14]

On the 17th, Biscoe sailed north past a row of small islands, now known as the Biscoe Islands. Magnificent mountains loomed behind them to the east. Convinced that these peaks were on something more significant than simple islands, Biscoe named the region Graham Land in honor of Sir James Graham, then Britain's First Lord of the Admiralty. Four days later, he landed on what today is called Anvers Island, the southernmost and largest landmass in the Palmer Archipelago, off the northwest coast of the Antarctic Peninsula. Because he mistakenly took it to be part of the mainland, he used this landing to justify claiming all Graham Land for Britain.

The *Tula* and the *Lively* finally reached the South Shetlands at the end of February. Biscoe then spent nearly a month in a largely futile hunt for seals—fur

THE SEALERS' AGE OF DISCOVERY: 1821–1839

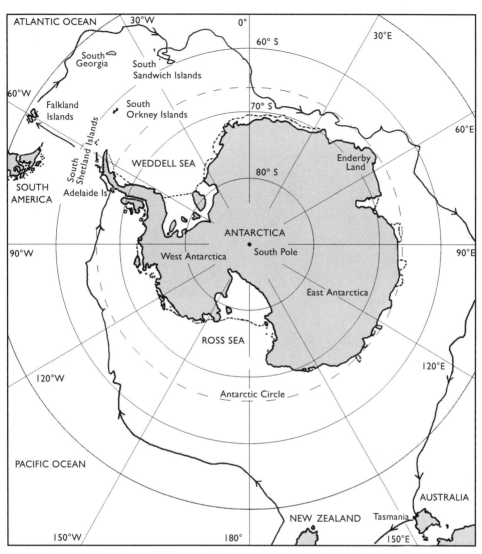

John Biscoe's 1830–32 high-latitude Antarctic circumnavigation — *Biscoe began his voyage in the Falklands, sailing directly from there to the South Sandwich Islands, then heading east toward the coast of East Antarctica. There he became the first to report sighting land along this coast, land that he named Enderby Land for his employers. After spending the 1831 winter in Hobart, he resumed his eastward sail, reaching the Antarctic Peninsula in February 1832. He completed the circumnavigation in the Falklands at the end of April 1832.*

or elephant, by now either would do. He gave up in April. When he reached the Falklands at the end of the month, he had completed the third high-latitude circumnavigation of Antarctica. His problems, however, were not behind him. Not only were there no seals, but several months after arriving, the *Lively* ran aground on a small islet. She was a total wreck. Eventually, Biscoe limped home with his scant cargo in the *Tula,* finally reaching England in February 1833.

Biscoe's voyage was a commercial disaster for the Enderby brothers, but Charles

Enderby, a founding member of Great Britain's Royal Geographical Society, was delighted with his captain's discoveries. Although the most significant was Enderby Land, Biscoe had also done important work in the Peninsula Region: Adelaide Island was a major discovery, and he had named and claimed a portion of the Antarctic Peninsula. The British were soon calling the entire Peninsula Graham Land.

The Enderbys, this time with government support, decided to follow up on Biscoe's work with another expedition in 1833–34. The plan was to start by exploring Graham Land, but the two expedition ships, the *Hopeful* and the *Rose,* met severe pack ice before they even reached the South Shetlands. The *Rose* was crushed and sank, the first ship known to fall victim to the Antarctic pack. After rescuing her crew, the captain of the surviving ship aborted the expedition.

The last major sealing expedition to explore the Antarctic went exclusively to East Antarctica. In February 1839, John Balleny discovered the Balleny Islands, the first land found south of the Antarctic Circle below New Zealand. Balleny's was a watershed voyage, the end of an extraordinary two-decade run of discovery by men going south for commercial reasons. The sealers' age of discovery was over.

Chapter 4

Three Great National Expeditions: 1837–1843

The sealers had left, but science still offered reasons to brave the ice. In 1830, the great German scientist Friedrich Gauss developed a theory that predicted the location of the Earth's magnetic poles. This was more than an academic question, for sailors knew that their compasses pointed to the magnetic rather than geographic poles. Knowledge of the location of the magnetic poles would allow a navigator to correct a compass reading and determine true headings. In 1831, Britain's James Clark Ross had reached and confirmed the location of the north magnetic pole, then on the Boothia Peninsula in the eastern Canadian Arctic. Fully testing Gauss's theory, however, required that someone do the same in the south.

In the 1830s, this scientific motive converged with national pride to inspire national Antarctic expeditions from France, the United States, and Great Britain. Although all three did their most famous work on the opposite side of the continent, each also spent a summer in the Antarctic Peninsula Region. The first to arrive were the French.

Dumont d'Urville Leads the French Into the Ice

Jules Sébastian-César Dumont d'Urville was 47 years old when he left France in 1837 in command of an expedition to the Antarctic and the South Pacific. Educated in the classics as well as the usual maritime subjects, he had concentrated his career in the French navy on survey and exploration. In 1819, he had been a junior officer on a voyage to the Mediterranean when his ship stopped at the Greek island of Melos. There a local peasant offered d'Urville a large statue he had found in his garden. The young officer immediately recognized it as an image of Venus and was eager to buy it. His captain refused, but when the ship reached Constantinople several days later, d'Urville appealed to the French ambassador and persuaded him that France should purchase the peasant's find. The government rewarded d'Urville's role in acquiring what is now called the Venus de Milo with a promotion and membership in the French Légion d'Honneur. Other voyages and promotions followed, including some that took him to the Falklands, Tierra del Fuego, and the South Pacific.

Early in 1837, d'Urville wrote to the French government proposing that they send him on a new exploratory voyage around the world. The centerpiece would be anthropological work in the South Pacific. Although King Louis-Philippe liked the idea, he made his approval subject to d'Urville's adding a voyage to the Antarctic that would attempt to better Weddell's record. Even though d'Urville had no experience with ice navigation, he accepted the challenge, writing, "I identified myself without hesitation with the king's thinking."[1]

As the time for departure neared, d'Urville found himself more and more excited about exploring the Antarctic, and his officers and crew joined him in his enthusiasm when the king, at d'Urville's urging, offered each man a reward of 100 French francs for reaching 75° S and 20 francs more for every degree farther south. This promise of a handsome monetary reward also may have helped the men overcome their early doubts about their commander. D'Urville wrote,

> At the commissioning of the ships, as they saw me walking slowly and heavily owing to a recent attack of gout, they had appeared surprised to learn that I was their commander, and some had even exclaimed naively, 'Oh, that old gaffer won't take us far!' From that moment on I swore that . . . that 'old gaffer' would show them something of the art of navigation such as they had never seen![2]

D'Urville sailed from France in September 1837 with 165 men aboard two men-of-war. Neither ship had received much preparation for the ice, nor had their open gunports been closed against the rough Antarctic seas. D'Urville sailed as captain of the larger of the two, the 380-ton *Astrolabe*. His second-in-command, Charles Jacquinot, was captain of the 300-ton *Zélée*.

After a month spent exploring the Strait of Magellan, d'Urville left Tierra del Fuego on January 9, 1838, and headed south for the Weddell Sea. To his surprise, ice floes began to dot the water as he neared 60° S, much farther north than he had expected to see them. And there were drifting icebergs studding the sea, made more dangerous by unremitting fog. Undaunted, d'Urville pressed on, but shortly after he passed 63° S, the pack, which had been growing more and more dense, completely blocked his way south. Weddell had found the sea almost entirely ice-free near here, but when d'Urville searched for a way through, all he found was more ice, a world that was "Austere and grandiose beyond words . . ." but also "inert, mournful and silent, where everything threatens man with annihilation."[3] He surrendered on January 25 and changed course for the South Orkneys. There, fog, snow, and rough seas frustrated every landing attempt. Nothing was going his way.

When the weather improved on February 2, d'Urville departed the South Orkneys for another attempt to push south. He made good progress for two days, but then the pack loomed ahead once more, this time just past 62° S. But now, he was determined to keep going. D'Urville wrote in his diary,

> I learn that their lordships, the sailors of the *Astrolabe,* just recently more than

fed up with my previous efforts, have suddenly become enthusiastic about the Pole and their only fear is that I may give up too soon. They need not worry. When I give up, none of them, I believe, will have any desire to push on any further [sic]![4]

D'Urville nosed his ships into a promising opening between the ice floes, and the *Astrolabe* and *Zélée* were soon smashing their way through. But then the ships reached a small pool of open water, completely surrounded by ice except where they had entered. D'Urville, who was beginning to question the wisdom of having taken his ill-prepared ships into the ice, tried to retreat. With the winds against him, it was impossible.

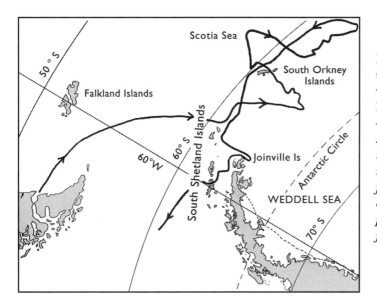

Dumont d'Urville's track during his voyage to the Peninsula Region — *D'Urville headed south in an attempt to better Weddell's record, but his farthest south, reached on his first attempt to penetrate the ice fell just short of 64° S.*

The next morning, the commander decided that he had to do whatever he could to get free of this trap. He began by having the men stand on the ice and literally pull the *Astrolabe* and *Zélée* through the little open water about them. From there, d'Urville was able to sail 3 or 4 miles before meeting more dense ice. Once again, the men went out onto the floes. As bad as things were, they soon grew worse: an iceberg began moving on a collision course with the *Astrolabe*. All the terrified men could do was watch helplessly as disaster bore down on them. But then the situation reversed completely. The iceberg stopped and the pack ice parted ahead of the ships. Unfortunately, d'Urville bowed to his officers' pleas to allow the crew a rest before sailing. He agreed to wait until morning, and by then, ice had again surrounded the ships. It was back to hauling on the ropes.

D'Urville finally escaped the ice on February 10, six days after he had so boldly entered it. A howling gale greeted him, replaced shortly by fog and snow as he resumed his quest for a way south. After four ice-blocked days, he gave up. But had he

The *Astrolabe* and *Zélée* in the ice of the Weddell Sea — *Painting by Ernest Goupil, the expedition artist (who would die during the following winter, at Hobart). (From d'Urville,* Voyage au Pôle Sud . . . , *1841–46)*

failed simply because it was a heavy ice year? He was unwilling to concede that possibility as he compared his experience with Weddell's report. Perhaps the Scotsman had had unusually favorable ice conditions, but d'Urville thought it equally possible that Weddell had exaggerated his achievement. Although the Frenchman was the only significant navigator to the region to express such doubts, his reaction was precisely what Weddell had anticipated, and defended himself against, when he had his men swear to the truth of his report.

If he could not explore the far south, d'Urville decided, at least he could survey in the South Shetlands. First, however, he detoured back to the South Orkneys. By February 20, his two battered ships were sailing along the northern coasts of the islands. D'Urville remained in the South Orkneys for three days doing a bit of charting and gathering a few geological specimens. It was useful work, but hardly the reason he had come south.

D'Urville reached the South Shetlands on February 25. After skirting Clarence and Elephant Islands, he steered his ships into uncharted territory, dipping down toward the eastern side of the Antarctic Peninsula. There he made some modest land discoveries, the largest a landmass he called Joinville Land. (It is, in fact, three islands, today known as d'Urville, Joinville, and Dundee islands.) The *Astrolabe* and *Zélée* reached 63° 27' S on March 5, near their farthest south for the entire summer. By now, d'Urville had a few land discoveries to his credit, had carried out physical and magnetic observations, and had done some meteorology and a little natural history. But in terms of his original objectives, the cruise was a failure. Although he might have tried to do more, he was depressed and suffering once again from severe gout. Many among the crew were also ill. It was time to head north.

D'Urville spent the next year and a half sailing the South Pacific, carrying out the exploratory and anthropological work he had proposed for the expedition in the first place. When he reached Hobart, Tasmania, in December 1839, he decided to return to the Antarctic. This time, he took his ships south of New Zealand in search of the south magnetic pole. D'Urville, who undertook this additional Antarctic challenge entirely on his own initiative, said he had to do it for the honor of France, because the Americans were in the field and the British were about to be. In mid-January 1840, he determined the general location of the south magnetic pole and discovered land along the coast of East Antarctica. D'Urville named the land (and a tuxedo-feathered penguin he had seen in the area) Adélie, after his wife, Adèle. These were the accomplishments that wrote d'Urville's name into Antarctic history. What he had done in the Peninsula Region paled in comparison. Still, that frustrating first season south had some value. If nothing else, it contributed to his later success, because it had prepared him for what he would face in the summer of 1839–40.

The French public and government greeted d'Urville enthusiastically when he reached home in November 1840. After a brief rest, he set to work preparing his expedition results for publication. Sadly, it fell to others to complete the job. On May 8, 1842, while returning from a spring outing to Versailles, d'Urville, his wife and son, and 56 others died in a blazing train crash. It was one of history's first fatal train disasters.

Charles Wilkes and the Americans Try to Better Cook

The summer after d'Urville's visit, the Peninsula Region administered another baptism in the ice, this one to the American national expedition. Officially called the United States South Seas Exploring Expedition, it is best known today simply as the Wilkes Expedition, after its commander, Charles Wilkes. This ambitious undertaking, the first government-sponsored American exploring expedition to venture outside North America, began in 1838 and did not return home until 1842. It encompassed surveying and scientific work in no less than the Atlantic Ocean, Brazil, Tierra del Fuego, Chile, the Pacific Ocean, Australia, New Zealand, the Philippines, the East Indies, the west coast of North America, and the Antarctic.

The Wilkes expedition emerged from lobbying that began almost as soon as the 1829–30 Pendleton-Palmer expedition returned home, but it had taken years of concerted effort, much of it by Jeremiah Reynolds and Edmund Fanning, before the government finally responded. By then, the expedition's objectives had ballooned far beyond the original proposals. Controversy affected many aspects of the planning, including selection of a commander. The first man appointed, a senior naval captain, resigned the post after rancorous disputes with the Secretary of the Navy. The U.S. government then considered several replacement candidates before the final choice fell to 40-year-old naval lieutenant Charles Wilkes. Wilkes was a man with a strong sense of mission, deeply interested in scientific matters, and he had the knowledge and ability the work required. Unfortunately, he had major flaws in matters of leadership. He was aloof, intemperate, and an extreme disciplinarian, and many of his officers and men would come to loathe him.

Wilkes left the United States in August 1838 with 440 men, including twelve civilian scientists, aboard six ill-chosen ships—the *Vincennes*, under Wilkes's own command, and the *Peacock, Porpoise, Sea Gull, Flying Fish,* and *Relief.* None had been strengthened against the ice, and three, like d'Urville's two ships, had open gunports. These were problems enough, but Wilkes soon realized that several of his vessels were dangerously unsound for any voyage. After an excruciatingly slow sail to Rio de Janeiro that consumed precious months, Wilkes had to spend weeks there making basic repairs.

The American fleet finally reached Orange Harbor, on the southern coast of Tierra del Fuego, on February 17. This would be Wilkes's base for his initial Antarctic work. Even though it was now too late to follow the original plan for the summer, he thought he could at least gain experience in the ice. To do that, he decided to send his ships on separate voyages. Wilkes ordered the 650-ton *Peacock* and the 96-ton *Flying Fish* to head southwest toward James Cook's 1774 record south location. He sent the 230-ton *Porpoise*, accompanied by the 110-ton *Sea Gull*, to explore the east side of the Antarctic Peninsula. The other two vessels, including his own *Vincennes*, would remain in the Tierra del Fuego area for survey and scientific work.

The *Porpoise*, with Wilkes aboard, and the *Sea Gull* were the first to leave. They sailed from Orange Harbor on February 25 and reached the South Shetlands on March 1. After two days surveying the known islands, Wilkes sailed south, on the same heading d'Urville had taken, toward the east coast of the Antarctic Peninsula. Icebergs soon blocked the way near where Wilkes made his first, and only significant, Peninsula Region discovery—three small bits of land he named the Adventure Islets. That evening, thick fog settled in. A heavy snowstorm followed. When the storm worsened two days later and ice began draping the ships' riggings, Wilkes ordered the *Sea Gull* to return to Orange Harbor, via Deception Island. The *Porpoise* would examine the more easterly of the South Shetlands before sailing north. Wilkes had been in the south less than a week and was already cutting his program short.

Wilkes soon abandoned even his truncated plan for the *Porpoise*, because his poorly clad men were suffering bitterly in the cold. The *Porpoise* reached Tierra del

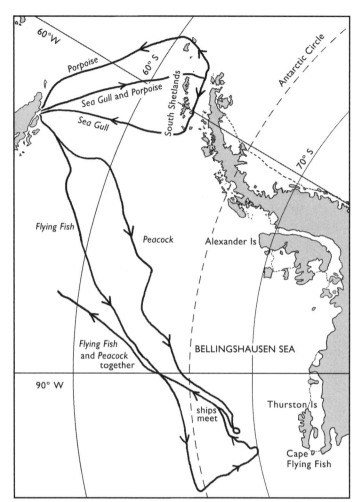

The Wilkes expedition ship tracks to the Peninsula Region in 1839 — *Two pairs of ships went south, the* Porpoise *and* Sea Gull *to the South Shetlands and the* Peacock *and* Flying Fish *west and south of the Antarctic Peninsula into the Bellingshausen Sea. The first two sailed south together but returned separately. The pair that went to the Bellingshausen Sea separated almost at once and sailed south on separate routes, meeting at last near their farthest souths. Following that, they sailed back north jointly until reaching 60° S, where they separated and continued their northern voyages independently.*

Fuego on March 14, less than three weeks after leaving. The *Sea Gull* lingered in the south only a bit longer. Her captain, Robert Johnson, spent a week anchored at Pendulum Cove in Deception Island's Port Foster. While there, the *Sea Gull*'s men searched unsuccessfully for the *Chanticleer*'s self-recording thermometers. Johnson departed Deception Island on March 17, leaving a record of his visit in a bottle at the foot of a flagstaff. He reached Orange Harbor five days later.

The other two ships made much more significant voyages, although they too experienced frightful weather and met far more serious problems with the ice than Wilkes had. Perhaps their captains felt an obligation to stick to their orders because Wilkes, the expedition commander, was not along. Wilkes could have changed the orders, but his officers had no such discretion. In any event, they persevered with their voyages much longer.

Wilkes had ordered the *Peacock,* commanded by William Hudson, and the *Flying Fish,* under William Walker, to sail westward as far as the Ne Plus Ultra of Captain Cook, in longitude 105° W, and from thence you will extend your researches as far to the southward . . . as you can reach. . . . [On your return] you will endeavour to get more and more to the southward, and to pass to the southward of the two small islands called Peter I. [sic] and Alexander . . . and then fall in with what Briscoe [sic] denominated Graham's or Palmer's Land (its proper American name). . . .[5]

In short, Wilkes had ordered them to go far south, late in the season, on a venture that was at least partially politically motivated.

A gale separated the ships on February 26, the first day out. Hudson, the joint commander of the two ships, spent about twelve hours looking for his consort before giving up and sailing on alone in the *Peacock*. His route entirely bypassed the South Shetlands and lay so far to the west that he was beyond sight of the Antarctic Peninsula. That the *Peacock* was grossly unseaworthy became clear less than a week out when the first of many days of ghastly weather hit. The resultant heavy seas proved fatal to a sailor who fell to the deck from the riggings. His shipmates buried him at sea on March 11.

The worst gale and wildest seas yet hit the *Peacock* a week later. The one consolation, Hudson wryly commented, was that the ice now coating the ship made her watertight for the first time on the voyage. Then dense fog settled in, this at a time when icebergs surrounded the ship. When Hudson reached 68° S, 90° W, on March 20, the fog lifted just long enough for him to see "an extended range of icebergs and field-ice in mass, representing a perfect icy barrier. . . ."[6] The fog thinned again briefly two days later, and Hudson saw icebergs all about him. Realizing how

The tiny *Flying Fish* — *Amid the icebergs of the Bellingshausen Sea. (From Palmer,* Thulia, *1843)*

hazardous the situation was, he began his retreat. Then thick snow replaced the fog as Hudson wove his way north. At last, on March 25, the skies cleared sufficiently for him to take his first sun sight in six days. The *Peacock* was at 68° S, 95° 44' W. It was here that Hudson finally found the *Flying Fish*.

Walker and the *Flying Fish* had also continued south after the ships separated. The *Flying Fish* was a far smaller vessel than the *Peacock,* and her ten-man crew experienced a dreadful voyage as the tiny vessel plowed through the heavy seas and storms. They also had to deal with an almost absurd danger—seas filled with whales, some of them larger than the ship. One swam so close that it bumped the vessel's side. By mid-March, three of the *Flying Fish*'s ten men were on the disabled list. Walker nonetheless sailed on. When clearing skies on March 20 revealed an iceberg-studded pack blocking the way south, he worked his way east along the ice edge until he reached 105° W at 67° 30' S. Walker crept south from there.

The *Flying Fish* was at 68° 41' S, 103° 34' W, when Walker found an opening in the pack. Excited, he headed southward in high hopes of passing Cook's mark. Those hopes, he wrote, were soon "blasted in the bud: it . . . became so thick we could not see at all."[7] Frustrated, Walker hove to, a fortunate decision because a raging gale soon descended on the ship. The result, per a contemporary account based on the crew's journals, was a ship

> beset with ice . . . and, as the hoar-frost covered the men with its sheet, they looked like spectres fit for such a haunt. . . . The vessel looked like a mere snowbank; every rope was a long icicle: the masts hung down like stalactites from a dome of mist; and the sails flapped as white a wing as the spotless pigeon above them. The stillness was oppressive. . . .[8]

Early in the morning of the 22nd, howling winds spiced with lightning illuminated a landscape of tightly packed icebergs all about the *Flying Fish*. That afternoon, when the ship was at about 70° S, 101° 16' W, just short of Cook's record in the vicinity, Walker thought he might be trapped. Then he spotted a place where the pack looked looser. He raised all the tattered sails the ship could carry and rammed her into the surrounding ice. It cracked, and the *Flying Fish* escaped. Walker remained nearby for two more days. Then the ice nearly captured him again. Once more, a desperate charge at the ice freed the ship. After this second narrow escape, Walker decided to retreat before his luck ran out. He met Hudson and the *Peacock* a day later.

Awful weather plagued the two ships as they headed north, but somehow, using the constant roar of the *Peacock*'s powerful horn, they managed to maintain contact. On April 1, the ships deliberately separated, the *Peacock* heading for Valparaíso while the *Flying Fish* sailed for Orange Harbor, carrying Hudson's reports to Wilkes.

Wilkes took his fleet, minus the *Sea Gull*, which was lost near Cape Horn, to Valparaíso and then across the Pacific, eventually reaching Sydney. In the summer of 1839–40, Wilkes, like d'Urville, sailed south in search of the south magnetic pole. The pole eluded him, but he did make scattered land discoveries in a 1,500-mile sail along the coast of East Antarctica. It was such an extensive stretch of apparent land that he became convinced he was seeing parts of a continental coast. He called it "the Antarctic Continent," becoming the first to formally use this name for the Southern Continent.[9] After leaving the Antarctic behind in March 1840, Wilkes went on to spend two years in the Pacific and along the west coast of North America. He finally reached home in mid-1842 with an immense body of survey and scientific results. These included thousands of specimens and artifacts that eventually became the foundation of the collections of the Smithsonian Institution in Washington, D.C.

History has paid much less attention to the Wilkes Expedition's 1838–39 season in the Peninsula Region than it has to his visit to the other side of Antarctica. Fair enough. The ships had sailed south far too late in the season and contributed little to the map. Sometimes overlooked, however, is the fact that the *Peacock* and *Flying Fish* had made remarkable voyages. Although they failed to reach Cook's farthest south, the tiny *Flying Fish* came amazingly close to her goal, and the *Peacock* did almost as well. Late in the season, their captains sailed these completely inadequate ships into one of the most difficult parts of the Southern Ocean. Today, the Walker Mountains and Cape Flying Fish, both on Thurston Island—the closest Antarctic land south of where the *Flying Fish* had retreated from the ice—honor their achievements.

As for Wilkes, he returned home to an uninterested public, a hostile Congress, and a court martial instigated by his junior officers. The charges against Wilkes included cruelty, scandalous conduct, oppression, illegal punishment, and, most serious of all, that he had lied about sighting land in East Antarctica on January 19, 1840. Although the court found Wilkes guilty on only one charge—that he had ordered excessive floggings seventeen times—his reputation was in tatters, as was that of the expedition. It was only in the 1930s that a later generation of Antarctic explorers confirmed Wilkes's East Antarctica sightings and placed the name Wilkes Land on the map. Similarly, it took years for his country to appreciate the expedition's other accomplishments, including work in the Pacific so good that the U.S. military used charts based on Wilkes's surveys for operations during World War II.[10]

After Wilkes left, the Peninsula Region was deserted except for William Horton Smyley, an American sealing captain, who hunted in the South Shetlands during both summers following Wilkes's brief visit. Although Smyley found few seals, his 1841–42 voyage was important. In February 1842, he sailed into Deception Island's Port Foster and came across the message that Johnson of the *Sea Gull* had left about his unsuccessful search for Foster's recording thermometers. Smyley decided to look for them himself. Unlike Johnson, he succeeded, but when he picked up the maximum thermometer, its indicator slipped. The minimum thermometer, which he picked up second and handled much more carefully, displayed a minimum reading of −5° F for the thirteen years it had stood there. This would be the lowest recorded temperature from the Antarctic until the *Belgica* expedition wintered south of the Antarctic Circle in 1898 and experienced cold down to −45° F.

Smyley had something else to tell the world about Deception Island. As he left, the no-longer-dormant volcano was erupting. Indeed, he reported that "the whole south side of Deception Island appeared as if on fire."[11] The island's name was more appropriate than anyone had realized. Not only was the wonderfully sheltered Port Foster hidden until one sailed through the narrow opening in the island's walls, but that seemingly safe, enticing harbor was also surrounded by a live volcano that could blow at any time.

James Clark Ross and the British Explore the Peninsula Region

The British national expedition, under the command of the same James Clark Ross who had reached the north magnetic pole in 1831, arrived in the Peninsula Region the summer after Smyley. Pressure from Britain's Royal Society had inspired the expedition, and the Society took an active role in preparing instructions for the scientific work. Ross's orders called for observations in many fields, including geology, zoology, botany, and hydrography, but his most important subject by far was to be a study of Southern Hemisphere terrestrial magnetism.

Ross's expedition differed from those of the French and the Americans in several significant ways. It was the only one of the three that focused specifically on the Antarctic regions, and Ross himself, 39 years old when he began the expedition, was a seasoned polar explorer, with years of Arctic experience preceding his Antarctic work. Unlike d'Urville, he had the respect of his men from the inception of the expedition, and in stark contrast to Wilkes, he retained that respect to the end. In addition, his ships were much stronger and better suited to the task than d'Urville's or Wilkes's ships had been. Finally, unlike the French and the Americans, Ross began his Antarctic explorations on the other side of the continent, with the result that his men already had two years of experience in the Antarctic ice before they reached the Peninsula Region.

Ross had left England at the end of September 1839 with two ice-strengthened vessels, the 370-ton *Erebus*, which he commanded himself, and the 340-ton *Terror*, each with a complement of 64 naval officers and crew. The expedition reached Hobart, Tasmania, in August 1840 after a slow voyage that included lengthy stops for magnetic observations at sub-Antarctic islands in the Indian Ocean. The *Erebus* and *Terror* left Hobart for the far south in late 1840. By early January 1841,

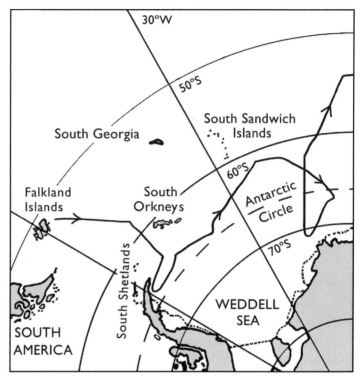

James Clark Ross in the Peninsula Region — *After leaving the Falkland Islands, Ross headed toward the east side of the Antarctic Peninsula, sailing south there until ice blocked his progress just short of 65° S. He then headed east across the Weddell Sea, following the ice edge, until he reached a place where he could resume his southward exploration. His farthest south point, at 71° 30' S, 14° 51' W, was only 45 miles from the low slopes of the northeastern coast of the Weddell Sea.*

Ross was pushing his ships into the pack. He broke through to open water four days later, into what we now call the Ross Sea. Ross saw his first Antarctic land on January 11. Continuing south, he passed Weddell's record latitude as he made and named discovery after discovery. His southernmost find was a towering, sheer wall of ice that was hundreds of miles wide. Ross wrote, "[We] might with equal chance of success try to sail through the Cliffs of Dover, as penetrate such a mass."[12] Today called the Ross Ice Shelf, this was the first major Antarctic ice shelf anyone had seen. After a winter in Hobart and Sydney, Ross returned to the Ross Sea for the 1841–42 summer. This time it took him more than 40 days, in contrast to the four of the previous summer, to force his way through the pack to the open water of the Ross Sea. The weeks spent fighting the ice left him so little time for exploration that his chief accomplishment this second summer was sailing farther west along the Ross Ice Shelf and slightly exceeding his previous farthest south, to a new record of 78° 10' S. He spent the 1842 winter in the Falklands.

Ross left the Falklands for the Peninsula Region in mid-December 1842. On the 24th, his men sighted their first iceberg for the season, about 50 miles northeast of the eastern end of the South Shetlands, and the ships were soon weaving their way through loose pack and scattered icebergs. Four days later, Ross sailed south past d'Urville's Joinville Land into the northwestern fringe of the Weddell Sea. There he made his first discoveries of the summer and began to dole out names, as he had in the Ross Sea. One went to a small conical mountain that Ross christened

James Clark Ross's *Terror* in the Erebus and Terror Gulf — *As painted by Robert McCormick, surgeon aboard the* Erebus. *The strangely shaped island on the right is probably Cockburn Island, a very distinctive bit of land discovered and named by Ross. (Illustration from McCormick,* Voyages of Discovery in the Arctic and Antarctic, *vol. I, 1884)*

Paulet Island. The name honored Captain Lord George Paulet, an officer friend of Ross's who had delivered provisions to the expedition in the Falklands. It was an obscure name for a seemingly insignificant landmass, one that nevertheless would become important at the beginning of the twentieth century. More discovery and christening filled the following days—Erebus and Terror Gulf, Cockburn, Seymour, and Snow Hill islands, and many more features.

January 5, 1843, saw Ross's first landing of the season. He and Francis Crozier, the *Terror*'s captain, went ashore on Cockburn Island and formally claimed the surrounding region. With them was Joseph Dalton Hooker, the young assistant surgeon/naturalist who would later become a renowned botanist. Hooker gloried in the nineteen species of tiny plants he found on the island.

Ross continued slowly south from Cockburn Island along the east coast of the Peninsula until January 9, when he met impenetrable ice just short of 65° S. Not only did the ice stop the ships, but it also surrounded and trapped them: they were beset, unable to move in any direction. Ross remained effectively imprisoned until January 17 when, in desperation, his men finally freed the ships after four exhausting hours of hauling and chopping a way through the floes.

The remaining days of January offered nothing but frustration, because the only openings in the pack led north rather than south. Ross finally gave up in early February and adopted a new strategy. He would follow the pack edge east until he located a way south. Ross crossed Weddell's southbound track at 40° W on February 14, but, he wrote, "under what different circumstances! He was in a clear sea: we found a dense, impenetrable pack. . . ." Ross knew that d'Urville's similar experience had led him to question Weddell's achievement. The British captain, however, reacted differently. He concluded that his countryman must have encountered an unusually open year and commented, ". . . we may rejoice that there was a brave and daring seaman on the spot to profit by the opportunity."[13]

Two weeks later, at 15° W, Ross at last reached a place where the pack edge turned south. The *Erebus* and *Terror* crossed the Antarctic Circle the next day at about 8° W. On March 3, Ross took a sounding at 68° 34' S, 12° 49' W. Because he had always hit bottom in less than 2,000 fathoms (12,000 feet) elsewhere in the Antarctic, he had only 4,000 fathoms of line prepared. When the line apparently failed to reach bottom, Ross thought he had made an important discovery, one that geographers would soon name the Ross Deep. Unfortunately, Ross's assumption of how his line had run out was in error. Instead of falling vertically, it must have drifted sideways with the deep-sea currents. The true depth here was about 2,200 fathoms, but for over half a century, Ross's erroneous sounding led scientists to seriously underestimate how near the Antarctic coast was.

Ross pressed on despite the lateness of the season. He crossed the 70th parallel the day after his sounding. Then, at 71° 30' S, 14° 51' W, the ships met impenetrable pack ice. Here Ross decided to end his exploration. He had no way to know that Antarctica's mainland coast was only 45 miles away to the south and east. But even without what would have been a great discovery, Ross was satisfied with what he had done in the Weddell Sea. If nothing else, he felt that the latitude his own ships had reached was high enough to vindicate Weddell.

When Ross began his retreat north, he encountered such gales, snowstorms, and thick pack ice that it took him four days to gain 70 miles. From then on, though, the only serious threats came from the icebergs that dotted the more or less open water. Ross recrossed the Antarctic Circle on March 11 for the last time on his epic three-year voyage.

The British national expedition had spent almost three months in the Peninsula Region, far longer than d'Urville or Wilkes, but Ross, too, had been frustrated by the ice and weather. Young Hooker, who had been so delighted with the plants he had collected on Cockburn Island, summed up what they had experienced: "It was the worst season of the three, one of constant gales, fogs, and snow storms. Officers and men slept with their ears open, listening for the look-out man's cry of 'Berg ahead' followed by 'All hands on deck!' The officers of *Terror* told me that their commander never slept in his cot throughout the season in the ice...."[14] Even with the frustration and harsh conditions, however, Ross had enjoyed major successes here. Not only had he made significant land discoveries off the east coast of the Antarctic Peninsula, but he had also penetrated deep into the eastern side of the Weddell Sea, a region that no other ship had visited.

There was one more matter of consequence. Ross wrote of seeing a "very great number of the largest-sized black whales...." They offered, he observed, "a valuable whale-fishery well worthy the attention of our enterprising merchants...."[15] Fifty years later, this report would inspire the first significant exploration into the Antarctic since that by Ross himself.

Ross returned home to a welcome much more akin to that enjoyed by d'Urville than the dark reception Wilkes received. Queen Victoria knighted him, and many of his men received promotions. History, too, has rewarded him, granting him an Antarctic status on a par with James Cook. Today the entire Antarctic map bristles with the names he contributed, as well as his own name, placed there by many others in his honor.

CHAPTER 5

Quiet Decades in the South; the New Hunters: 1844–1896

D'Urville, Wilkes, and Ross all returned home with reports of substantial continuous coastlines in the Antarctic, the best evidence yet that there was indeed a continent in the south. Even so, people in the mid-nineteenth century saw little reason to pursue the matter. What the national expeditions had seen was an infertile icescape, of little value to anyone, especially when so many other matters demanded world attention: economic recessions in Europe and the Americas; the almost obsessive search for Sir John Franklin, lost in the Arctic in 1846 while seeking the Northwest Passage; wars, including the Crimean War, the U.S. Civil War, and the 1870–71 Franco-Prussian War. . . . All these and much more absorbed the attention of world governments, as well as any funds that might have gone to renewed Antarctic exploration.

From 1844 to 1893, a mere six ships are known to have sailed south of the Antarctic Circle, all but one outside the Peninsula Region. Only two of these voyages were historically important. The first was made by an American sealer, Mercator Cooper. On January 26, 1853, he landed on the northern coast of Victoria Land, in the Ross Sea Region. This is the earliest known continental landing other than those on the Antarctic Peninsula. The British naval/research ship H.M.S. *Challenger* made the second significant voyage. She spent several months in the far southern reaches of the Indian Ocean as part of a four-year, worldwide oceanographic expedition, and on February 16, 1874, made history as the first steamship to cross the Antarctic Circle. More substantively, the *Challenger*'s deep-sea dredging in the area brought up rocks that geologists later determined had originated on continental land. This was the first scientific evidence that there was indeed an Antarctic Continent.

As for the Peninsula Region, people largely deserted it for nearly a decade after Ross left. The South Shetlands saw a very limited renewal of fur sealing in the early 1850s, and then things quieted down again for almost twenty years. By 1870, the fur seal population had recovered sufficiently for a few hunters to find a profitable cargo, particularly when they supplemented it with elephant seal oil. Although this effort never approached the intensity of the 1820s, a smattering of sealers worked the South Shetlands every summer from 1871–72 through 1880–81. Then, the fur seal population once more devastated, the hunters left. This time it would be nearly 80 years before anyone saw more than a solitary representative of these animals in the South Shetlands.

Eduard Dallmann and the *Grönland*

Among those who had hunted the South Shetlands in the 1870s was a 43-year-old German whaler/sealer named Eduard Dallmann. His 1873–74 voyage to the Peninsula Region was a pioneering one in two key ways. Although the *Challenger* was the first steamship to cross the Antarctic Circle, Dallmann's vessel, the *Grönland*, a sailing ship with a 95-horsepower auxiliary engine, inaugurated the age of steam in the Antarctic a few months earlier, in mid-November 1873. Equally significant, Dallmann was the first to sail to the Antarctic regions with instructions to investigate Ross's report of Antarctic whales. He began with what turned out to be a disappointing month sealing and searching for whales in the South Shetlands. Then, like the more enterprising sealers of the 1820s, he decided to look farther afield.

At the end of December, Dallmann headed south along the west coast of the Antarctic Peninsula, into an area bearing little resemblance to the existing sketchy charts. He was just north of 65° S when he spotted an opening in the mountains to the east. It seemed to cut deep inland to the north as it separated the islands of the Palmer Archipelago (then unnamed and thought to be part of the mainland) from Biscoe's Graham Land. Dallmann, who speculated that it might be the western end of a channel through the Antarctic Peninsula to the Weddell Sea, named it the Bismarck Strait. It was, in fact, the southern end of what we now know as the Gerlache Strait. More discoveries and christenings followed—Booth, Krogmann (now Hovgaard), and Petermann islands, among others—before the *Grönland* turned back just short of the Antarctic Circle.

Dallmann was back in the South Shetlands by mid-January. He spent the rest of the summer hunting there and at the South Orkneys. On March 1, just before quitting the south, he landed on King George Island and erected a post at Potter Cove with a copper plaque memorializing his voyage. Today it is designated a historic monument under the Antarctic Treaty, the oldest such monument anywhere in the Antarctic. And the pioneering steamship *Grönland*, another reminder of Dallmann's voyage, resides today as a museum piece in her home port of Bremerhaven.

The *Grönland* voyage was a modest commercial success because of the elephant seal oil and the few fur seal skins it garnered, but it failed as a whaling effort since the only whales Dallmann saw were too fast for his nineteenth-century ship. His discouraged sponsors did not try again. But the voyage had a far more important return than its slim profit from hunting. The rough survey Dallmann had made along the Peninsula's west coast was the first from this area since Biscoe in 1832, and his discoveries appeared almost at once on German maps.

Others besides Dallmann and his employers were considering Antarctic whaling at virtually the same time. In 1874, David and John Gray of Peterhead, Scotland, published a pamphlet based on Ross's account of the whales he had seen. The Grays wanted to generate support for sending whaling ships to the Peninsula Region, the area where Ross had made a particular point about whaling prospects. For years, however, nothing happened.

In the meantime, a few individuals called for a return to Antarctic exploration. One such was Matthew Fontaine Maury, superintendent of the U.S. Hydrographic Office, who began his campaign in the late 1850s. When civil war consumed American attention and Maury faded from the scene, a German, Georg von Neumayer, replaced him as the world's chief promoter of Antarctic exploration. Others took up the cause in the 1870s, and interest gradually grew, motivated for some by the commercial possibilities, for others by the potential for valuable scientific and geographical work. But as with the Grays' proposal, it took years for their efforts to bear fruit.

Government-sponsored scientific expeditions from France, the United States, Great Britain, and Germany did join elephant sealers working at Iles Kerguelen and elsewhere in the Indian Ocean in 1874–75. They came to these locations on or near the Antarctic Convergence to establish temporary bases from which to observe the transit of Venus. But other than these and the limited visit of the *Challenger*, the only important government expedition to the south between 1846 and 1896 was an undertaking by a German group that spent a year on South Georgia. They were there as participants in the 1882–83 International Polar Year (IPY), an international program of coordinated scientific observations in the two polar regions. Eleven countries took part, with fourteen significant bases, twelve of them in the Arctic. Of the two Southern Hemisphere bases, the German South Georgia effort was the only one south of the Antarctic Convergence. (The other one was a French base located in the south of Tierra del Fuego.)

THE GERMANS ON SOUTH GEORGIA

The German IPY expedition, led by Karl Schrader, reached South Georgia on August 16, 1882, aboard the *Moltke*, a German naval vessel. Four days later, the ship steamed into Royal Bay on the southeast coast of the island. Schrader had found his home on the shore of what he would name Moltke Harbor.

Everyone was soon at work establishing the base, and on September 3, Schrader's eleven-man team moved ashore with their livestock and pet dog. The *Moltke*'s captain and officers joined them briefly in their new home, and together all the men hung a portrait of Kaiser Wilhelm I in the living room and drank toasts to His Majesty's health. Then the contingent from the *Moltke* returned to the ship and sailed away.

The Germans gradually settled into life at South Georgia. Food consisted of the provisions they had landed, augmented by whatever they could obtain locally. They baked fresh bread weekly, enjoyed beef and mutton from the livestock they had brought, and ate elephant seal as well as local birds' eggs in season. Once, they tried leopard seal steaks. It was Karl von den Steinen, the expedition doctor and zoologist, who offered up the leopard seal. He wrote, "It met with the approval of the referees who did not suspect its origin. The dish had a terrible chocolate-brown colour, and after I had explained what it was it was never repeated."[1]

The major scientific event in the first months was the December 6 observation of the transit of Venus. The weather cooperated, and the Germans obtained excellent results. They had been very lucky, because the weather was frequently nasty. Indeed, a violent gale blasted the base almost as soon as the buildings were finished, whipping up the snow on the ground and driving icy crystals into the huts through the tiniest cracks in the walls. Many more storms followed. The worst one of their entire stay assaulted them right in the middle of summer.

The official IPY scientific program of astronomy, meteorology, geomagnetism, and tidal observations formed the core of their work, but when the men had time, they also studied South Georgia itself. Some explored their immediate vicinity on short day trips, and in late October, five men went farther, hiking a few miles into the mountains behind the base. Others focused on natural history. Von den Steinen particularly enjoyed observing the birds and the few elephant, Weddell, and leopard seals he came upon. Hermann Will, the botanist, concentrated on local botany and geology. He also attempted to grow rye, barley, and wheat, an effort that failed miserably because of the frequent snowstorms. Another scientist's garden plot produced a bit of watercress but little else. South Georgia, all concluded, was a poor place for farming.

The men found the penguins to be a special delight. In May, von den Steinen brought a king penguin chick back to the station. Two more captured king chicks joined him in June. The men named these birds after the Bible's three wise men, and a seaman made leather corsets for them

> with holes at the sides for their 'arms' to stick through and with a lacing arrangement at the back. Wearing these harnesses . . . the little fellows were tethered by a cord, which ran along a low telegraph wire that we didn't need. When they got the urge to get away they would lean into their harnesses in unison and, like horses harnessed to a mired wagon would try with all their might to pull the astronomical observatory down.[2]

Much to everyone's regret, the bird they named Caspar [sic] died almost immediately. Balthazar followed him in August.

The German IPY expedition's three pet king penguin chicks — *In their leather corsets. This illustration makes clear that the three were chicks, not adults. (From reproduction in von den Steinen, "Zoological Observations..., part 2," 1984. Originally published in the expedition scientific reports, 1890)*

The navy returned for Schrader's team in early September. They sailed away from South Georgia on the 6th, just over a year after beginning their stay on the island. Before closing the doors for the final time, Schrader placed notes in German, English, French, and Spanish in each hut, explaining the origin of the buildings and requesting that subsequent users take good care of them. As for the surviving king penguin, he accompanied the men. Von den Steinen wrote, "[The captain] had a comfortable poultry cage secured on deck for my 'son,' as Melchior was generally known among the officers."[3] Sadly, this accommodation was an inadequate substitute for the bird's South Georgia home. He died shortly after the ship reached Montevideo.

The German IPY expedition was highly important in spite of its limited field of study. This first scientific party to spend a year south of the Antarctic Convergence had rich results to show for their time ashore. In addition to the official IPY program observations, their achievements included the local surveys, studies of South Georgia's fauna and flora, and the earliest photographs of South Georgia—in fact, the first known from anywhere in the Peninsula Region. (Unfortunately, none of these photos were ever published.) Although little remains of the station buildings today, a number of names the expedition contributed to the map do, permanent reminders of the work of eleven pioneer winterers.

Antarctica's quiet years began to lift in the summer of 1892–93. In a sense, history was repeating itself. Just as commercial interests had played a central role in exploring the Peninsula Region in the 1820s, so would they again in the early 1890s. This time, however, it was whalers, not sealers, who led the way.

QUIET DECADES IN THE SOUTH: 1844–1896

THE DUNDEE FLEET: FRUSTRATED WHALERS AND SCIENTISTS IN THE SOUTH

Although the pamphlet the Gray brothers had published in 1874 generated much discussion and even a few false starts, nothing concrete developed for years. Finally, in 1892, another Scotsman, Robert Kinnes of Dundee, took the first step. A whaling company owner, he decided to send an expedition south. The four ships he sent—the *Balaena, Active, Polar Star,* and *Diana*—were veterans of the Arctic whaling fleet, stout sailing vessels with auxiliary steam engines. Kinnes's original plan was for a purely whaling venture, with no scientific or exploration component, but the Royal Geographical Society, excited over this first Antarctic-bound expedition to leave the British Isles in decades, persuaded him to add a modest scientific program. The whalers would keep meteorological records, and the ships' medical officers would undertake natural history studies. An eager competition for the few physician/naturalist positions ensued.

The key doctors selected were William Speirs Bruce, aboard the *Balaena,* and Charles Donald, sailing on the *Active.* Both were trained physicians, but each, especially the 25-year-old Bruce, was much more interested in the voyage's scientific possibilities and the opportunity to visit the Antarctic. Another nonwhaler expedition member was Bruce's friend W. G. Burn Murdoch, who signed on as an assistant surgeon to accompany Bruce. A professional artist with no medical training whatsoever, Murdoch sailed south with a prepaid commission in his pocket to write a book about the voyage.

The four ships left Scotland in September 1892 to great fanfare. Indeed, fourteen young Dundee boys were so excited that they stowed away, all but two unsuccessfully, in hopes of going along. (The twelve caught in time were put ashore in Scotland. The other two boys, not found until the ships were at sea, officially joined the crew rosters in the Falklands.) The *Balaena* reached the Falklands first, on December 8. There her captain, Alexander Fairweather, learned that the *Jason,* a Norwegian whaling ship commanded by Carl Anton Larsen, was heading to the same area as the Dundee fleet.

Three days later, the *Balaena* and the *Active,* the second ship to reach the Falklands, sailed for the south. By December 18, they were skirting the pack edge toward the Erebus and Terror Gulf, off the northeast coast of the Peninsula. The colorful and fantastical ice forms enchanted Bruce and Murdoch, but scenery was the only rewarding thing in sight. Ross had written of seeing right whales here, the species historically taken in the Arctic. Instead, the only whales the Dundee men spotted were blues, finbacks, or humpbacks—animals too fast for hunters in whaleboats. The men thus turned to seals for a cargo, but that too was disappointing. Instead of elephant or fur seals, all they found were the much less profitable Weddells and crabeaters.

The whalers were unhappy about finding no right whales, and Bruce and

Dundee whaling fleet vessels — *Surrounded by icebergs as the crew members hunt seals on ice floes. (From Murdoch,* From Edinburgh to the Antarctic, *1894)*

Murdoch, aboard a ship whose captain was unsympathetic to anything but hunting, were equally frustrated. Murdoch wrote, "If we lay hold of glacier rocks or birds' skins we raise a whirlwind of objections, and an endless reiteration of the painfully evident truth that 'this is no a scientuffic expedeetion [sic].'"[4] Both he and Bruce yearned to explore the new world about them, a place "where we could lay the chart over leagues of undiscovered land." But instead, ". . . blubber is apparently to be the only interest, and we steer away [from that land] . . . in search of it. . . . We are in an unknown world, and we stop—for *blubber* [Murdoch's emphasis]."[5]

Bruce, Murdoch, and Donald got together on Christmas Day. They compared notes on their scientific observations, and all three "bewailed the utter commercialism of the expedition."[6] Donald also took advantage of the outing to try using barrel staves as skis. Unfortunately, they simply sank under his great bulk. But despite this comical failure and his other complaints that day, Donald was actually enjoying a relatively satisfying voyage. Unlike the *Balaena*'s Captain Fairweather, the *Active*'s captain, Thomas Robertson, was personally interested in exploration and supported Donald's scientific work.

The Dundee whalers spied another sail on December 26. Initially, they took it for their own truant *Polar Star*, the one ship of the fleet yet to arrive. Instead, the ship was Larsen's *Jason*. In the days and weeks that followed, the Scots and Norwegians worked in the same areas, saw each other often, and occasionally cooperated. The *Polar Star* finally joined them in mid-January.

As the days wore on with no right whales in sight, the Scots began to speculate that Ross might have misidentified the whales he had seen. The seas teemed with whales, but they were all the wrong kind, including a humpback they tried to catch. Donald wrote, ". . . and a most exciting chase it led us. . . . On being struck, the whale ran the five lines in the first boat straight out and got free. It was again struck, four additional harpoons and six rockets being fired into it. In spite of all this it escaped after a thirteen hours' battle...." [7] The Dundee whalers were simply unequipped to take the whales they met.

During the second week of January, the *Active* chanced on a strait that ran right through the southern part of d'Urville's Joinville Land. Captain Robertson named it Active Sound, after the ship, and christened the newly cut-off southern landmass Dundee Island. Here they enjoyed several beautiful days before a storm forced the ship onto what Robertson dubbed Active Reef. The *Active* lay there for six hours as the men heaved ton after ton of their hard-won sealskins overboard to lighten the vessel so she could float free. Other dangers faced the ships as they hunted in the northwestern Weddell Sea. The ice nearly trapped them several times, but somehow, aided by their steam engines, they always managed to break free. Steam had added a completely new dimension to Antarctic work.

The Dundee fleet quit the south in mid-February. Kinnes, the expedition sponsor, was understandably disappointed when his ships arrived in Scotland four months later. His whalers had seen hundreds of whales and taken none at all. There was a slim profit from sealing, but that was it, far from the bonanza a successful whaling voyage would have returned. Kinnes decided not to pursue the matter.

The scientific returns were also limited. Not only had that work been an afterthought, but several of the captains, especially Fairweather of the *Balaena* with whom Bruce sailed, were reluctant to support even that slender effort. Still, there were some scientific accomplishments. Bruce and Donald's work resulted in better

"Ship and Ice, Midnight, 1892. Lat. 64° 23' S, Long. 56° 14' W" — *One of Donald's two photos published in 1894, the first known published photos of land and ice from this deep in the Antarctic. The quoted position puts this Dundee whaling fleet vessel in the south of the Erebus and Terror Gulf. (Photo and caption from Donald, "The Late Expedition to the Antarctic," 1894, plate facing p. 62)*

understanding of seals and penguins, and the meteorological observations were good. They had also discovered Active Sound and Dundee Island and produced a more accurate chart of Erebus and Terror Gulf. Further, the two photographs Donald included with his 1894 article about the expedition in the *Scottish Geographical Magazine* were the first published of land anywhere below the convergence. And Murdoch's commissioned book, *From Edinburgh to the Antarctic,* which appeared in 1894, proved a charming and delightfully illustrated narrative.

The expedition's other significant contribution was the beginning of Bruce's love affair with the Antarctic. But nine years would pass before he found a way to return, this time with his own expedition, devoted solely to science. Robertson, the Dundee whaler captain most sympathetic to scientific work, would be with him.

THE NORWEGIANS ARRIVE: CARL ANTON LARSEN AND THE *JASON*

The Norwegian expedition that had hunted with the Dundee whalers was the first from that country to brave the Antarctic ice. Christen Christensen, a major Norwegian whaleship owner, had sent the venture south because he too had read the Gray brothers' report. But unlike the promoters of the Dundee whalers, Christensen encouraged his captains to explore.

The man whom Christensen chose to captain his ship, 32-year-old Carl Anton Larsen, would become a pivotal figure in the Antarctic whaling story in later years. Profit was certainly his first priority, but like Weddell 70 years earlier, he was also interested in science and exploration. He and Christensen were thus an excellent match. Larsen's ship, the 495-ton *Jason,* was a veteran Arctic sealer equipped with a 60-horsepower auxiliary steam engine.[8]

Larsen reached the South Shetlands in late November 1892, several weeks before the first of the Dundee whalers. He began by coasting south along the east side of the Antarctic Peninsula, reaching 64° 40'S, 56° 30' W, before pack ice blocked the way. He then backtracked and landed on Seymour Island, a few miles to the north. There, to his great excitement, Larsen discovered fossil wood, the first found so far south.

When Larsen met the Dundee whaling ships in Erebus and Terror Gulf on December 26 as related above, he told the Scots of his exploratory voyage south, his fossil find, and his total catch to date of 500 seals. The bad news was that he, too, had not seen a single right whale. By his own choice, Larsen worked in the same area as the Dundee fleet for the rest of the season. He did manage to take two bottlenose dolphins—two more cetaceans than the none-at-all the Dundee whalers obtained, but otherwise Larsen had no more whaling success than the Scots had.

The Norwegian newspapers declared the trip a disaster when Larsen reached home in early June: unlike the Scottish venture, the *Jason* voyage had actually lost money. Larsen, however, had a much less negative view. His cargo, of course, was disappointing, but he had gathered information about the potential whaling

QUIET DECADES IN THE SOUTH: 1844–1896

grounds, and he had some valuable exploratory and scientific results. Christensen, who felt the same way, was willing to try again. It probably helped that Larsen had collected souvenirs for his employer. These included a stuffed penguin and seal pup and the skull of a leopard seal, all of which Christensen put up in his home's entrance hall.[9]

Thus Christensen, unlike Kinnes, sent a follow-up expedition in 1893–94. This time there were four ships. Larsen, again as captain of the *Jason,* was in charge. The other vessels were the *Castor* under Morten Pedersen, the *Hertha* captained by Carl Julius Evensen, and a supply ship that would remain in the Falklands. Larsen's small fleet sailed from Norway in mid-August 1893. After calling at the Falklands, the three whaling ships headed south separately.

The *Jason* reached the South Shetlands in early November. On November 18, Larsen again landed on Seymour Island. Although he found more fossils, there were no seals about, and after a futile two weeks looking for animals in the area, Larsen headed farther south along the east side of the Peninsula. This year, open

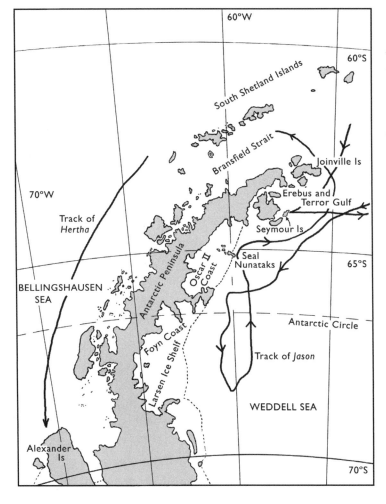

Tracks of Carl Anton Larsen's *Jason* and Carl Julius Evensen's *Hertha* in 1893–94 — *Larsen's* Jason *penetrated the Weddell Sea along the east coast of Peninsula to 68° 10′ S, far beyond where any previous ships had gone, becoming the first to see significant stretches of land and the ice shelf that would later be named for Larsen. Evensen's* Hertha *went even farther south, to 69° 10′ S, where he became the first to see Alexander Island since Bellingshausen's discovery of that land in 1821.*

water beckoned him forward beyond 65° S, into a region that no other ship had penetrated. On November 30, he spied mountains to the southwest. The next day, just past 66° S, he saw more land. Larsen named this major discovery King Oscar II's Land (now Oscar II Coast). He wanted to explore farther, perhaps even take his ship to the coast and go ashore, but he resisted the temptation because he knew his primary objective was obtaining a cargo.

Even if he could not take the time to investigate his land discovery, exploring in search of seals and whales was very much a part of his orders. Larsen thus continued south, hugging the coast of the Antarctic Peninsula as closely as possible, observing and recording numerous deep indentations in a massive ice shelf that extended out from the base of the mountains. (Others later named it the Larsen Ice Shelf in his honor.) Larsen crossed the Antarctic Circle on December 3. Three days later, he reached an estimated 68° 10' S. Here, at last blocked by dense pack, Larsen reversed course. Sailing north, he made a new discovery: land that fog had obscured on the southward sail. Larsen named it Foyn Land in honor of Svend Foyn, the Norwegian whaler who had invented the harpoon gun in the 1860s.

On December 11, Larsen discovered several small volcanic islands just off the ice shelf near 65° S. A small party landed on the offshore ice, and Larsen and a sailor donned skis to cross to what Larsen named Christensen Island (now Christensen Nunatak). Norway, in the person of Larsen, thus claimed the first successful ski trip in the Antarctic. Near here, Larsen at last found seals, animals in such enormous numbers that they "lay in places so closely packed that we had to make circles in order to advance. It was a delightful sight to see those masses of animals. . . ."[10] He could finally amass a cargo, and to recognize this, he named the island group the Seal Islands (today, Seal Nunataks).

While Larsen was exploring the east side of the Peninsula, Evensen made his own important voyage down the west coast in the *Hertha*. He crossed the Antarctic Circle on November 9, passed Adelaide Island a day later, and reached 69° 10' S before ice blocked him on November 21. The next day, Evensen sighted Alexander I Land. He was the first to see it since Bellingshausen in 1821. Unfortunately, the Norwegian had little to say about his historic sighting. Jean-Baptiste Charcot, a Frenchman who would explore this region a few years later, wrote, ". . . though the estimable and kindly Evensen is a daring and skillful captain, geographical questions seem to interest him very little. . . . [All] that I could get out of him about Alexander I Land was: 'Very high and fine mountains, plenty of icebergs!'"[11]

The three ships, including the *Castor*, which had not done any serious exploratory work, returned to the Falklands to re-coal in late January 1894. After a second voyage south that yielded little and a final call at the Falklands, Larsen took his ships to South Georgia for a few days. It was his first exposure to this island that would later play a prominent part in his life. On April 20, the men of the *Castor* and *Hertha* visited the abandoned German IPY station. Afterward, the *Hertha* harpooned a right whale—the first one the Norwegians had seen in two summers—in Moltke Harbor. Despite the fact that they lost it before they could take it aboard, the experience gave the Norwegians hope for future Antarctic whaling success.

The Norwegians headed for home in early July with over 13,000 sealskins, primarily crabeaters, and 1,100 tons of seal oil. Although the expedition was another complete failure as a whaling venture, the exploration results were the most significant from the Antarctic since Ross's, especially those from the *Jason*'s voyage along the east side of the Antarctic Peninsula. Larsen had penetrated these usually ice-clogged waters nearly 200 miles farther south than anyone before him, discovering not only large tracts of land, but also the first major Peninsula Region ice shelf.

Christensen was delighted with Larsen's exploration success. Still, it had been a second disappointing commercial season, and the shipowner decided to end his Antarctic whaling efforts for now.

No one made a significant visit to the Peninsula Region for several years after Larsen's second *Jason* voyage. In 1894–95, however, a final whaling reconnaissance effort headed south, this time to the Ross Sea. This Norwegian expedition, sponsored by Svend Foyn and led by Henrik Bull aboard the *Antarctic*, made several important landings in the Ross Sea Region in January 1895. The first was on the Possession Islands. There the men found lichens, the first plants discovered south of the Antarctic Circle. A few days later, seven men landed at Cape Adare on the northern coast of Victoria Land. This landing was widely regarded at the time as the first on the main Antarctic Continent. But as a whaling effort, Bull's voyage fared no better than the Dundee whalers or Larsen had.

For the time being, the poor commercial results of the four whaling reconnaissance voyages discouraged the hunters. Their geographic discoveries and scientific findings, however, excited those who had been promoting the resumption of Antarctic exploration. When the British Royal Geographical Society hosted the Sixth International Geographical Congress in London in mid-summer 1895, two sessions dealt with the Antarctic. Shortly before the conference ended, Professor von den Steinen, the doctor/naturalist for the 1882–83 German South Georgia expedition, presented a resolution to the effect that

> . . . the exploration of the Antarctic regions is the greatest piece of geographical exploration still to be undertaken. . . . [The] Congress recommends that the scientific societies throughout the world should urge, in whatever way seems to them most effective, that this work should be undertaken before the close of the century.[12]

The resolution passed unanimously, laying the foundation for a renewal of Antarctic exploration.

Chapter 6

De Gerlache and the First Antarctic Night: 1897–1899

A 31-year-old Belgian naval lieutenant, Adrien de Gerlache, led the first exploratory effort to head south after the 1895 International Geographical Congress, but the origins of his expedition actually predated the Geographical Congress. De Gerlache had had Antarctic fever for years and had even applied to join an Australian/Swedish expedition proposed for the early 1890s. That expedition never sailed. De Gerlache then mounted his own effort, but it took him years to raise the necessary funds and organize matters.

In 1895, de Gerlache traveled to Norway. There, he joined a Norwegian voyage to east Greenland and to Jan Mayen Island, north of the Arctic Circle between Greenland and Norway, to gain experience in the ice. He also found his expedition ship, an old but strong 110-foot Arctic whaler with a 150-horsepower auxiliary steam engine. He purchased her in 1896 and renamed her *Belgica*. (The expedition would thus become known to history as the *Belgica* Expedition after the ship's name, a practice that became traditional for the Antarctic expeditions of the early twentieth century.)

De Gerlache found assembling his expedition team a major challenge. The ship's officers were the easiest. Most of them, including Georges Lecointe, the second-in-command, were from Belgium. The one non-Belgian was a 25-year-old Norwegian, Roald Amundsen. Recruiting scientists was much more difficult. Although several Belgians applied for positions when de Gerlache first announced his plans, organizational delays discouraged all but one—de Gerlache's good friend Émile Danco, who would be the expedition's geophysicist and magnetic observer. Eventually, de Gerlache looked outside Belgium. The geologist, Henryk Arctowski, and the assistant meteorologist, Anton Dobrowolski, came from Poland. The naturalist, Émile Racovitza, was Romanian. As for his ship's doctor, de Gerlache hired and lost several men before he took Frederick Cook, an American. Cook, who had spent a full year in northern Greenland in 1891–92, was the only expedition member who had previously wintered in the polar regions. De Gerlache had even more problems assembling an acceptable crew. The composition of this group, drawn from Belgium and Norway, kept changing right up to the day he left civilization for the south.

The *Belgica* party was thus a polyglot mix, without a single common language. For the most part, the officers and scientists spoke French or German, and the crew spoke Norwegian, French, or Flemish. De Gerlache tried to make a virtue out of necessity by portraying it all as a grand experiment in internationalism.

De Gerlache planned to sail to the Antarctic Peninsula early in the 1897–98 summer. After exploring there, he would go on to the Ross Sea, where he and three others would winter in northern Victoria Land. They would attempt to reach the south magnetic pole in spring. The *Belgica* would winter at Melbourne and return the following summer to collect the shore party.

The *Belgica* expedition left Antwerp, Belgium, on August 16, 1897, to great fanfare. Engine problems, however, forced de Gerlache to abort the departure, and then, the day after the *Belgica* limped into nearby Ostend for repairs, two sailors quit. De Gerlache managed to replace them locally, but he had a more serious problem on his hands. The doctor had also resigned. This led to a last-minute cable to New York offering the job to Cook. The American, who had earlier applied unsuccessfully for the position, jumped at the chance.

Adrien de Gerlache at the time of the expedition.

De Gerlache finally left Ostend in late August and headed for Rio de Janeiro, where he picked up Cook. The next stop was Montevideo. Here de Gerlache dismissed his cook for insubordination. The new man he hired would have a very short tenure, because he fell ill the day after the expedition left Montevideo. The *Belgica* reached Punta Arenas, Chile, then a boomtown of some 5,000 souls, on December 1. There, more crew troubles ensued. Not only did de Gerlache have to let the new cook go, he also fired five sailors. That created new problems, because there were no acceptable replacements for any of these men in Punta Arenas. De Gerlache turned the galley over to the ship's steward, Louis Michotte, who swore

The *Belgica* at Antwerp, August 15, 1897 — *With flags flying proudly on the day before sailing for the great adventure. (Both from de Gerlache,* Quinze Mois dans l'Antarctique, *1902)*

that he knew how to cook, and reluctantly accepted the fact that the *Belgica* would sail south the first season seriously undermanned.

Then delay followed on delay, beginning with two weeks loading supplies at Punta Arenas. At this point, de Gerlache changed plans. He would devote his first season to work along the Antarctic Peninsula, thus deferring his efforts in the Ross Sea area to the following summer. This decision made sense, given the shaky manpower situation, but de Gerlache had another motive as well. It meant that he could afford the days needed to take advantage of Argentina's offer of free coal at Ushuaia, a tiny military and convict station on the Beagle Channel. The coaling consumed another two weeks. Then, the evening after leaving Ushuaia, the *Belgica* grounded on a reef in the Beagle Channel. The men freed her with help from the residents of a nearby house, but not before they had jettisoned most of their on-board water to lighten the ship. That cost de Gerlache a stop at Staten Island to replace the water, a diversion that lost another week.

Finally, on January 14, 1898, the nineteen men of the *Belgica*—eleven officers and scientists and eight seamen—reached the Drake Passage and headed south. They sighted the South Shetlands on January 20. The ship grounded again the next day. This time, the men freed her without difficulty, and even though she hit several more rocks almost immediately, she somehow escaped unscathed.

One of the men was not as fortunate a day later. January 22 began with a storm. By afternoon, huge seas were sweeping everything movable on deck overboard. Then everyone heard a piercing cry. A Norwegian sailor, 21-year-old Auguste-Karl Wiencke, who had been dangling over the *Belgica*'s side on a rope while cleaning the scuppers, had plunged into the water after losing his grasp on the line. Everyone rushed to join the desperate efforts to save him. Wiencke managed to grab the rope they threw down, but his freezing hands could not grip it tightly enough for his shipmates to pull him aboard. Since it was impossible to lower a boat in the wild seas, Lecointe, with a line tied about his waist, leapt into the water to help. Just as he was about to seize Wiencke, the young sailor lost his hold on his own line and drifted away before his helpless shipmates' horrified eyes.

Wiencke's death was a devastating introduction to the Antarctic. Still, it was no reason to abandon the expedition. The next day, the *Belgica*, her flags flying at half-mast, sailed for Hughes Bay, a few miles down the west coast of the Peninsula. Here, where John Davis's sealing gang had made the first landing on the Antarctic Continent in 1821, de Gerlache began the first scientific survey of the region. In the evening, de Gerlache, Racovitza, Cook, and Arctowski went ashore on a small island. This initial experience of stepping foot on Antarctic ground was magical. An enchanted Cook wrote, "The scenery, the life, the clouds, the atmosphere, the water—everything wore an air of mystery."[1]

Many more landings followed. The second, on the 24th, was more significant

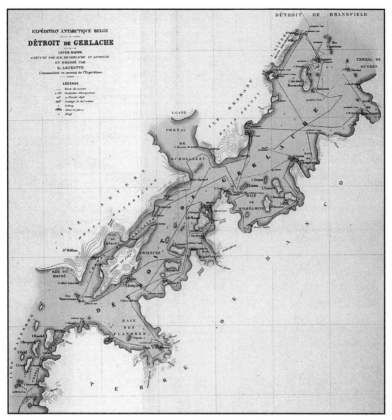

Lecointe's detailed chart of the *Belgica*'s explorations in the Gerlache Strait — *Nearly all detail on this chart was a product of the* Belgica's *surveys. Note that the coasts of land to the west (left) are shown as undefined, a reflection of the fact that no one had formally surveyed this land (although sealers, including Biscoe and Dallmann, had sailed along and seen the west coasts of these islands, which the* Belgica *expedition cut off from the mainland by its discovery of the Gerlache Strait). (From Lecointe,* Au Pays des Manchots, *1904)*

than the first. This time, a large group went ashore with a boatful of scientific paraphernalia, and Racovitza, the expedition naturalist, quickly claimed the expedition's first important scientific discovery—tiny, nearly microscopic, insects, which, Cook wrote, Racovitza "hailed with as much delight as if he had found nuggets of gold."[2]

That afternoon brought more excitement as de Gerlache resumed his survey of Hughes Bay. When the fog that had been obscuring the land lifted, he could see two large gaps in the mountains along the sides of the bay. One led northeast, separating Trinity Land from the Peninsula proper. The other pointed southwest.

The *Belgica* left Hughes Bay on January 27 via the southwest opening. De Gerlache wondered, "Was it really a way through? And where would it lead? Into the Pacific, as its initial direction suggested, or towards [the Weddell Sea]. . . , in the South Atlantic? . . . It was impossible to say whether it was a strait or just a fjord." The excitement of entering this unknown world was intense, a time when, "Despite our fears we were experiencing that joy and those particular emotions that come over true mariners when

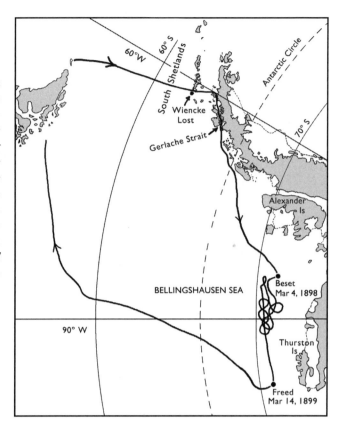

The *Belgica* track during her time in the south, January 1898 to March 1899 — *After his productive weeks along the northern Antarctic Peninsula, de Gerlache took the Belgica farther south, to where she became beset in the Bellingshausen Sea on March 4, 1898, and began a helpless drift that lasted for months. The ice finally freed her to return to civilization on March 14, 1899.*

the bow of their ship first cuts through virgin waters."[3]

The men made as many as two or three landings a day during the latter part of January and early February. Some of the landings were so brief that they frustrated the scientists, who wanted to linger and explore. Then, on January 30, de Gerlache decided to make a weeklong excursion on a mountainous 33-mile-long island just to the west of the mysterious strait. All the scientists and officers wanted to go, but some had to stay with the ship. De Gerlache finally settled on a shore party of himself, Amundsen, Arctowski, Cook, and Danco.

The week ashore on what de Gerlache would later name Brabant Island was difficult and frustrating for these men, the first to attempt land travel in the Antarctic. Pulling their overloaded supply sledge was a struggle, and on the second day, Danco crashed into a hidden crevasse. Fortunately, his skis caught on the sides and saved him from a possibly fatal fall. Farther up the slope, another, much broader, crevasse stopped them cold. Near the end of their stay, a warm storm savaged their tent and swamped their campsite with meltwater. Unsurprisingly, all five pioneer Antarctic travelers were greatly relieved to depart the land when the *Belgica* returned for them on February 6.

It would be 85 years before anyone else spent as much time on Brabant Island as de Gerlache's team had. Then, from 1983 to 1985, an ambitious British group would remain through two entire summers and a winter. One member of the wintering team would be 22-year-old François de Gerlache, Adrien's grandson. Unlike his grandfather, he would conquer the crevasses and reach the mountain summits.[4]

The 1898 shore party had much to report about their experiences. So did Lecointe. While de Gerlache's group had been camped ashore, he had sailed the length of the possible strait to where it opened into the Pacific. The channel thus cut off land to the west that contemporary charts showed as part of the Antarctic Continent. De Gerlache named this great discovery the Belgica Strait (later renamed by others the Gerlache Strait). He christened the newly cut-off islands the Palmer Archipelago. Three of the bigger islands in the group received titles honoring Belgian provinces—Brabant for the island camped on, Anvers and Liège for the other two. De Gerlache baptized another large island Wiencke in memory of the lost sailor.

On February 8, the *Belgica* entered a narrow branch of the Gerlache Strait between Anvers and Wiencke Islands. De Gerlache christened this scenically magnificent passage, which enthralled all aboard, the Neumayer Channel in honor of the German who had worked so hard to promote Antarctic exploration. Cook, by now the official photographer in addition to his medical duties, spent hours taking photos. He wrote, "As the ship steamed rapidly along, spreading one panorama after another of a new world the noise of the camera was as regular and successive as the tap of a stock ticker."[5] Cook's splendid work here and elsewhere would make the *Belgica* the first Antarctic expedition with an extensive photographic record.

Four days later, at the southern end of the Gerlache Strait, the men made their final shore landing. Then de Gerlache headed south through the spectacular Lemaire Channel. His choice of name for this glorious and now-famed strait had a link to a much different climate. It was a tribute to Charles Lemaire, a Belgian explorer of the African Congo.

De Gerlache had already accomplished a great deal when he

The sledging trip on Brabant Island — *Laboriously hauling the overloaded sledges up the icy slopes. Note the skis and snowshoes, both of which the men used, and the first "modern" tent pitched by explorers in the Antarctic. (Both from Cook,* Through the First Antarctic Night, *1900)*

exited the Gerlache Strait. In addition to discovering this strait, he had made twenty landings and had spent a week on the first land exploration in the Antarctic. His results obliterated much of the detail on the existing charts: the *Belgica* had sailed right over locations indicated as land, and her men had gone ashore where charts showed seas. But de Gerlache wanted to do more. He decided to push farther south.

As he sailed south, he attempted to approach the mainland coast, but dense offshore ice blocked the way. Then the pack ice closed in around the *Belgica*. De Gerlache was unconcerned. He had a steam engine to supplement his sail power, and was confident that he could force his way to the open water he could see a few miles away. And so he soon did, and continued on south. On the evening of February 15, the men could see a star for the first time since leaving Tierra del Fuego. It was a message that the bright Antarctic summer nights were waning. The star also had a more practical significance: Lecointe used it to fix their position. His dead reckoning had indicated that they were very near the Antarctic Circle, and the nautical sighting confirmed his estimate. That very night, the *Belgica* crossed the Antarctic Circle heading south.

De Gerlache sighted Alexander I Land late the next day. It was a magnificent sight standing "out superbly with its mighty glaciers, gleaming yellowish-white against the deep blue of the sky. . . ."[6] But despite the continuing excitement of the voyage, some of the scientists and officers, concerned that autumn was approaching, began debating what ought to be done. Should they retreat, or risk being trapped in the ice for the winter? De Gerlache, himself noncommittal, continued south, and by evening, they lost sight of Alexander I Land. It was the last land the men of the *Belgica* would see for thirteen months.

From February 17 to 28, de Gerlache sailed west along the edge of the pack. Several times, the ice briefly trapped the *Belgica*. On the 23rd, during one of these periods beset, de Gerlache asked his men how they felt about wintering in the ice. Everyone expressed reservations. De Gerlache heard them out, and then promised to turn north to work along the Peninsula and in the South Shetlands before returning to Ushuaia. Four days later, however, he resumed the push south.

As the day began on February 28, the pack to the east and south looked invitingly open. De Gerlache wrote after the fact, "Whether we would get through the pack-ice or whether we would be stopped; whether we would succeed in escaping in time in order to avoid having to over-winter, or whether we would be beset, it was our duty, it seemed to me, at least to try."[7] He pushed into the ice, and by March 4, the *Belgica* was 90 to 100 miles deep into the pack. Unable to progress any farther, de Gerlache decided to retreat. He was too late. By the time he had raised steam, the ice gripped the ship too tightly for the *Belgica*'s small engine to force an opening.

The *Belgica* was stuck, beset in the Bellingshausen Sea at 71° 22' S, 84° 55' W. So long as a solid floe remained firmly about the ship, she would simply drift, trapped but safe. If her encircling floe were to crack, however, pressure from storm-driven convulsions in thousands of tons of sea ice could easily crush her. Loss of the ship would be an almost certain death sentence for all on board. No one in the outside world had any idea where they were—trapped about 300 miles from the closest land they knew of, and well over 1,000 miles from the nearest other human beings. Rescue was a virtual impossibility.

De Gerlache was one of the few who faced the situation positively. To him, "We were about to become the first to spend a winter in the Antarctic pack, and this fact alone promised plenty of data to collect and phenomena to study. Was this not what we had asked for?"[8] Despite their initial fears, most of the scientists and officers gradually adopted de Gerlache's point of view, more or less. They were soon walking about on the ice and thinking about the scientific work they could do under the circumstances. The crewmen, for their part, were busy making the ship snug for the coming winter. The situation, however, carried dangers of a sort that no one had anticipated. These men were completely unprepared mentally for the psychological problems of a winter that would swallow the sun for months.

The first weeks before the winter night fell had their cheerful times, and the men were inventive in finding ways to differentiate the days. They watched the calendar closely to ensure that they missed no legal holiday, from any country. They also observed every significant birthday possible, all their own plus those for anyone else of note. (Many later expeditions would follow the same practice.) In April, Lecointe, Racovitza, and Amundsen organized a "beauty contest" with magazine illustrations as contestants. The contest was extremely elaborate, with splendidly complicated rules and intense electioneering for several days before the actual voting took place. When done, the majority had chosen two of the 464 entrants as the world's most beautiful women.

The sun was so low from late April on that Lecointe had to rely on the stars to calculate their position. The men really hungered for this information, even though all it meant was identifying their momentary location in an ice-covered sea where currents and winds were carrying the *Belgica* along a drunken, corkscrewing route north and south of about 71° S. Cook commented, "[When] we obtain the observations which fix our position accurately in this lonely world of desolation, a kind of boyish rejoicing runs along the line of men on the decks. . . . [This] knowledge seems to bring us nearer home because it offers us something tangible with which to make comparisons. . . ."[9]

The sky cleared sufficiently on May 16 for Lecointe to determine their position once again. They were at 71° 35' S, 89° 10' W, and their Antarctic night had begun. If their latitude remained unchanged, it would be 70 days before they saw the sun again. If they drifted north, it would be sooner; if south, later. There was no way to predict.

Until the loss of the sun, most of the men were reasonably healthy, though not

The *Belgica*, beset in the ice — *Frederick Cook's photo taken by moonlight, May 20, 1898, using a 60-minute exposure. (From Cook,* **Through the First Antarctic Night,** *1900)*

because they were enjoying a nourishing diet. In fact, there had been complaints about the food for months. Many developed a particular distaste for the canned and preserved meats, which they disparagingly called embalmed beef. They tried penguin meat, but most thought that was even worse. Cook commented,

> No one seemed to eat the penguin steaks with any kind of relish.... It is rather difficult to describe its taste and appearance.... The penguin, as an animal, seems to be made up of an equal proportion of mammal, fish and fowl. If it is possible to imagine a piece of beef, an odoriferous codfish, and a canvas-back duck, roasted in a pot, with blood and cod-liver oil for sauce, the illustration will be complete.[10]

The most serious problem with the food, however, was neither de Gerlache's provisioning nor the possibilities the local fauna offered. It was the lack of a competent cook. Michotte, the steward turned chef, had more enthusiasm than aptitude. De Gerlache wrote, "His culinary efforts, in which to be sure imagination was certainly not lacking, in so far as he was capable of putting together the most disparate of ingredients, were in general hardly successful.... The only thing he was always good at was soup—though it must be added that all he had to do was to heat it up without even needing to add salt."[11] This was not a trivial matter, given the importance of food in their confined daily lives. As later expeditions would learn, the difference between appealing seal and penguin steaks and unpalatable ones depends crucially on the cook. For the men of the *Belgica,* these vital fresh meats were awful as cooked by Michotte, and most of the men refused to eat them. Scurvy, the dreaded disease of sailors long at sea, was the inevitable result, because the rest of their food, processed and preserved as it was, contained little or no vitamin C.

Everyone grew weaker and more lethargic as the light faded. Before long, they all seemed to be suffering from something. The scientists and officers tried to fight the epidemic by planning an ambitious work program for the weeks before the sun returned. Instead, as darkness increased, they all did less and less. The crew members were a bit better off, because they had essential day-to-day tasks to occupy their time, but they too felt the strain. In the evenings, now distinguished largely by the clock, the men tried to divert their thoughts by reading and playing cards. Music came from the sailors who could play the accordion and stories from those with creative imaginations or colorful pasts. No matter what they did, however, it was impossible to escape thinking about their helpless situation.

Frederick Cook, as doctor, was concerned with the health of all the men, but Émile Danco was his most seriously ill patient. The geophysicist had a pre-existing heart problem and had been complaining of shortness of breath even before the sun set. He might have been all right under normal circumstances, but life on the *Belgica* was hardly normal. On June 1, Danco took to his bed and stopped eating. He died June 5. His grieving shipmates, now down to seventeen, buried him at sea through a crack in the ice.

When Danco died, most of the other men were showing signs of what Cook called polar anemia, probably the first signs of scurvy combined with polar-night-induced depression. Danco's death, however, was a pivotal moment, and as the dark days wore on, the men grew increasingly despondent. Even the ship's cat seemed affected. He died in late June. Cook, busy with his medical duties, was in the best shape. Amundsen also seemed to be coping better than most. But everyone suffered to some degree.

In early July, Lecointe, who up to then had been one of the healthiest men on board, joined de Gerlache, who had already taken to his bunk, on the sick list. At this point, the moral leadership fell to Cook, and the effective command and responsibility to Amundsen, the ranking semi-healthy officer. Both men moved aggressively. Amundsen examined the equipment and realized that most of the crewmen's clothing was inadequate for the winter cold. Rifling a store of red blankets, he had them cut up and sewn into loose suits. They were effective for warmth, although "when the men appeared on deck with them they certainly produced a bizarre and theatrical effect."[12]

Cook attacked the health situation in several ways. He placed the sickest men close to a stove for an hour, stressed warm clothing and dry conditions, and, probably most important, changed their diet. He prescribed milk, cranberry sauce, and fresh penguin and seal meat. Cook tried his treatment first on Lecointe, telling him that he would be out of bed in a week if he followed it. Somewhat to Cook's surprise, the treatment worked as promised.

All the men began recovering their health under Cook's care. Just as important for morale, the sun was on its way back. Lecointe managed his first nautical fix in several weeks on July 21 and excitedly told the men that they might glimpse the sun the next day. This news, Cook wrote, sent "a thrill of joy from the cabin to the forecastle."[13] Shortly before noon the next day, the men fanned out onto the ice to watch. The sun did not disappoint. At noon, as their hungry eyes stared greedily at the horizon, "a fiery cloud separated, disclosing a bit of the upper rim of the sun." Cook continued, "For several minutes my companions did not speak. Indeed we could not at that time have found words with which to express the buoyant feeling of relief, and the emotion of new life. . . ."[14]

The men's spirits and hopes rose with the sun. The scientists resumed their work, and de Gerlache started planning the expedition's future. Once they escaped the ice with the arrival of summer, he would go on to Victoria Land and sledge to the south magnetic pole. In the meantime, he would prepare, using a long trip over the pack ice to test the equipment and gain sledging experience. He decided to start with a trial trip to a tabular iceberg a few miles from the ship, choosing Amundsen, Cook, and Lecointe from the many aboard who volunteered to go. It seemed relatively simple, a short excursion to a nearby iceberg. In fact, these three men were setting out on a dangerous adventure over a difficult and shifting surface that could separate them from the ship and even make it impossible for them to find her on their return.

The three travelers set off eagerly on July 31 pulling a 270-pound sledge. Their confidence in their recovery from the physical maladies of the winter night soon faded. But the real difficulties began when they reached pressure ridges in the ice separated by valleys covered by ultra-dry sandlike snow that made it nearly impossible to draw the sledge. Moreover, the iceberg that had seemed so near was, in fact, more than 16 miles distant. They stopped for the day in mid-afternoon. Their tent, a new one Cook had designed during the winter, worked very well, but their cooker was another matter. The device took hours to heat their dinner.

After the night on the ice, they set off again. A wide lead—an open channel of water between the floes in the pack—soon stopped them. It was, Cook wrote, "a great polar river in a mid-polar sea of ice."[15] When the lead still blocked them the next morning, they turned back for the ship. But where was the *Belgica*? They could only hope they were heading in the right direction. They camped for a third night when it grew too dark to travel. The next morning, fog pinned them down. When the skies cleared in the afternoon, much to their surprise they saw the *Belgica*. She was only a mile away, and men were streaming their way to help. Unfortunately, another lead lay between them and their approaching shipmates, and a fourth night on the ice ensued. The next morning, after three men from the ship reached them, the sledge party abandoned most of their equipment so that they could make their way back over the difficult hummocked route their shipmates had found for them. By afternoon, all were safely back aboard, but it had been a narrow escape.

August proved to be a month of disappointment after the excitement of the returning sun. The men again began weakening physically. One crewman went insane, and several others nearly so. The weather was stormy and cold, the sun seldom seen even though it was rising higher and higher with each passing day. Indeed, they experienced the coldest temperature of the entire expedition, -45° F, in early September.

The one positive note was that life was gradually returning to the pack ice. The burgeoning population of seals and penguins about them meant more fresh meat for the table. And these animals also gave the men something new, something alive, to look at in the bleak ice desert about them. One returned form of animal life, however, was much less welcome. Rats had boarded ship when the *Belgica* was at Punta Arenas. Now, with the cat dead, they were thriving. Indeed, they were much healthier than the men. Not only healthy, they were also bold and noisy, especially at night. Although the men fought this plague with everything they could think of, the rats, de Gerlache wrote, "smelled out all our ruses and carried on insolently snapping their claws at us."[16] They remained undefeated for the rest of the expedition.

In September, de Gerlache had the ship's engine overhauled, more to encourage the crew than for any practical purpose, since they were still completely frozen in. He also began experimenting with their tonite, an explosive supposedly more effective on ice than dynamite. The resulting explosions were exciting but entirely ineffective.

The year 1899 opened with the *Belgica* still trapped in the ice. By then, the men were feeling desperate, dreading the thought that they might have to spend a second winter beset. That made the leads they could see in the distance, open enough to sail through if only they could reach them, all the more tantalizing. Cook proposed a plan to help nature clear the way, and de Gerlache agreed to try it, mostly because he thought it might give the men a sense of hope. Cook's idea was to dig trenches from the bow and stern toward the open water. This might, he thought, create a flow of water that would weaken the ice sufficiently that it would crack and release them. The men finished the trenches by January 12, but the project failed, because the water that flowed in during the warmest part of the day simply re-froze at night.

Next, they tried a new, much more ambitious, plan. They would cut a channel from the stern all the way to the closest lead, nearly 3,000 feet away. This was a huge task, by far their most strenuous physical work in months. Michotte, the only man exempted from the cutting, worked the hardest of all, feeding the ravenous appetites the effort produced. By January 30, they had only another 100 feet of ice left to cut. Then the ice cracked and shifted, closing their hard-won channel. Weeks of exhausting effort seemed to have been wiped out by a casual whim of the ice.

On February 4, just as the men were resigning themselves to the horrible thought of a second winter, the *Belgica* began to roll gently. A slight, almost imperceptible, swell was moving the floes about. A week later, cracks appeared in the floe surrounding the ship. De Gerlache promptly ordered the boilers fired up so they could start the engine as quickly as possible, should the ice offer them an opening. The next day, February 12, they were all prepared when their manmade channel reopened and widened briefly. Then it closed again.

It was February 15 when the ice at last relented sufficiently for them to begin their escape. That day, the men's channel opened just enough for them to push through to the lead they had been trying to reach for over a month. From there, de Gerlache forced his way to another lead, and then to another and another. But the Antarctic pack was a stubborn jailer. The next evening, when the *Belgica* was about 12 miles north of her winter prison, she found herself once more beset. The ice would hold the ship fast here for nearly a month. This time, however, there was more hope. The men could see the open sea at the edge of the pack less than 10 miles away.

The ice finally released the *Belgica* on March 14, 1899. She was at 70° 45' S, 103° W, about 335 miles from where she had entered the pack ice a year earlier. The total drift, though, had been 1,700 twisting and turning miles across a net 18° of longitude and less than a degree of latitude. In the last days of besetment, the men had been just over 100 miles from land, their closest approach of the entire drift.

The *Belgica* was free at last, and no one had to persuade de Gerlache to leave the Antarctic immediately. He reached Punta Arenas on March 28, 1899, and his men rushed ashore as soon as the ship was secure. It was an unnerving yet wonderful experience to step onto land after so many months. Cook wrote,

> When we mount the first hill we sit down and watch and wait to see if it, too, does not move like the hills of ice upon which we have rested so long. . . . A few of the sailors who came ashore remained on the beach, kicked about in the sand, and tossed pebbles. . . . [They] continued to play in the sand for hours with the delight of children at the seashore.[17]

After several weeks in Punta Arenas, de Gerlache, too broke to buy coal for his engine, sailed the *Belgica* home.

Despite the beset winter and abandonment of most of the original program, the expedition actually achieved much scientifically. The meteorological records from the year in the ice were particularly significant. There were also those very fruitful three weeks along the Peninsula. De Gerlache published his findings in a series of scientific volumes, and several expedition members wrote popular accounts of their experiences. The first of these works to appear was Cook's *Through the First Antarctic Night . . .* , published in 1900. De Gerlache followed in 1902 with his own *Quinze Mois dans l'Antarctique* (*Fifteen Months in the Antarctic*).

There was one more result, human rather than scientific. The expedition had introduced Roald Amundsen to the Antarctic. The lessons he learned on the *Belgica* would serve him well in 1910–12, when he would lead the first expedition to reach the South Pole. Amundsen, however, would be the only man from the *Belgica* to return to the Antarctic. Although de Gerlache's interest in Antarctic exploration persisted, the closest he would come to returning himself would be in 1903, when he signed on as adviser to Jean-Baptiste Charcot's first expedition (to be described in Chapter 9). He would leave before the expedition reached the south. De Gerlache, however, did lead three voyages to the Arctic aboard the *Belgica*. Money problems defeated Arctowski's attempt to organize his own southern expedition. (He would return in spirit when Poland named a scientific station on King George Island for him in 1977.) Cook did go back to the Polar Regions, but to the Arctic, where he had been before the *Belgica* expedition. Sadly, the doctor who had served so nobly in the south would become notorious for his widely discredited claim of reaching the North Pole in 1908.

De Gerlache's expedition inaugurated a flurry of exploratory work in the south. The years from 1897 to 1917, a period historians call Antarctica's Heroic Age, would see fifteen exploratory expeditions, including the *Belgica*, from eight different countries head south.[18] Six would come from Great Britain, two each from France and Germany, and one each from Australia, Belgium, Japan, Norway, and Sweden. (The United States did not join in largely because American interest in the polar regions was more focused on the Arctic, especially during the early years of this period.) These efforts would open an entirely new chapter in the Antarctic story. Rather than leaving after a summer of exploring coasts from aboard ship, the expeditions of the Heroic Age would spend a full year or more investigating Antarctica's land and ice. More than 600 men would participate in these undertakings. As pure numbers go, this figure is not particularly impressive. Indeed, this headcount roughly equals the total that had sailed south with d'Urville, Wilkes, and Ross. But in terms of man-days spent in the south, there is no contest. And more significantly, over 360 of these men, including those of the *Belgica*, would spend at least one winter in Antarctica.

The next major Antarctic exploratory effort after the *Belgica* was Carsten Borchgrevink's 1898–1900 *Southern Cross* expedition to Cape Adare at the northern edge of the Ross Sea, thousands of miles distant from the Peninsula Region. A ten-man party spent the 1899 winter there in a small hut, the first wintering on land south of the Antarctic Circle. In the summer of 1901–02, three more exploratory expeditions set out for the south. The British *Discovery* party, led by Robert Falcon Scott, also sailed to the Ross Sea, wintering in McMurdo Sound. Among other things, this group made the first long trips into the Antarctic interior, including one on which Scott, Edward Wilson, and Ernest Shackleton sledged to 82° 16' S on the Ross Ice Shelf. The German *Gauss* expedition led by Erich von Drygalski spent the 1902 winter frozen in the ice aboard their ship off the coast of East Antarctica. And then there was a Swedish expedition. This one sailed to the Peninsula Region.

Chapter 7

Nordenskjöld's Saga of Survival: 1901–1903

Nils Otto Nordenskjöld, the man who would lead the next exploring expedition to the Peninsula Region, was a 32-year-old Swedish geologist who already knew something about living in or near the polar regions. Not only did he have personal experience from expeditions he had led to southern Patagonia in 1895–97 and to east Greenland in 1900, but polar exploration was also in the family blood. In 1879, Otto's uncle, Baron Adolf Erik Nordenskiöld, had been the first to sail the Northeast Passage, through the ice-filled seas of the Arctic Ocean across the northern shores of the European and Asian continents. The Baron had also considered leading an Australian-sponsored expedition to the Antarctic in the late nineteenth century—the very one that de Gerlache had hoped to join, but that had died stillborn.

For years, the younger Nordenskjöld had been thinking about an Antarctic expedition of his own. Reports of Carl Anton Larsen's fossil finds on Seymour Island in the 1890s focused his interest on the east side of the Antarctic Peninsula. He decided to take a ship as far south down this coast as possible. There he would land to establish a base where he and five others would live for the next year to carry out a scientific program, especially in geology, and explore the surrounding country. In the meantime, the ship would sail north for the winter and return the next summer to collect the six men who had wintered at the base.

Nordenskjöld recruited Larsen, the man who knew the area best, as his ship's captain. The vessel, the *Antarctic,* was also a south polar veteran. Henrik Bull had used her for his 1894–95 Ross Sea whaling voyage. Like the *Belgica,* she was a sailing ship with an auxiliary steam engine. Most of the ship's officers and crew were Norwegians, because Larsen selected them, but the scientific participants were primarily young Swedes. Johann Gunnar Andersson, an old friend of Nordenskjöld's, was second-in-command, though only for the second summer because he first had to meet academic commitments in Sweden.

The *Antarctic* left Sweden in mid-October 1901. Nordenskjöld sailed first to England and met with William Speirs Bruce, the Dundee whaling expedition veteran who was now planning his own exploring venture for the following year. The two men discussed scientific cooperation and mutual relief should either need it.

Nordenskjöld reached Buenos Aires in mid-December. There he picked up two more expedition members—Frank W. Stokes, an American artist, and José Sobral, an Argentine naval officer. Nordenskjöld had already agreed to the Argentine meteorology office's request that Sobral come along as a government representative. Now, however, he learned that the Argentines wanted their man to join the wintering party. Despite initial reservations, Nordenskjöld agreed. In return, Argentina promised to help the expedition in any way it could. It was a welcome commitment, though not one that appeared particularly important at the time.

As for the 21-year-old Sobral, he had been told only three days earlier that the navy had selected him to go. He was to take part in magnetic, oceanographic, and meteorologic work, and at the end of the expedition was to turn all the data he collected over to the Argentine government. And that was all he was told. Not only did he have no polar experience, he did not speak a word of Swedish, nor did he own any cold-weather clothes. And when he frantically rushed about Buenos Aires in search of something to wear in the Antarctic, he quickly learned that stores, rich with summer garments, had little to offer. He collected what he could, but, he later wrote, "With the exception of underwear, everything was perfectly useless."[1]

De Gerlache had planned to man-haul his sledges, but Nordenskjöld intended to use sledge dogs. Tragically, all but four of the fifteen Greenland huskies he had taken aboard in Sweden had died from the stress of passing through the tropics en route to Buenos Aires. Thus, after leaving Buenos Aires on December 21, 1901, Nordenskjöld detoured to the Falklands to find replacements. There, to his disgust, the Falkland Islanders took advantage of the situation, and Nordenskjöld, bitterly disappointed in what he was offered, purchased only four dubious and grossly overpriced sheepdogs.

When the expedition reached the South Shetlands on January 10, 1902, Nordenskjöld began his work with a few days exploring the northwest side of the Antarctic Peninsula. The experience infected the *Antarctic*'s men just as it had those aboard the *Belgica*. Using words that echoed de Gerlache's sentiments as he sailed these same waters, Nordenskjöld wrote that it was a time of "feverish eagerness, evoked by our voyaging amidst unknown surroundings where none knew what surprises the next minute might produce."[2] The main work of the expedition, however, was elsewhere. Nordenskjöld reluctantly tore himself away and rounded the Peninsula to voyage south in search of a winter home.

It was January 15 when the *Antarctic* passed a small bay near the tip of the Peninsula. Nordenskjöld suggested to Larsen that it looked like a good place for a future depot. Opposite the bay was the north end of an unnamed sound lying between the Peninsula to the west and Joinville Island to the east. Nordenskjöld headed south through this channel, christening it Antarctic Sound, after his ship. The small island that Ross had named Paulet in the summer of 1842–43 was just beyond its southern

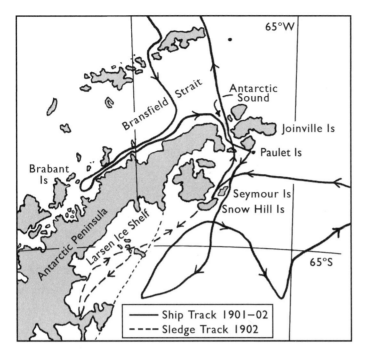

Routes of the *Antarctic* in 1901–02 and of Nordenskjöld's 1902 September–November sledging trip — *Nordenskjöld began his work in the south by surveying and exploring with the* Antarctic, *concluding with the ship landing the wintering team at Snow Hill Island. In the fall of 1902, he and two others made one major sledge trip, across the Larsen Ice Shelf to the Jason Peninsula and back to Snow Hill Island.*

opening. When the scientific party and several crewmen landed there, the richness of the wildlife astounded them. Tens of thousands of Adélie penguins crowded the rocky beach, and hundreds of seals lolled about on the offshore ice. This day was more significant than anyone could know at the time: Nordenskjöld's men had seen two locations—Paulet Island and the small bay that Nordenskjöld had pointed out to Larsen—that would mean much to them the following year.

The next day, the *Antarctic* pushed through the ice to Seymour Island. Not only had Nordenskjöld promised Bruce that he would leave an information cairn there, but he also wanted to establish a provision depot for himself. In addition, he was considering the island as a winter base site. Nordenskjöld located a good place for his message cairn and the depot, but otherwise Seymour Island was a disappointment: the only fossils the eager geologist unearthed were little different from those that Larsen had collected here nearly a decade earlier. Ignorant that through sheer bad luck he had looked in the part of the island with the most limited fossil variety, Nordenskjöld ruled out Seymour Island as a wintering site.

Dense pack ice halted the *Antarctic*'s southward progress just past 66° S. This would prove to be her farthest south for the entire expedition. Since it was only the third week of January, Nordenskjöld decided to follow the ice eastward in search of a southward opening in the pack. This voyage into virgin waters was scientifically useful, but that was all it achieved. Nordenskjöld gave up at the beginning of February and turned back to find a place to winter.

On February 12, he stood on deck gazing at the place that would be home for six men for the next year. It was a rocky shore on Snow Hill Island, just south of Seymour Island. Nordenskjöld had finished choosing his companions only a few days earlier. Four of the men in the shore party had been definite from the start—Nordenskjöld himself, Anders Bodman for meteorology and magnetic work, Sobral because of the promise made to Argentina, and a sailor, Ole Jonassen, chosen for his Arctic experience. Nordenskjöld had been recruiting the doctor, Erik Ekelöf, since leaving Sweden. Ekelöf now agreed to stay. When Stokes, the American artist and original sixth man, pulled out at the last minute, Nordenskjöld replaced him with a sailor, 18-year-old Gustaf Åkerlund.

Everyone unloaded feverishly for a day and a half, and then the *Antarctic* left the wintering men at their new home. She began her return to civilization with a detour south to lay depots in support of Nordenskjöld's planned spring sledging trip, a project that would investigate the lands Larsen had discovered in 1893–94. Perhaps the ship would call at Snow Hill Island on her way back north, perhaps not. The six men left behind, "deserted and alone . . . , who were to be the first settlers on that desolate strand," at once got to work making their prefabricated wooden hut habitable.[3] They moved the kitchen gear inside on the 17th. That evening, they ate their first meal in their new home, a small but cozy building with one central room, a tiny kitchen, and three equally minute bedrooms.

When the *Antarctic* did return, on the 21st, Larsen came ashore briefly and reported that the ice had defeated his efforts to establish the southern depots. Then he departed for the north. The Snow Hill team and their eight dogs were now truly on their own until the ship returned in the spring.

Larsen's inability to establish the depots meant that the wintering party would have to lay their own depots for the spring sledging. On March 11, Nordenskjöld, Sobral, and Jonassen set out across the ice with the four huskies and one of the Falklanders to place a depot as far south as possible. The men were fit and eager, and the dogs, even the sheepdog, pulled well, but problems plagued the trip. As a result, the depot the group managed to establish was only a few miles from Snow Hill Island.

Once back at base, the men found ample work to occupy their time, because Snow Hill Island—a place with a wealth of fossils—was rich in its scientific possibilities. They also finished the hut. When done, they had pictures on the walls, embroidered tablecloths, cushions, and red-checked window curtains. Sadly, kitchen smoke and mildew soon attacked their decoration efforts. The result was more than a cosmetic problem: everything on the walls became a sticky mess, and moisture running down those walls gathered on the floor and froze, creating huge mounds of ice that gradually swallowed up bits of their gear.

Winter announced itself emphatically at the beginning of May. From then until spring, the men endured storm after storm. During a particularly bad one, Nordenskjöld noted in his diary,

> Someone of us has compared our house in this storm to a railway train.... The shaking, which is so severe that the water in a basin on the table trembles as if there were an earthquake; the rattling in the kitchen-range damper; the howling and booming in all keys; the door which is opened and slammed-to again, letting in each time the winter cold and a thick cloud of condensed vapor—everything reminds one vividly of ... an express-train rushing along a line which is not too solidly constructed.[4]

Because their base, at 64° 15' S, was north of the Antarctic Circle, the sun rose every day. Still, the winter days were very short, and when storms assaulted them, it was dark outside even when the sun was theoretically up. They treasured their rare nice days.

Mid-August saw the first signs of spring, when animal life began to return. That meant welcome fresh meat for both men and dogs. Spring also meant that it was time to begin the sledging program. Nordenskjöld, Sobral, Jonassen, and five dogs set out over the sea ice on September 30 for a long exploring trip south. By the end of the second week, they could see the southern coast of Larsen's King Oscar II Land. It was clearly more substantial than a simple group of islands, a key question Nordenskjöld had been eager to answer. On October 18, they sledged onto the Antarctic Peninsula mainland at 65° 48' S. All were anticipating exciting discoveries ahead, but instead the southbound trip had to end abruptly the following day when Jonassen nearly broke his arm. The return, slowed by both Jonassen's injury and unrelenting storms, was difficult, and near the end they were racing starvation. The three exhausted men and their equally worn-out dogs finally staggered into the Snow Hill Island hut late on November 4. They had traveled more than 400 miles in just over a month, inaugurated serious sledge travel in the Peninsula Region, and mapped the coast of the Antarctic Peninsula between Snow Hill Island and their farthest south.

The six men now settled down to wait for the *Antarctic*. In the meantime, Nordenskjöld made several short trips to Seymour Island. There, to his great delight, he discovered large leaf fossils, clear indicators that the Antarctic Continent had once enjoyed a much warmer climate. (These fossils were some of the earliest evidence for the theory of continental drift, which would be first proposed in 1912 by the German Alfred Wegener.) Equally exciting, Nordenskjöld also came upon fossil bones from an unknown species of penguin considerably larger than the emperor. Seymour Island *was* a geologist's paradise, a place where later generations of geologists, following Nordenskjöld's lead, would make more and more significant finds.

The days before Christmas passed rapidly, and Nordenskjöld's men celebrated the holiday itself with a great feast. After that, time began to drag, but in early January they could see open water to the north. They were sure the *Antarctic* would arrive any day. They were wrong.

Where *was* the *Antarctic*? She had reached Ushuaia in Tierra del Fuego on

March 4, 1902, after a stormy trip. Larsen had then sailed to the Falklands to pick up Andersson (Nordenskjöld's friend, who now joined the expedition as second-in-command) before continuing to South Georgia in mid-April. The *Antarctic* men spent almost two months there, the scientists working about the island while Larsen and the crew hunted elephant seals.

After wintering in the Falklands and spending spring at Tierra del Fuego, Larsen headed south on November 4, 1902, to pick up the Snow Hill Island party. He met the pack ice five days later, at 59° 30' S, much farther north than it had been the previous summer. The ice stopped the *Antarctic* for two full days before Larsen managed to ram her forward a bit. Then the ice closed in again, and a heavy storm on the 17th compounded the problems. Six difficult days later, Larsen at last reached the South Shetlands. The scientific staff then spent a week and a half in the islands and the northern Gerlache Strait, making landings to collect specimens, surveying the area, and correcting errors on de Gerlache's pioneering chart of the area.

The scientists ended their work on December 5, and Larsen set a course for Snow Hill Island. It was a glorious summer day as everyone shifted to thoughts of seeing their old friends again. They had brought along mutton and wild geese from Tierra del Fuego for a celebration dinner and planned to decorate Nordenskjöld's cabin with evergreen beech branches from Patagonia. It was not to be.

When the lead the *Antarctic* was steaming through in Antarctic Sound ended, Larsen retreated and headed for the longer route to the east of Joinville Island. Ice blocked that way as well. Worse, the pack trapped the ship. As the *Antarctic* spent days drifting north of d'Urville Island, Larsen and Andersson prepared a backup plan that they would implement as soon as the ship escaped the drift. Their idea was to land Andersson and two other men on the northern tip of the Peninsula. These three men would then sledge to Snow Hill Island over the sea ice. If the *Antarctic* herself failed to reach Snow Hill Island before a specified date, Andersson would take his two men and Nordenskjöld's six back to his starting point and wait for the *Antarctic* to return. But first they had to escape the ice, and in the meantime, all they could do was wait. Larsen described Christmas aboard the ice-bound ship as a day "darkened by gloomy apprehensions"—a marked contrast to the happy mood that day at Snow Hill Island, where six men excitedly anticipated the *Antarctic*'s imminent arrival.[5]

On December 29, the ice at last opened sufficiently for Larsen to regain control of the ship and sail her into the small bay Nordenskjöld had pointed out the previous January. That evening, Larsen landed Andersson, Samuel Duse, the expedition cartographer, and Toralf Grunden, a Norwegian sailor. The three men wasted no time once on shore. First, they organized a cache of supplies, enough for nine men for two months. It was insurance in case they had to bring the wintering party back. They then strapped on their skis, harnessed themselves to their sledge, and set off for Snow Hill Island. It was about 70 miles away, across frozen sea and snow-covered land.

None of these men had any polar sledging experience, and they found the first short day on the trail difficult, the second more so. But despite their inexperience, they made reasonable progress until January 4. Then they met open water, a cruelly ironic development. Ice-covered seas had stopped the *Antarctic,* and now the *lack* of ice blocked the sledging party. With no way to go forward, their only choice was to return to what Nordenskjöld would later name Hope Bay in honor of their experience there.

Andersson's party regained Hope Bay on January 13, and settled down to wait for the *Antarctic.* The first days passed happily enough, with Duse and Grunden busy with camp work while Andersson studied the local geology. Like Nordenskjöld, he collected a wealth of important fossils. But as the days turned to weeks, a forced wintering, something they had once discussed only as a remote possibility, "grew gradually to a threatening certainty."[6] The men at Hope Bay now faced the same question as those at Snow Hill Island. Where was the *Antarctic*?

After dropping Andersson's trio at Hope Bay on December 29, Larsen had taken the *Antarctic* for another attempt to reach Snow Hill Island. This time, he advanced a bit farther around the north coast of d'Urville Island before the ice again trapped the ship. On January 9, 1903, the ice began to do more than simply imprison the *Antarctic.* It was now squeezing her sides and doing real damage. When the pressure finally eased off, the men cleared the ice to free the rudder. To their horror, they saw that the ice had ripped it away: it was nowhere to be seen. And the pack was still in control, carrying the grievously battered *Antarctic* south past Joinville Island with her stern resting on a floe.

Three weeks later, the ice under the stern broke, and water flooded in through gashes in the hull. The men worked the pumps for more than a week while they tried desperately to repair the damage. On February 13, Larsen at last accepted that

The *Antarctic* beset in the ice, about to sink, and then she goes, her flag still flying — *Photos by Larsen taken while on the ice after abandoning ship. (From Nordenskjöld et al.,* Antarctic: Två År Bland Sydpolens Isar, *1904)*

the *Antarctic*'s wounds were fatal and moved provisions and equipment—as much as time allowed—onto the surrounding ice. Then the men gathered below for one last time before each went on deck to say an emotional good-bye to the ship. Finally, they raised the Swedish flag and abandoned ship. When everyone, including the terrified ship's cat, was off, the desolate group stood on an ice floe and watched the *Antarctic* sink.

Twenty men and one cat now stood on an ice floe, their only hope of survival a 25-mile trek over the pack ice to Paulet Island, the nearest land. After a night spent preparing, they piled their gear into their small boat. Then they put their backs into hauling the heavily laden boat—essential for crossing open water between the floes—over the ice, in and out of the water, struggling, battling their way for days with rafted floes, pressure ridges, and leads. They finally reached Paulet Island on February 28. It had taken two weeks of exhausting effort to gain this barren, circular island, only about a mile in diameter. It was about 40 miles from Hope Bay, where three men looked for a ship that would never come, and 65 miles from Snow Hill Island, where six others waited and wondered.

The Antarctic ice had splintered one expedition into three, three groups so close that each could see where the others were if only they knew where to look. Afterward, Nordenskjöld would remember,

> During the second year I was often to cast longing glances towards [Hope Bay and Paulet Island], while wondering if some short communications to us might not be lying there; but in my boldest flights of imagination I never supposed that, during that long winter, each of them [was] the dwelling-place of a division of our expedition. [7]

But none of the three groups knew was where the other two were. The only thing that all 29 men knew by the end of February was that they would be spending an unplanned winter in the Antarctic.

The six men at Snow Hill Island were the best prepared. They had a sturdy hut and a stock of emergency second-year provisions they had landed the previous summer. That, with seal and penguin meat for food and blubber for fuel, should see them through. These men were even relatively relaxed about the situation. Once they realized that they faced a second winter, Nordenskjöld wrote, "No one complained, no one showed any signs of fear, but from that moment we spoke no more of relief. When we mentioned the future it was but to consult on the best means of preparing for, and employing, our second winter."[8] Despite this sensible frame of mind, it was a difficult time. These six men had been living together for many months, and they began to wear on one another. Time took its toll on their hut as well. When the men ran out of candles, they devised a seal-blubber lamp. It worked, but, Nordenskjöld observed, "The lamp lights up the room almost too well, for it is not pleasant to see its clear gleam fall upon walls covered with sticky

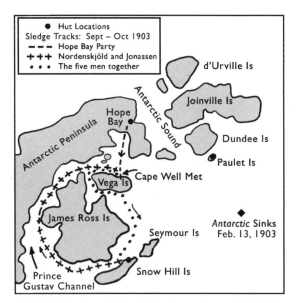

The three huts during the winter of 1903 — *The locations of the huts where Nordenskjöld's three marooned parties spent the winter were nearly in sight of one another, if the men had only known where to look. Hope Bay was about 40 miles from Paulet Island, 70 from Snow Hill. Paulet Island, in turn, which was about 25 miles from where the* Antarctic *sank, was approximately 65 miles from the hut on Snow Hill Island. The tracks here show the routes taken by the men from Hope Bay and by Nordenskjöld from Snow Hill Island that resulted in their meeting at Cape Well Met in October 1903.*

cardboard. . . . The poor pictures are black and damp; all articles of iron are rusty, and the bedclothes are falling to pieces. . . ."[9] Still, these were minor concerns compared to those the other two groups faced.

The problem for the three men at Hope Bay was sheer survival. They had two crucial needs: shelter and food. Both had to come almost entirely from the meager resources their environs offered. When they finally accepted the reality that no ship was coming, Andersson wrote, "[We] made a complete alteration in our manner of living, changing hastily from enjoying perfectly civilized fare [from the depot supplies] to supporting ourselves almost exclusively on the products of the land around us."[10] They killed all the penguins and seals they could, finding just enough through the fall and winter to ward off starvation.

Because their tent was too frail for the winter, they built a stone hut. The tent, erected inside, provided an insulating wall, and penguin skins carpeted the rocky floor. Heat and light came from penguin and seal blubber burned in an improvised stove and lamp. Both devices produced black, greasy smoke that soon coated the men and their gradually rotting clothing. These three men used everything they had. Ends of rope, bits of wood, empty tins—all were carefully hoarded and used to sustain a life far from the level of comfort that Nordenskjöld described so gloomily.

Three men, living in enforced togetherness in appallingly Spartan conditions. Two Swedes and a Norwegian. A scientist, an educated officer, and an ordinary seaman. They shared the work equally and made a determined effort to be considerate of one another. Sundays were their great festivals, all three meals using some of their precious depot provisions. The first Sundays of the month were the best. On these days, they allowed themselves a small sip of gin each from Duse's pocket flask. They declared extra drink days on their three birthdays and on May 17, the Norwegian national day. With the work of survival to keep them busy, the days often passed quickly. Evenings were more difficult. With not a single book to their name and desperate for anything to read, they sometimes devoured the labels on their food tins as hungrily as they did the contents. At other times, they chatted or told each other stories. Much of the time, however, they simply existed "in a desert of intellectual nothingness...."[11]

The twenty men at Paulet Island differed from the three at Hope Bay in two significant ways. They were a larger group, and they knew what had happened to the *Antarctic*. Having more men was a mixed blessing. Although it meant added hands to share the work and a wider variety of personalities for all to interact with, there were equally more mouths to feed and bodies to shelter. And knowing the fate of the ship also had its downside: these men were the only ones without even a slim hope that the *Antarctic* would appear the following summer.

Differences, yes, but the Paulet Island group faced the same essential survival problem as their friends at Hope Bay: wresting food and shelter from the Antarctic. The food situation on Paulet Island was dire. Larsen's men had even fewer civilized luxuries than the Hope Bay three to relieve the monotony of their seal and penguin diet, and worse, the local food resources were problematic. When the men first reached the island, they found that most of the penguins had already departed for the season. Only a few thousand remained, in marked contrast to the seemingly uncountable masses there in January 1902. And what of the seals they remembered? They, too, were mostly gone. Larsen estimated that they would need 3,000 to 4,000 penguins to survive the winter, but they garnered only 1,100 in their first large hunt. Starvation was thus a constant threat. Fortunately, just enough wandering penguins and seals chanced by to see them through. A nearby freshwater lake was such a welcome water source, one not requiring scarce fuel to melt, that the men easily overlooked the rotting penguin bodies paving its bottom.

Building a hut for twenty men was also a challenge. When finally complete, their stone home, built of basalt slabs and roofed with sails and seal skins, provided a tightly crowded space for men who had to spend day after stormbound day inside. And they soon found that they had built the hut on a foundation of foul-smelling penguin guano.

Having many hands to build their home and collect a stock of food became a disadvantage once they had completed these tasks. Only one man had a real job. He was the cook, Axel Andersson, who "stayed in his kitchen, day in and day out . . . half choked by the nauseous smoke from the blubber."[12] For the others, boredom soon became a serious problem. Carl Skottsberg, the 21-year-old biologist, wrote, "How we spent our every-day life during the . . . winter months can be realized by taking the description of one day—any one day."[13] The few books they had managed to rescue from the *Antarctic* were a godsend.

Slowly, the days, and then months, passed at Paulet Island. All but one of Larsen's men remained remarkably healthy, given their living conditions. The exception was a young seaman, Ole Christian Wennersgaard. He coughed violently, and with no medical supplies or doctor amongst them, his companions could do little for him. Wennersgaard died on June 7, probably from heart disease exacerbated by the primitive living conditions. Ekelöf, the expedition doctor, who was now at Snow Hill Island, had known on leaving Sweden that Wennersgaard had heart problems, but had not thought it serious enough to disqualify him. Now the young sailor was gone. His grieving friends buried him in a snowdrift outside the hut, there to wait until spring when they could dig him a proper grave. When spring came, they buried him in the rocky ground with a cairn to mark the spot.

Not only did the men at Snow Hill Island have more comfortable living conditions, they had a scientific

Three Huts in the Antarctic — **Top:** *The Paulet Island hut, home for twenty men.* **Center:** *The Hope Bay hut, where three men lived.* **Bottom:** *Nordenskjöld's Snow Hill Island hut, home to six men. (All from Nordenskjöld et al.,* Antarctic: Två År Bland Sydpolens Isar, *1904)*

program to keep them occupied. When spring came, Nordenskjöld decided to make one long sledge trip before a relief ship—be it the *Antarctic* or one from somewhere else—could be expected to arrive. He and Jonassen headed north on

October 4. Another party was already on the trail. They were traveling south.

The men at Hope Bay had had ample time to consider their situation during the winter. One thing was clear. If the *Antarctic* had been lost with all hands, no one would know where they were, and their only hope of rescue lay in sledging to Snow Hill Island. They were ready to go on September 20. To leave, to do something to help themselves. The excitement at the thought made the storm that pinned them down for another week all the harder to bear. While waiting, Duse engraved a board with the words "J. G. Andersson, S. Duse, T. Grunden, / from S. S. *Antarctic* / wintered here 11/3–28/9 1903."[14] They tied the board to a tent pole when they finally left on the 29th. Andersson also put a short note in a bottle and placed the bottle under the board. The note—in English as was the board's message—provided a brief account of their stay and their plan to travel to Snow Hill Island.

Bad weather plagued the Hope Bay men for days, but the traveling surfaces were excellent, and they made good progress when not tentbound by storms. They reached the north shore of Vega Island, roughly midway between Hope Bay and Snow Hill Island, on October 9. The next morning, Andersson took a day off from the trail to allow Grunden to rest a frostbitten foot. Duse and Andersson used the time for cartography and geology. It was remarkable. On a trip undertaken to rescue themselves, with limited food and inadequate equipment, aware through hard experience that storms could delay them for days, they chose to spend time pursuing scientific work. They knew they were close to Snow Hill Island, so why rush. Andersson decided to tarry another day or two.

Meanwhile, Nordenskjöld and Jonassen, the Snow Hill Island duo, were slowly sledging north. They camped for the night of the 11th just north of the very same Vega Island where the Hope Bay men had been for the past two days. The next morning, the two men discussed their plans. Nordenskjöld felt that they should head for Paulet Island. First, however, they would explore their immediate vicinity, in particular "a well-marked, dark and prominent headland which attracted my attention each time I looked in its direction. It was as though a premonitory feeling told me that something important and remarkable awaited us there."[15]

Something was: the Hope Bay men were sledging toward that same headland. While stopped for a midday meal, they saw seals on the ice. Or were they something else? Andersson wrote, "'What the deuce can those seals be, standing up there bolt upright,' says one of us. . . . 'They are moving,' cries another. A delirious eagerness seizes us. A field-glass is pulled out. 'It's men! It's men!' we shout."[16] Duse fired his pistol as a signal. Then he and Andersson took off to intercept the men, whoever they were.

Nordenskjöld wrote of what he saw at the same time:

> I imagine for a moment that I catch a glimpse of something of an unusual appearance, but pay no further attention to it, when Jonassen speaks again:

'What's that strange thing there close by the land?' I glance thither and say: 'Yes, it looks like men, but it can't be, of course; I suppose it is some penguins!' and continue to march onwards. But Jonassen says at once: 'Hadn't we better stay so that you can see what it is?' For the third time I look. . . . I take my field-glass. My hand trembles a little when I put it to my eyes, and it trembles still more when the first look convinces me that it is really men . . . [We] hurry shorewards at a run. . . . I soon hear a faint cry. . . . I do not answer, for the matter is as yet all too mystical for me, and I can now see so much that I mark the strangeness of the figures that are coming towards us. . . .

Two men, black as soot from top to toe; men with black clothes, black faces. . . . Never before have I seen such a mixture of civilization and the extremest degree of barbarousness. My powers of guessing fail me when I endeavour to imagine to what race of men these creatures belong. They hold out their hands with a hearty, 'How do you do?' in the purest English. 'Thanks, how are you?' was my answer. 'Have you heard anything of the boat?' they continue. 'No!' 'Neither have we!' 'How do you like the station?' 'Oh, very well in every respect.' Then comes a moment's pause, and I puzzle my brains without result. They are members of the *Antarctic* Expedition, but still they know nothing of the vessel. A dim idea comes into my mind that I ought to ask who they are, and why they are here.[17]

Andersson, realizing that Nordenskjöld had not recognized them, had deliberately spoken in English as a joke. Then, his game over, he revealed their identities. Once all

"Two men, black as soot from top to toe. . . ." Andersson and Duse meet Nordenskjöld at Cape Well Met — *Samuel Duse's drawing of the event. (From Duse,* Bland Pingviner och Sälar . . . , *1905)*

four men recovered from the shock, they skied to where Grunden was cooking. They spent the night together at the headland that all agreed to name Cape Well Met.

Nordenskjöld led his enlarged party straight back to Snow Hill Island the next morning. They reached the hut on October 16, the second anniversary of the day the expedition had left Sweden. That night, nine men, rather than six, celebrated the date with a splendid banquet. The pièce de résistance was a roast emperor penguin. It was the first time they had had this bird, a real rarity in the area, and the lone bird had been carefully photographed and studied before being killed several days earlier in anticipation of the anniversary feast.[18]

The Hope Bay men's arrival boosted morale at Snow Hill Island and reinvigorated the interest in scientific efforts. Over the next month, while the nine men waited and wondered when, and from where, relief might arrive, they spent their time cheerfully working.

November 8, 1903, began as a routine day. Two of the nine men had gone to Seymour Island to collect penguin eggs. Since Nordenskjöld expected them back that afternoon, he thought nothing of it when someone reported men walking toward the hut. Soon, however, everyone was outside wondering why there were four men approaching, not two. Even more curious, the strangers looked clean and civilized. The egg-gatherers resolved the mystery when they excitedly introduced Julián Irízar, captain of an Argentine relief ship, and one of his officers. Nordenskjöld wrote, "For the second time within a month we stand face to face with one of those moments when one's whole world of sense seems to resolve itself into a mist. . . ."[19]

Irízar's surprise arrival was the result of a chain of events set off by the *Antarctic*'s failure to return at the end of the 1902–03 summer. The Argentine government, true to its pledge to help Nordenskjöld, had immediately swung into action. They had ordered Irízar, their London naval attaché, to obtain equipment and supplies in Europe and then to return home to command a search expedition. Among the many whom Irízar had consulted in London was Ernest Shackleton, recently back from a year with Scott's *Discovery* expedition to the Ross Sea. It was Shackleton's first brush with Peninsula Region matters. (Twelve years later, he would have a personal interest in the stores Irízar had purchased.) For an expedition ship, Argentina had upgraded a small naval vessel, the *Uruguay*, to prepare her for the Antarctic. Sweden had simultaneously mounted her own rescue effort. She used a veteran Arctic whaler, the *Frithjof*, and chose Olof Glydén, an experienced polar sailor, to command her.

The *Frithjof* had reached Buenos Aires on October 30. There, Glydén had learned that Irízar, who was officially cooperating with him, had already left. The *Uruguay* would wait at Ushuaia for the Swedes, but only until November 1. When preparations delayed the *Frithjof* in Buenos Aires, Irízar had sailed south alone. He had reached Seymour Island on the 6th and begun searching for Nordenskjöld, using information Larsen had provided while in the north the previous summer. Irízar's hunt had ended when he spotted the tent belonging to the egg-gatherers.

Nordenskjöld told Irízar that he could leave Snow Hill Island in two days. But there was still the question of the missing *Antarctic* and her twenty men. Should they search, or first go north to consult with the authorities? No decision had been reached when Irízar returned to the ship after dinner. Then the dogs began howling. The man sent to investigate reported that six people were approaching, probably from the *Uruguay*. But when someone else went out to greet the visitors, those inside the hut heard "wild ear-piercing cheers, mingled with shouts of '*Larsen!* [Nordenskjöld's emphasis] Larsen is here!!'" Nordenskjöld wrote, "[We] have experienced so much during the last few days that nothing can seem impossible to us; but still, I can scarcely believe my ears. There must be some mistake; it must be the day's unrest that has made one of us give a form of reality to his wishes. But I hurry out like the rest, and the next instant all doubts are vanished."[20]

Larsen had realized that he must reunite with one of the other groups if his men were to be rescued. Thus, on October 31, leaving thirteen men behind to await his return with help, he and five others had set out in their small boat for Hope Bay. They found the place deserted when they arrived on November 4. But there was the stone hut, a desperate shelter so like their own, and Andersson's note saying that his group had gone to Snow Hill Island. Larsen's only choice was to follow. After waiting out a storm, he and his companions took to the boat again, and they reached Snow Hill late in the evening of the very day the *Uruguay* arrived.

On November 10, Nordenskjöld wrote a note to Glydén, left it in a bottle on the kitchen table, and closed the hut. Then fifteen expedition men and their dogs sailed away with the Argentines to collect the thirteen men on Paulet Island. When the *Uruguay* reached the island early the next morning, the Paulet contingent was sound asleep. They woke quickly, however, when Irízar blew the ship's whistle. Young Skottsberg wrote of their reaction, "Arms wave wildly in the air; the shouts are so deafening that the penguins awake and join the cries; the cat, quite out of her wits, runs round and round the walls of the room; everybody tries to be the first out of doors, and in a minute we are all out on the hillside, half-dressed and grisly to behold."[21]

Before leaving Paulet Island, Irízar established a provision depot using the supplies brought from England, and the *Antarctic* men erected a cross atop the cairn over Wennersgaard's grave. A brief stop at Hope Bay to pick up scientific collections brought Nordenskjöld's expedition in the south to an end.

The *Uruguay*'s voyage north was one last challenge for Nordenskjöld's men, a stormy passage that nearly wrecked the crowded ship before she reached Cape Horn. Once there, however, the problems were behind them. Irízar made two stops before he reached Buenos Aires. The first was at the Año Nuevo Observatory off

Staten Island. There he left Nordenskjöld's dogs, a gift from the Swedes to the Argentine navy. After the second stop, to send a wireless message announcing the successful rescue, he reached Buenos Aires on December 1. The Argentine navy delayed the official arrival for a day so that the ship could be repaired, thus to shine in the welcoming festivities. And wild festivities they were when the *Uruguay* officially arrived in Buenos Aires on December 2. Thousands of people crowded the docks to greet her, and hundreds of thousands thronged the streets that the men passed through after leaving the ship.

Nordenskjöld's amazing story captured the public imagination, and Nordenskjöld, Andersson, Larsen, and Skottsberg jointly produced a popular narrative, *Antarctic: Två År Bland Sydpolens Isar (Antarctica: Or Two Years Amongst the Ice of the South Pole)*. The book appeared in 1904 and was quickly translated into several other languages. The more important product of this expedition, however, was its handsome scientific harvest, published from 1904 to 1920 in six massive volumes.

Larsen was the only senior member of the expedition who would return to the Antarctic, initially to South Georgia in 1904–05. Nordenskjöld's plan to return in 1914–15 on a British-Swedish expedition fell victim to World War I.

And what of the *Frithjof*? Despite knowing that the *Uruguay* had already sailed south, Glydén continued with his mission. He reached Snow Hill Island on December 3, a day after the *Uruguay* landed its passengers to cheering crowds in Buenos Aires. The next day, a small group of Swedes visited the hut and found Nordenskjöld's note. What might have been a significant voyage had become a historical footnote, but Fridtjof [sic] Sound, a small channel on the southwest side

The Paulet Island hut in January 1995 — *A ruin of tumbled stone walls where Adélie penguins nest, their downy chicks resting on the remains of where men once lived. (Author photo)*

of Antarctic Sound, does honor Glydén's effort.

Other names on the Antarctic map also commemorate the expedition. Several are particularly evocative, among them Cape Well Met, Hope Bay, Wennersgaard Point, Antarctic Sound, and Uruguay Island. But Nordenskjöld's expedition left more than names in the Antarctic. All three expedition huts and Wennersgaard's cross and cairn are now historic monuments under the Antarctic Treaty. The Snow Hill Island hut, now carefully restored and lovingly preserved by Argentine archaeological teams, is in the best condition. Recovered artifacts from the expedition are on display inside for the benefit of tourists and others who come to visit during the summer. What remains of Andersson's Hope Bay shelter is mostly a reconstruction, another Argentine effort. As for Larsen's Paulet Island structure, the roof is gone, but substantial remnants of the original walls remain. For years, penguins nested in these ruins during summer, but in 2007, Argentine teams began efforts to fence the birds out and started work restoring parts of the hut.[22]

CHAPTER 8

Bruce and the *Scotia*, Bagpipes in the South: 1902–1904

Nordenskjöld's men had had company in the Peninsula Region in the winter of 1903. The other residents were the members of William Speirs Bruce's *Scotia* expedition, and unlike the marooned Swedes, they had intended to be there.

Bruce had longed to return to the Antarctic for years following the Dundee whaling expedition. Instead, chance offered him scientific positions in the Arctic, including several summers on Spitsbergen and a full year in Franz Josef Land, north of Siberia. As a result, Bruce was among the most experienced polar region scientists in Great Britain when he began organizing his Antarctic venture in 1900.

The *Scotia* expedition was proudly and, at times, defiantly, Scottish. Indeed, Bruce later wrote, "While 'Science' was the talisman of the Expedition, 'Scotland' was emblazoned on its flag."[1] One unfortunate result of this expressed national pride was that the British government, which granted £40,000 to Robert Falcon Scott's *Discovery* venture, dismissively declined to support "this purely Scottish expedition."[2] Virtually all the expedition's very tight budget would come from Scottish sources, most of it from the textile-manufacturing Coats brothers, James and Andrew. In early 1902, Bruce purchased an elderly 140-foot-long Norwegian Arctic whaler with auxiliary steam engines, selected in part because she was cheap. When he took the vessel to Scotland, the maritime architect who inspected this bargain declared her a disaster, saying that the best thing to do would be to "fill her with stones and . . . sink her."[3] Bruce instead found the money for repairs. He then proudly rechristened his ship the *Scotia*. To the surprise of many, she proved to be an excellent Antarctic expedition vessel.

With the exception of an Englishman, David Wilton, a zoologist who had been with Bruce in Franz Josef Land, the ship's entire original company came from Scotland. The scientific staff, six including Bruce himself, all had experience working under snow and ice conditions. The officers and crew were similarly well prepared, almost all of them whalers who had sailed the Arctic ice for years. The captain, Thomas Robertson, also had Antarctic experience, as captain of the *Active* on the

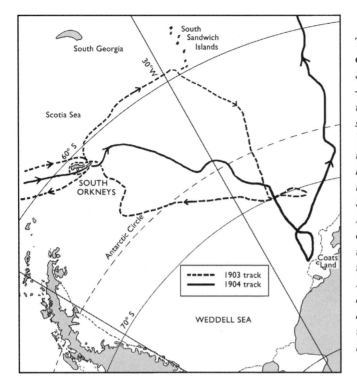

The tracks and explorations of the *Scotia*, 1903 and 1904 — *Bruce's original plan was to winter with his ship on the coast of the Weddell Sea. When ice blocked his way, he retreated and spent the 1903 winter in the South Orkneys, at Laurie Island. At the beginning of summer 1903–04, he initially went north, then returned south, first to Laurie Island, then sailed deep into the Weddell Sea, at least reaching the coast, where he discovered what he named Coats Land.*

Dundee whaling expedition.

The core of the expedition's work was to be oceanography and meteorology, but Bruce also intended to do some geographic exploration. His plan was to sail south into the eastern side of the Weddell Sea until he found land. There he would establish a base. The ship would leave for winter oceanographic work and return the following summer to retrieve the wintering party.

The *Scotia* sailed from Scotland in mid-November 1902 and reached the Falklands on January 6, 1903. There the staff began their scientific work while Bruce completed preparations for the south. By now, however, he had changed his plans. He would still explore the eastern Weddell Sea, but instead of leaving after dropping off a wintering party, the *Scotia* would remain to serve as the base. Everyone would winter in the south.

On January 26, Bruce left the Falklands with 34 men and 8 sledge dogs. He met the ice pack a week later, just east of the South Orkneys. Most of the scientific staff made a hurried landing in these islands on February 4 to gather a few specimens. Then it was back to the *Scotia* to look for a winter home. Bruce had already started daily oceanographic soundings, work that his men had quickly realized was extremely dangerous when carried out on a pitching deck in icy seas. As the *Scotia* sailed eastward, skirting the northern edge of the pack along the 60th parallel, the effort yielded a major discovery. It was part of a lengthy underwater ridge, today called the Scotia Ridge (or Scotia Arc), that loops eastward from Tierra del Fuego all the way to the South Sandwich Islands and then west to the South Shetlands and Antarctic Peninsula. (South Georgia, the South Sandwich Islands, and the South Orkneys are all high points along this ridge.)

Bruce had almost reached the South Sandwich Islands when the pack edge began slanting south on February 15. Three days later, he excitedly crossed the Antarctic Circle in a virtually ice-free sea. The pack, however, soon reappeared. Several days of zigzagging advanced the *Scotia* to 70° 25' S, 17° 12' W, where the ice briefly trapped the ship. Once free, Bruce quickly retreated north. A frustrating week followed as he battled the ice, trying to find any way south. There was none, and after the pack again temporarily held the *Scotia* captive several times, Bruce conceded defeat. Although willing to allow the *Scotia* to freeze in near a coast where he could work on shore, he was determined to avoid spending a winter drifting helplessly as *Belgica* had done.

With the far south no longer an option, Bruce decided to winter in the South Orkneys. He knew that these rugged islands offered good opportunities for productive work, because no other scientists had spent significant time there. Bruce also saw another positive aspect to the location: the *Scotia* would be able to come and go during the winter for oceanographic work. The men of the *Scotia* saw the South Orkneys for the second time on March 21 and began searching the uncharted coasts for a safe harbor. After four days spent creeping along in snow squalls and near-zero visibility, Bruce found his home on the south coast of Laurie Island. He named the place Scotia Bay.

The political status of the South Orkneys was hazy in 1903. Although George Powell had claimed the islands for Great Britain in 1821, the British government had never followed up. Bruce might have made a new claim based on occupation, but he felt that his private, Scottish, expedition lacked a government mandate. After much discussion, the *Scotia*'s men declared the South Orkneys a no-man's land. John Fitchie, the third mate, did have an idea. He thought the islands would make a splendid penal colony where "in summer the convicts would be employed in housebuilding, and in winter he would keep them busy shoveling snow off the glaciers."[4]

Bruce dropped anchor in Scotia Bay on March 25, 1903, fully intending to come and go during winter. But the Antarctic had other ideas. Pack ice drifted in and clogged the bay three days after he arrived. Within a week, the ice had frozen solid, imprisoning the *Scotia* in her namesake bay. The one compensation was that the ice cover provided a convenient way to travel the quarter mile to shore.

His year in Franz Josef Land had convinced Bruce of the value of fresh meat. He thus directed the cook to serve seal and penguin three times a week. Bruce's men were well aware of the *Belgica* party's low opinion of penguin meat. Because their own reaction was so different, they concluded that the only possible explanation for Frederick Cook's utterly negative comment on the taste of penguin was that the *Belgica*'s cook was incompetent. R. N. Rudmose Brown, the expedition botanist, wrote,

> I really must protest against this caricature; and while I admit [that penguin meat's] dark red-black appearance is unusual, and that it cannot be compared in taste to anything I know, I yet must bear testimony to its excellence as a food. . . . I think it would be worth while [sic] to establish penguin rookeries on many of the barren rocks off the western isles of Scotland, and so introduce a new and delicious food to the inhabitants of this country. . . .[5]

The 1903 winter at Laurie Island was a happy one. The men had good food, comfortable quarters aboard the snug *Scotia,* and more than enough work to keep them busy. In addition, Bruce and Wilton, both experienced skiers, taught everyone to ski. This sport, a great novelty to these Scottish whalers, soon became a favorite recreation. They also celebrated every holiday and birthday possible. The mid-winter party was their biggest. The crew had an especially good time on that occasion because the cold had had an unanticipated effect on the barrels of Guinness-donated ale: most of the water in the ale had frozen, and the liquid the men poured out was almost pure alcohol. Soon, Brown wrote, "Sounds of revelry in the fo'c'sle rapidly became pandemonium. . . ."[6]

Although the men lived aboard the *Scotia* during the months in the South Orkneys, Bruce also had them build structures on shore—a survey cairn, a magnetic observation hut, and, requiring the most work, a house. This last would provide shelter should something happen to the ship. Its chief purpose, however, was to house the meteorological party Bruce planned to leave behind when the *Scotia* returned to South America to refit in the spring. By the time the house was complete in September, more than 100 tons of local rock had been used to construct the 14-foot-square building. Bruce christened it Omond House, in honor of Robert Traill Omond, an Edinburgh meteorologist who had been an enthusiastic supporter of the expedition from its earliest planning.

Once he realized that the ice had trapped the *Scotia*, Bruce altered his winter scientific program to focus on the South

In the laboratory on the *Scotia* — *From left to right, Bruce, Pirie, and Wilton. (From Three of the Staff,* The Voyage of the "Scotia," *1906)*

Orkneys themselves instead of oceanography. The first field team set out on July 28 to begin surveying Laurie Island. It was a successful trip, but Bruce cut it short and recalled the trip leader, John Pirie, the expedition doctor, on August 5. Allan Ramsay, the *Scotia*'s 25-year-old chief engineer, was dying.

Ramsay had been ill for months, a recurrence of heart problems he had experienced in the Falklands. Unfortunately, he had concealed his condition at the time because he had so wanted to continue with the expedition. Ramsay died on August 6. Two days later, on a rare beautiful day, his 33 companions carried his body to shore as Gilbert Kerr, the piper, played the bagpipes. They buried Ramsay at the foot of a mountain Bruce named Mount Ramsay in his memory.

Three Heroic Age expeditions to the Peninsula Region. Four deaths. Three of the dead had had health problems before sailing south but were nonetheless determined to go. The *Belgica*'s Danco and *Antarctic*'s Wennersgaard had died when extreme living conditions exacerbated their already frail condition. Ramsay, however, had been living in much better circumstances. Perhaps the cold of the south and the strain of the expedition had sealed his fate. In any event, he, like Danco and Wennersgaard, had made his choice. Now he too was gone.

The cross on Laurie Island marking Allan Ramsay's grave at the foot of Mount Ramsay. (From *Three of the Staff*, The Voyage of the "Scotia," 1906)

Ramsay's death saddened everyone, but work went on. In the meantime, winter gales raised Bruce's hopes that the ice in Scotia Bay would break up and allow the ship to escape to the open water that he could see just a few miles away. Then the winds dropped, and by the end of August, ice again blanketed the sea.

The ice might be stubborn, but spring was on its way, heralded first by the return of the seals. On September 10, the men captured a Weddell seal pup, and Bruce gave her the run of the ship. He was delighted when he "got her to speak into the phonograph. The result was remarkably successful."[7] It was the first time anyone had recorded an Antarctic seal voice.[8] Sadly, persuading her to take tinned milk from a bottle was far less successful, although she was willing to suck at anything else, especially the men's boots and trousers. She was so endearing that everyone missed her deeply after she went into convulsions and died on the 20th.

The seals had returned at the end of August. When the penguins joined them in the second week of October, Laurie Island became a naturalist's delight. The scientists were delighted to have the penguins to study, and everyone welcomed the new source of fresh meat. Bruce also tried filming the returning fauna with his movie camera. To his frustration, this effort was less successful than his recording of the mewling voice of the seal pup. He did get about 50 feet of film of a penguin rookery, but that was it. Still, this first movie of *any* length from the Antarctic was another historic achievement.[9]

Despite the advent of spring, ice continued to grip the *Scotia*. At the end of October, the crew decided to sacrifice the "jinker" who was bringing this bad luck. First, they brought out a life-size dummy to be tried at court. It was clearly modeled on the botanist, Brown, of which the man himself wrote, ". . . what made the semblance stronger was that the effigy was clothed in an old suit of mine, which I had unwittingly given away to help in the manufacture of the victim." Captain Robertson quickly pronounced the effigy guilty and imposed a sentence of burning. Then, Brown wrote, "Out on the [ice] . . . I had the unique experience of setting fire to myself after pouring half a gallon of spirit over myself to aid the conflagration. It was a very enjoyable evening."[10] The Scotia Bay ice was unimpressed. It held firm.

Bruce and five others were in the field when the ice finally freed the *Scotia*. On November 23, it suddenly broke up and moved out of Scotia Bay without warning. Pirie set off early the next morning to tell Bruce the news, and what ensued was rather farcical. Pirie had left the *Scotia* saying he would be back by 11 a.m. When he reached Bruce's camp at about 8:30 a.m., the whole party immediately headed back to the ship, arriving just in time to see the *Scotia* steaming out of Scotia Bay. Captain Robertson had taken the 11 a.m. time literally, and when Pirie had not shown up by then, Robertson decided to take the *Scotia* round the island to collect Bruce. Despite the camping party's shouts, the ship sailed on. Bruce's group then backtracked to their camp, arriving in ample time to greet the *Scotia*, only to find that heavy swells made it impossible for them to reach the ship. And so Robertson returned to Scotia Bay to wait for Bruce to come back overland.

The *Scotia* left Laurie Island on November 27. Six men remained behind to continue the scientific work until Bruce returned for a second summer's work. Robert Mossman, the meteorologist, was in charge. With him were Pirie, three crewmen, a scientific assistant, and one of the sledge dogs, to serve as a pet.

Bruce reached the Falklands five days later. The local residents greeted him warmly, and the men of a British naval vessel brought round a pile of current newspapers and magazines. Bruce's men were most interested in the news of the other Antarctic expeditions—that Scott's *Discovery* was frozen fast at Ross Island, with two ships going to her relief; that Drygalski's *Gauss* had returned after her year frozen in the ice off the coast of East Antarctica; and that Nordenskjöld's *Antarctic* had disappeared. Bruce was personally concerned about Nordenskjöld's situation because of his mutual support agreement with the Swede. He was thus relieved to learn that the *Uruguay* had already sailed south. A mail boat arrived several days later with a report of Nordenskjöld's rescue. That news ended Bruce's brief thoughts about joining the search. Since the mail boat was en route to Buenos Aires, he seized the opportunity to hitch a ride. The slower *Scotia* followed the next day.

The *Scotia* party spent nearly a month in the Argentine capital while the ship was being repaired and Bruce wrestled with money matters. The expedition was broke, placing the second summer's work in jeopardy, not to mention the fate of the six men still on Laurie Island. Bruce wired desperate appeals home for money. Once more, the Coats family came through.

Argentina had asked to participate in Nordenskjöld's expedition. Now Bruce offered the Argentines another opportunity in the south. He proposed that they take over the Laurie Island meteorological station. Although Bruce had not consulted with the British government before making the offer, he did advise British officials in Buenos Aires of his actions. They wrote to the Foreign Office in London, and the gist of the official response, when it finally arrived in late March, was that the British had no objections: they thought the South Orkneys were worthless. Argentina, in contrast, had jumped at the chance.

The *Scotia* left Buenos Aires on January 21, 1904, with three Argentines for the Laurie Island base aboard. By then, Mossman and his group of Scots had been at Scotia Bay alone for more than two months. When February opened with no ship in sight, they began to worry that something bad might have befallen the *Scotia*. If she had been lost before reaching civilization, no one would know that there was anyone on Laurie Island. Would they have to find some way to rescue themselves? Was there, in fact, a way? When the ship at last arrived on February 14 after a detour to the Falklands, the relieved shore party ran up the flag to show all was well. Then, Pirie wrote, "Not to seem too anxious, we returned to lunch and allowed the visitors to seek us out. . . ."[11]

Mossman was agreeable to remaining for another year as head of the observatory, under the auspices of the Argentine Meteorological Office. Bill Smith, the crewman who had served as cook for the summer, also agreed to stay. The other four men and the dog rejoined the *Scotia*. Although the three Argentine newcomers immediately raised their national flag, none suggested that this meant an Argentine claim to the islands. They had, however, brought postage stamps and a specially designed postmark, "Orcadas del Sud. Distrito 24," for the benefit of stamp collectors.[12] The first cancellations carried the date February 20, 1904. Some were on Falklands stamps, others on Argentine.

The *Scotia* left the South Orkneys for her second Weddell Sea cruise on February 22. The ice was much more open this time, and by late evening on the 28th, the ship was across the Antarctic Circle at approximately 28° W.

March 3 was a beautiful day. An early morning sounding found the water depth was only 1,131 fathoms. Robertson, surprised by this abrupt drop from the 2,500 or more fathoms they had been seeing prior to this, hurried to the crow's nest, and there was the explanation. "Land ahead," he shouted down excitedly to the men below. A sighting here was completely unexpected, because Ross's 1843 soundings that had found the "Ross Deep" had implied that any land must be much farther away. Bruce wrote of what they saw, "There were great stretches of ice cliff, possibly 100 or 150 feet high at the highest. . . . The surface of the ice-sheet appeared to rise gently towards the interior. . . . This went on indefinitely, until the whole merged itself into the sky . . . one great field of ice, extending as far as one could see. . . ."[13] Bruce named his discovery Coats Land for the brothers who had made the expedition possible.

The weather deteriorated as Bruce sailed south along Coats Land the next morning. By the time the skies finally cleared on the 6th, ice was forming in the open lanes in the pack at an alarming rate. The excitement of discovery, however, had infected Bruce. Abandoning his usual caution, he ordered Robertson to continue south. That day, they traced the ice cliff for 150 miles.

The wind rose during the night, and by morning it was howling. The new engineer, taken on in Buenos Aires because of Ramsay's death, lacked ice experience and thus did not realize how important it was to maintain steam while in the ice. When the winds shoved the *Scotia* into a gluey slurry of ice, the engine was unavailable to push her through into open water and she was trapped. By the time the winds finally calmed on the 9th, they had driven the *Scotia* and her surrounding ice into a small bay in the ice shelf at 74° 01' S, 22° W, just short of Weddell's record set over 200 miles to the west. Even though ice gripped the *Scotia*, Bruce saw no reason to be idle. The staff continued scientific work and, when time permitted, everyone enjoyed himself on the ice. Kerr even played his bagpipes for emperor penguins who had wandered up to investigate the ship. The birds were indifferent to his efforts.

By the 12th, ice covered the water outside the bay, and the prospect of a winter beset in the Weddell Sea loomed. Then Robertson spotted a small crack in the surrounding ice. A few minutes later, the pack began breaking up, and Bruce raised the flags in anticipation of a quick departure. The Royal Scottish banner waved on the foremast, while the Union Jack and a silk Saint Andrew's cross—crafted by Bruce's wife—flew at the ship's bow. The excitement was premature. Even though the ice beyond them had splintered, the *Scotia* remained firmly embraced by an encircling floe. The men tried all sorts of stratagems to escape: blasting; everyone jumping simultaneously on the ice; running together from one side of the ship to another. . . . Nothing worked. Bruce wrote, "Nature smiled at our puny efforts, leaving us still embedded, with water all round us, chips breaking off everywhere, except where the ship was."[14]

The *Scotia*'s bagpiper, Gilbert Kerr, plays for an emperor penguin — And forces the penguin to stay by tying him in place. Many have experimented with music and penguins, and most have found, as Kerr did, that the birds pay little attention. (From Three of the Staff, The Voyage of the "Scotia," 1906)

Then nature began to play with them. The floe broke up, releasing the ship. Now she floated free, but the tantalizing open lanes about her had closed. The *Scotia* was simply in a different trap. The next afternoon, the pack loosened, and after desperate hours of ramming, the *Scotia* advanced 2 or 3 miles closer to the open water. She finally reached it a day later.

It was March 14, and Bruce was still deep in the Weddell Sea. But despite the lateness of the season, he wanted to satisfy one more question before sailing north. Did the Ross Deep actually exist? A sounding on March 23 confirmed doubts that he had had for years. Within a few miles of Ross's position, he found the depth to be less than 2,700 fathoms.

Four days later, the *Scotia* crossed the Antarctic Circle corkscrewing madly in a wild storm. Bruce completed his expedition work with a brief visit to Gough Island, a small isolated bit of land at 40° S in the southern Atlantic Ocean. He then continued on to Cape Town, arriving May 5. There Bruce gave Reuters a brief note and sent an equally spare message home: "Have made bathymetrical survey 4400 miles of ocean, sounding and trawling in 2660 fathoms: left Mossman and four others in Scotia Bay, South Orkneys: discovered great ice barrier 74 degrees south, 17 to 28 degrees west: visited Gough Island: experienced bad weather: arrived good condition: all well. Bruce."[15] It was truly important news, but hardly written in a fashion to excite public interest.

The years following the *Scotia* expedition were difficult for Bruce. The expedition's outstanding achievements received little recognition, especially in England; raising money to publish his scientific results was an ongoing battle, with no help at all from the British government; and the feeling that his venture was so underappreciated hurt him deeply. Unfortunately, his expedition had experienced insufficient drama to excite the British public. Bruce, though, was also a poor publicist for his work. When he became seriously ill with influenza shortly after reaching Scotland, he abandoned the idea of writing an expedition narrative of the type the public expected from Antarctic explorers. He had never really wanted to do it anyway. Instead, Brown, Mossman, and Pirie took up the task. It was thanks to them that *The Voyage of the "Scotia,"* officially written by "Three of the Staff," appeared in 1906. Bruce did write the introduction.

The public may have taken little note of Bruce's work, but it is not forgotten on the Antarctic map. Today, the chart features the Scotia Sea, Scotia Arc, Scotia Bay, Mount Ramsay, Coats Land, and several places that others named for Bruce himself. The ruins of Omond House, now honored as an historic monument under the Antarctic Treaty and preserved by the Argentines, also serve as a reminder of Bruce's very fruitful expedition.

The *Scotia* expedition officially ended when Bruce reached Scotland on July 21, 1904. Robert Mossman and Bill Smith, however, were still on Laurie Island, living through their second Antarctic winter. This turned out to be a much less happy year for the two Scots. They were ill matched with their Argentine companions; their Omond House accommodations were far less comfortable than the *Scotia* had been; and there was insufficient work to keep everyone busy. It did not help that Omond House nearly collapsed in early March when gale-whipped waves undermined its walls. Smith, an incurable optimist, was the most cheerful among the five men, but even he sometimes commented that "Life's too — [sic] slow for a funeral."[16]

The relief ship arrived at the end of December, bringing with her five new men for the base and a prefabricated wooden building to replace Omond House. This structure, modest as it was, served notice that a new era had begun in the Antarctic regions: humans had come to stay—year-round. Argentina has fulfilled that promise. Since taking over from Bruce in 1904, she has occupied Laurie Island without interruption, even in time of war.

As for the relief ship, she was the same *Uruguay* that had rescued Nordenskjöld, now with a new captain, Don Ismael F. Galíndez. Mossman was soon aboard, arranging for departure and learning that they would not be sailing directly back north. Another expedition, Jean-Baptiste Charcot's *Français* venture, was now wintering in the Peninsula Region, and Galíndez had orders to head south to pick up messages that Charcot had said he would leave.

The *Uruguay* left Laurie Island for the south on New Year's Day 1905. Galíndez sailed first to Deception Island. Finding no message there from Charcot, he left one of his own, in case the Frenchman were to show up. Then he continued to Wiencke Island. Unfortunately, he missed the cairn Charcot had left there, and difficulties with the ice ended a brief attempt to search farther south. As a result, when the *Uruguay* arrived in Buenos Aires on February 8, Galíndez reported that Charcot was apparently lost.

Bruce was the only man from the *Scotia*'s scientific staff who even attempted to return south, but chronic money problems defeated him. (After he died in 1921, however, a whaling vessel transported his cremated remains to South Georgia. On Easter Sunday 1923, the South Georgia magistrate boarded a whale catcher to carry Bruce's ashes out to sea, where, following a simple ceremony, he scattered them in the Southern Ocean.) A few of the *Scotia*'s crew would come back to the south as whalers. Otherwise, the sole returnee to Antarctica with a link to the *Scotia* expedition was the *Uruguay*. She would make many more relief voyages to the South Orkneys and farther south. The Argentine government eventually awarded her an honorable retirement and preserved her as a museum at a dock in Buenos Aires. Along with Robert Falcon Scott's *Discovery* and Roald Amundsen's *Fram*, the *Uruguay* is one of only three surviving ships from Antarctica's Heroic Age.[17]

CHAPTER 9

Charcot and the *Français* Explore the Antarctic Peninsula: 1903–1905

Jean-Baptiste Charcot, the 36-year-old Frenchman who commanded the expedition that Galíndez reported lost, was a complete newcomer to the Antarctic. He also had little experience leading a scientific expedition. Although he was an experienced sailor and an excellent navigator, all his prior voyages had been as a gentleman yachtsman. His formal training was as a physician, because his father, a wealthy and distinguished neurologist, had wanted his son to follow him professionally. But Jean-Baptiste's heart was elsewhere. After his parents died in the mid-1890s and left him a substantial inheritance, the younger Charcot gradually abandoned medicine and spent increasing time doing what he really loved, sailing his yacht to remote places like Iceland and the Faroe Islands, high in the North Atlantic. In 1901, Charcot learned that four expeditions were heading for the Antarctic and began to think about going south himself. He discussed the idea with Adrien de Gerlache, the *Belgica* expedition leader, in January 1903 and found him enthusiastic. Indeed, de Gerlache agreed to come along as an adviser. Several months later, the world heard that Nordenskjöld was lost. That settled it. Charcot would go south and join the search.

Charcot knew that he wanted to explore and to carry out a scientific program. Other than looking for Nordenskjöld, however, he had no idea where he was going to work in the south. Despite this ill-defined program, the French scientific establishment embraced the idea of a French Antarctic expedition. Charcot was prepared to finance the venture himself to the extent he could, but his personal fortune was insufficient to fund a major Antarctic expedition, even after he had sold his most prized possession, a painting by the eighteenth-century French master Jean-Honoré Fragonard. The Paris newspaper *Le Matin* launched a subscription drive on his behalf, and the French government made a helpful grant, but money was still tight. In the end, the scientific staff participated as unpaid volunteers, the two naval officers lent by the French navy sailed with only base pay, and the crew—virtually all of whom had been with Charcot on his yacht voyages—received nothing extra for this risky expedition. Charcot himself wore three hats. He was leader, ship's captain, and doctor.

The expedition's ship, the *Français*, was custom built in just five months. She was a 245-ton vessel, roughly 150 feet long, built more for strength than comfort. Significantly, Charcot economized by equipping her with a used 125-horsepower steam engine. There was one very un-French thing about the ship's stores—Charcot stocked little wine because he disapproved of heavy wine drinking with meals. Instead, he supplied the ship with a large library.

The *Français* left France on August 25, 1903, with 21 men, including de Gerlache. Charcot stopped en route at Madeira and met with Captain Glydén of the *Frithjof* to discuss plans for the Nordenskjöld search. Something much more important, however, happened here: de Gerlache and two of the scientists informed Charcot that they were quitting the expedition. All three left when the *Français* reached her next port, in Brazil. De Gerlache's public explanation was that he missed his fiancée. The truth was less politically palatable. He had secretly been dubious about Charcot's vague plans from the beginning and had come along only because Charcot had pleaded with him. The first stage of the voyage had added fatal doubts—about both the ship and Charcot's leadership—to his earlier concerns.[1]

De Gerlache's departure posed a major problem. Could Charcot, who had had no ice experience, lead an Antarctic expedition by himself? He talked with André Matha, the senior of the two naval officers aboard and the expedition's second-in-command, with J.-J. Rey, the meteorologist, and with Paul Pléneau, the photographer. All were eager to continue. Charcot then gathered the crew and offered each man the opportunity to return home. All wanted to go on. Although he could not replace de Gerlache's Antarctic experience, Charcot could at least replenish his scientific staff. He wired home for replacements.

When Charcot reached Buenos Aires on November 16, he learned that the *Uruguay* had sailed weeks earlier to rescue Nordenskjöld. The news actually came as a relief, because Charcot could not have gone south himself at this point. Not only did he have to wait for his new scientists to arrive, but the *Français* had had to be towed to Buenos Aires from Montevideo after the geriatric engine had broken down. Repairs were essential. The Argentine government, as eager to aid the French as it had been to assist the Swedes, provided all the repair help needed. It also arranged to transport expedition equipment to Ushuaia aboard a government vessel and to provide free coal when Charcot arrived there to collect his equipment.

Two weeks later, the *Uruguay* returned with Nordenskjöld's party. Now Charcot was free to concentrate on his own expedition. After consulting with Nordenskjöld and Larsen, as well as with Bruce, who had just arrived from the Falklands, he decided to work on the west side of the Antarctic Peninsula. There he would explore the west coast of the Palmer Archipelago, continue south to Adelaide Island, and then try to reach Alexander I Land, last seen in 1898 by de Gerlache, but never surveyed since its discovery by Bellingshausen in 1821.

Nordenskjöld's arrival afforded Argentina another way to help Charcot: the navy offered him the dogs Nordenskjöld had given to Irízar. Charcot had never considered using sledge dogs, perhaps because his adviser, de Gerlache, had had none, and no one aboard the *Français* knew how to work with them. Still, he accepted. He would figure it out when he had to. The *Uruguay*'s captain contributed another animal. He was a pet pig named Toby, about whom Irízar said to Charcot, "He has been our mascot; I hope he will be yours!"[2] As a final gift, the Argentine government promised to send a ship south the following summer to pick up any messages left at Deception and Wiencke islands.

The two new scientists—Ernest Gourdon, a geologist, and J. Turquet, a biologist—showed up in early December. They were completely unknown quantities to Charcot, and he commented that neither looked much like his idea of an explorer. Still, they were there and he would take them. One additional new man joined the expedition in Buenos Aires. In an echo of the *Belgica* expedition, Charcot fired his unsatisfactory original cook and replaced him with a local hire. Fortunately the similarity to the *Belgica* experience ended there, since the new cook, a man with a mysterious past named Rozo, proved to be an excellent choice. So, too, were the new scientists, especially Gourdon, who quickly overcame Charcot's doubts.

The *Français* left Buenos Aires on December 23 with twenty men, Toby the pig, and a cat. Charcot sailed first to Tierra del Fuego for surveying and scientific work, stopping en route to pick up the dogs at Staten Island. After completing his work in the region, he made one last stop, at Ushuaia, to collect his equipment and Argentina's donated coal.

Charcot at last departed Tierra del Fuego on January 27, 1904. The ensuing Drake Passage crossing was so rough that nearly everyone aboard was seasick, but the men forgot their queasy stomachs when they saw their first iceberg, on February 1. Then the mist cleared, and they could see the whole coast of Smith Island, "a scene which was so excessively beautiful in its sinister grandeur that one almost felt a sort of pain in the contempla-

Jean-Baptiste Charcot and most of the scientific staff of the *Français* expedition — *In front, from left to right, J.-J. Rey, Charcot, André Matha; in rear, left to right, Paul Pléneau, J. Turquet, and Ernest Gourdon. (From Charcot, Le 'Français' au Pôle Sud, 1906)*

tion of such magnificence."[3] Despite his delight with the landscape, Charcot did not linger in the South Shetlands, and he completely bypassed Deception Island, where he had said he might leave a message. Instead, he immediately pushed south along the west coast of the Palmer Archipelago.

The magical quality of the voyage came to an abrupt end when the recently repaired engine failed again. Relying on his sails for the time being, Charcot examined several possible stops for repairs before mooring the ship to the fast ice in Flanders Bay, just north of the Lemaire Channel, on February 7. There the crew spent eleven grueling days laboring over the engine. Others, however, had a better time. The staff pursued scientific matters, and Charcot allowed the delighted dogs to run free. He, himself, spent one day skiing toward the end of the bay with Matha. This was his first time on skis, and, he wrote, "I got into a fearful tangle, inevitably ending up by falling over . . . [and at] times, it seemed to me that I had become one of those magazine puzzles: 'Today's problem. . . .'"[4] Matha had equal difficulties. Worse, he refused to put on his sunglasses and spent the next day in agony from snowblindness. These were definitely polar neophytes, learning their craft the hard way.

By February 19, the crew had repaired the engine well enough to continue. For the next few days, Charcot cruised about the Palmer Archipelago and the Gerlache Strait. De Gerlache's survey had been the first serious work here, but inevitably, he had made errors—hardly surprising given the intricate maze of islands, channels, and bays he had encountered. Charcot thus was sailing in an area still ripe for discovery. Fall, however, was on the way, and he had yet to find a winter base site. He also needed to leave the promised message cairn on Wiencke Island. In

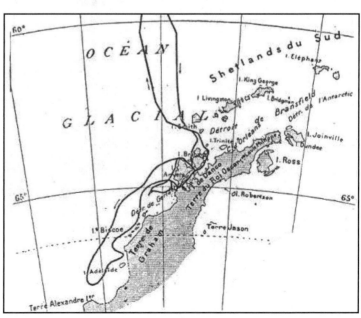

The 1903–05 track of the *Français* expedition in the south — From Charcot's Le 'Français' au Pôle Sud, *1906, this map shows the expedition's route in summary, related to the geography of the region as Charcot knew it.*

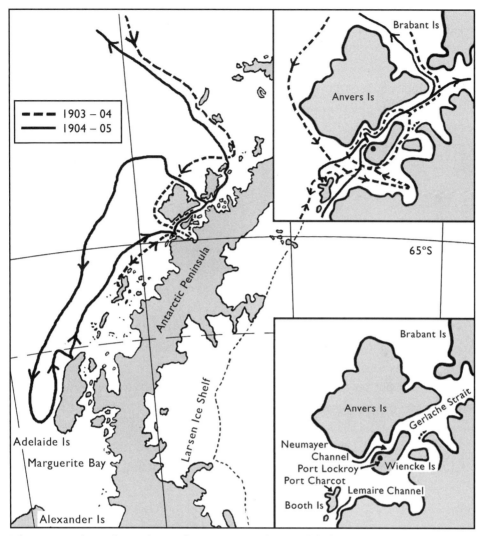

The *Français*'s track on the modern map — *This simplified version of Charcot's journey through the complex maze of waterways between the islands of the Palmer Archipelago and farther south shows the route of the Français for the summer explorations of 1903–04 and 1904–05. It was just off Adelaide Island, where the 1904–05 track jogs sharply to the right, that the Français ran aground. The small inset maps provide detail for the portions of the voyage in the southern part of the Palmer Archipelago and just to the south of there, including showing the locations of Port Lockroy—discovered by Charcot and used as a haven to repair his battered ship—and of Port Charcot, on Booth Island, where Charcot spent the winter of 1904.*

sailing around the island, he discovered a small inlet just off the Neumayer Channel with a wonderfully well-sheltered anchorage. Charcot named the location Port Lockroy after the French Minister of Marine. It was perfect as a wintering site, with two exceptions: not only was it farther north than Charcot had hoped to establish his base, but the excellent shelter it promised, although good for comfort, was a problem for useful meteorological work.

Charcot built his cairn, and everyone enjoyed his first experience of a penguin rookery. Then they continued on to use the last days of summer to explore and search for a winter home. Charcot's first objective was to sail through the Lemaire Channel to investigate an eastward opening, just to the south, that de Gerlache had thought might lead to the Weddell Sea. The Lemaire Channel, however, was full of ice, far too heavy for the *Français*'s small engine to penetrate. Charcot next tried to go around the islands on the west side of the channel. To his frustration, the ice was nearly as thick there.

He then retreated to a bay on the north side of Booth Island, the landmass forming the west side of the Lemaire Channel.[5] While waiting for the ice to open, he spent his time investigating the area and soon concluded that it offered a good winter site. Its only drawback was that it was nearly as far north as Port Lockroy. The next day, February 23, he again tried to force his way through the Lemaire Channel. When he found the ice still impossibly dense, he gave up after a few miles and returned to the small cove on Booth Island. Then an engine boiler burst. Another day lost to repairs.

On the 25th, Charcot left Booth Island once more. Although the weather was gorgeous, the sea, still choked with ice, was much less friendly. When the *Français* finally broke through west of Booth Island, Charcot was too far out from shore to see the coast south of the Lemaire Channel. He reached his farthest south for the season, at 65° 58' S, two days later. There the pack was so thick and the feeble engine in such dreadful shape that he decided to return to Booth Island for winter.

It was March 3 when Charcot anchored the *Français* at Booth Island, in the small bay now known as Port Charcot. Although he was only at 65° 4' S, not nearly as far south as he had hoped to winter, he was at the farthest limit of the *Belgica*'s Antarctic Peninsula explorations. That made the location a promising venue for scientific work. A few days after arriving, Charcot stretched an anchor chain across the entrance to the narrow inlet to keep the pack ice out. He relied on the shallow water where the ship lay to protect her from icebergs.

The *Français* was to be their winter home, but the staff would also be doing scientific work on shore. Charcot set the dogs to work pulling sledges loaded with the materiel for the shore installations. At the same time, he banished them from the ship, because they had been such a nuisance aboard. Set free on a small island in the bay where he thought they were safely confined, the dogs soon escaped, gleefully attacked snoozing seals, and created havoc in a penguin rookery. A horrified Charcot had the dogs tied to a line near the ship.

The scientific work ashore began as soon as the men finished unloading. Many of the instruments proved inadequate, but Libois, the carpenter-stoker, was particularly inventive at improving matters. Among other things, he transformed used provision tins into floats, buckets, stovepipes, tubs, and even flower pots for the

The northern end of the Lemaire Channel — *The scenically spectacular strait between the Antarctic Peninsula mainland, on the left, and Booth Island, on the right. Port Charcot, where the* Français *expedition spent the winter, is on the northwest coast of Booth Island, out of sight beyond the promontory on the right side of Booth Island as shown here. (Author photo)*

geologist, Gourdon, who had brought a gooseberry bush and tried, unsuccessfully, to raise asparagus.

Charcot's men settled down for the winter very cheerfully. They ate well because the cook, Rozo, a man who padded about the deck in his slippers, understood the art of cooking seal and penguin and baked fresh bread three times a week. Fancy rolls appeared on Sundays. And Charcot worked hard to banish boredom from the long winter evenings. He organized classes for the men, and for everyone there was that well-stocked library. Some of the men discussed the news in the months-old newspapers. Two even studied the racing form, picking winners from races long since lost and won. There were also the usual parties in honor of every possible occasion, and when weather permitted, Charcot organized ski and sledge contests.

Although Charcot did all he could to keep his men happily occupied, there were limits to what he could do to make them comfortable aboard the small, cramped *Français*. Keeping the ship warm during winter was also difficult. Charcot's own cabin, the farthest from a reluctant stove, was the coldest of all. His one compensation was that several seals took up residence beside the ship and soothed him to sleep at night with their snores.

The cold also posed difficulties for scientific work ashore. So too did the nearly constant wind that plagued them during the long winter nights. Charcot wrote,

> We [would] . . . set out, equipped with what we considered our best lantern, protecting it with care against the wind; and just as we were about to use it, a gust would blow it out. As it was no use thinking of taking matches in the storm and snow, we must needs return [to the ship] to relight the lantern, and . . . we often had to make the journey three or four times over.[6]

Charcot had only one major health problem to deal with during the winter. Matha, the second-in-command, fell ill in mid-July, perhaps with scurvy, perhaps with heart problems. Charcot was unsure. He knew, however, that Cook's treatment on the *Belgica* had succeeded. Thus he prescribed hours naked in front of a red-hot stove in addition to the fresh meat that all the men were already eating.

Matha's illness concerned Charcot not only because he was worried about the man, but also because it could affect the spring sledging program. He was still eager to investigate de Gerlache's possible strait south of the Lemaire Channel, and he now planned to sledge there over the sea ice. A week after Matha fell ill, Charcot

The animals of the *Français* — Top Left: *The dogs donated by Argentina investigate a Weddell seal at Port Charcot.* **Top Right:** *Toby the pig, "notre ami Toby," peeks through an interior cabin window.* **Bottom:** *Two of the staff work aboard ship during winter, accompanied by the ship's cat. (All from Charcot, Le 'Français' au Pôle Sud, 1906)*

and two others skied down the frozen-over Lemaire Channel to find a place for a provision depot. Three days later, leading a group of six men with two dog teams, he established the depot at the chosen site, at the southern end of Hovgaard Island, just south of Booth Island.

The depot party returned just in time. That night, a raging storm slammed the ice in Port Charcot against the *Français*'s hull. Charcot hurriedly evacuated the ship, the crew carrying the still-ill Matha ashore in his sleeping bag to a hut on shore. The rest of the men huddled in tents as everyone wondered if the ship would survive. To their relief, the storm relented early in the morning, and the *Français* was safe. For then, that is. Another violent storm a few days later led to an equally frightening time on shore as the Antarctic weather attacked their home.

Matha's health improved, and men and ship survived the storms. Not so the sledging plans. When gales repeatedly destroyed the sea ice almost as soon as it formed, Charcot realized that his spring trip would have to be by boat. Ironically, that meant waiting until the sea ice broke up even more.

When the weather finally warmed in November, the men found that they preferred the cold, because the damp air made everyone moody and irritable. But when Charcot set them to work collecting penguin eggs, the resulting exercise and fresh new cuisine restored their good humor. Moreover, the penguins, the source of the eggs, provided delightful entertainment. Charcot fancifully speculated about what the penguins might think of these creatures who were raiding their nests:

> What were these strange beings doing, looking like giant penguins, sometimes beneficent, sometimes hostile, who could do such amazing things? . . . The legends would pass from one generation of penguins to the next, and perhaps some forgotten tool, some ancient empty tin, some pieces of wood, preserved as idols, would be shown to the free-thinking penguins who would shrug their shoulders and, in spite of all, would still express their doubts. However other penguins from distant regions of the Antarctic . . . would nod their heads and tell how in their part of the world too, prodigious miracles had been performed; that there too, almost identical legends prevailed. . . . And who knows whether in this icy world, so previously peaceful, penguin wars might not break out to decide whether de Gerlacheism, Scottism, Brucism, Nordenskjöldism, or Charcotism should triumph![7]

Watching penguins was good fun, but what Charcot really hungered to do was resume his exploration program. On November 24, he and four others set out in a heavily loaded whaleboat. Five days later, they stood on the mainland coast south of the Lemaire Channel. The previous three days had been a difficult struggle across the pack ice, sometimes too loose to walk across but still often too dense to row through. At times, they had had to pull the boat out and haul it over the ice. Then, when an opening appeared, it was back into the water to row a bit. Their reward was another challenge, a strenuous climb to the top of 2,900-foot-high Cape

Charcot converses with Adélie penguins at Port Charcot. *(From Charcot,* Le 'Français' au Pôle Sud, *1906)*

Tuxen. A wonderful view repaid their effort, but Charcot still could not see into de Gerlache's possible strait. Perhaps, he thought, an island he spotted 4 miles out in the pack might provide a better vantage point. This became his next goal.

After leaving a cairn at the foot of Cape Tuxen, Charcot's team set out for the island and quickly found themselves back in the same battle with the ice. They finally reached Charcot's target on December 2. It turned out to be a collection of small islets (today called the Berthelot Islands). The view from the 650-foot summit of the largest bit of land there gave him his answer. Nothing on the opposite shore even suggested a strait.

On December 5, after an equally difficult return trip, the travelers were home at the *Français*. Matha, long since recovered from his mysterious winter illness, had been working hard in Charcot's absence to prepare the ship for departure. There was much to do, because hundreds of feet of winter ice now lay between the *Français* and open water. Charcot tried blasting and sawing the ice, a nearly useless effort until strong winds came to his aid in mid-December. Meanwhile, the crew continued trying to restore the unreliable engine to health.

One event dampened everyone's spirits in the last days at Port Charcot. Toby, the pig, died. Charcot wrote, "[We] were all very upset; we really loved him. Poor old pig! What a strange existence his had been!"[8] On one occasion during winter, Toby had swallowed at least half a dozen fishhooks when he stole some fish. Although Charcot's emergency surgery had saved him then, he had fallen inexplicably ill at the beginning of December. Now, though the crew had devotedly fed him spoonfuls of condensed milk for nearly three weeks, he was gone. His human companions buried him alongside a canine friend who had died earlier.

Just before leaving on Christmas Eve, the crew of the *Français* celebrated with a gramophone recital for the penguins. A cardboard Christmas tree hung with tinsel and toys, a gift from Charcot's sister, enlivened the wardroom, and flags and Chinese lanterns decorated the crew's quarters. At midnight, everyone gathered, lit candles, and set plum puddings ablaze.

The *Français* sailed away from Port Charcot on Christmas Day 1904 bound, Charcot hoped, for new discoveries. He left the huts, a stock of provisions, and a whaleboat on shore just in case he had to spend another winter there.

Charcot began with a detour north to Port Lockroy to update his cairn. Then he turned the *Français* south, into a battle with storms and icebergs that went on for days. When the weather finally improved on January 11, 1905, Charcot wrote, "The pack-ice was ahead of us, and the rapidly clearing blue sky revealed a wonderful spectacle. Never was the great calm after a storm more complete, more perfect, more impressive, and yet more radiant."[9] Beautiful yes, but that pack was impenetrable. It was maddening, because Charcot could see Alexander I Land, looming up far to the south, as well as magnificent mountains to the east, and the ice rendered both places unreachable. He coasted along the ice edge for almost two days looking for a way through to Bellingshausen's discovery, but it was hopeless. As he gave up and turned north, he promised himself that he would try again.

On January 15, a few miles north of where the chart placed Adelaide Island, Charcot discovered an opening in the ice that led him closer to the mountains he had seen several days earlier. Because the map showed Biscoe's island to be only 8 miles long, Charcot thought he had made a major discovery—never-before-seen continental land to the east of Adelaide Island. He excitedly named it Loubet Land for French President Émile Loubet, an enthusiastic expedition supporter. Then disaster struck. Charcot wrote, "Suddenly . . . we felt a terrible impact, the masts trembled and bent to the extent that we feared that they might come down and the boat seemed to climb almost vertically, with an ominous crack. . . ."[10] The *Français* had hit a reef because Charcot had naively assumed that the nearby icebergs meant deep water. That would be true if they were floating, but not if they were grounded, as they were here. *These* icebergs were indicators of shallow water, not deep.

Water surged in at the bow. With the engine nearly useless, the men desperately worked the pumps by hand while the carpenter applied an emergency patch to the hole in the hull. Charcot knew the damage was too serious to permit further exploring, and renewed storms increased the danger to the crippled ship as she retreated north. At the end of January, Charcot finally reached the relative safety of Port Lockroy. There he examined his hull and found the damage to be even worse than feared. The crew spent ten days at Port Lockroy making the best repairs they could before Charcot took the still-leaking *Français* back to civilization.

On March 5, the *Français* anchored at the small Patagonian port of Puerto Madryn, several hundred miles down the coast from Buenos Aires. Here Charcot heard of what had taken place in the world in his absence. The most important event was the Russo-Japanese War, begun after he had reached the ice. Little had happened in the Antarctic regions, but there was the voyage of the *Uruguay*, sent as Argentina had promised she would, to look for Charcot's messages. Now he learned that she had returned with a report that his expedition was lost, and that there had been much concern about his fate. Indeed, the telegram he sent from Puerto Madryn announcing his safe return reached France just in time to head off a rescue effort.[11] And there was sad news on a personal front: his wife, unhappy with his long absences from home, had decided to leave him.

Charcot arrived at Buenos Aires on March 14 to an enthusiastic welcome, and the Argentine government offered to fund all needed repairs for the *Français*. They turned out to be considerable. When she was dry-docked, the engineers found her so severely damaged that they were amazed she had reached Buenos Aires before sinking. Still, she was repairable. When the work was complete, Charcot sold her to Argentina to service the Staten Island and Laurie Island meteorology bases. The Argentines renamed her *Austral* and used her as the Laurie Island relief ship for the 1905–06 and 1906–07 seasons. Then, in December 1907, she grounded on a reef in the Río de la Plata just as she began a voyage to establish a meteorological station at Charcot's base on Booth Island. This time, she was a total wreck. So were the plans for the base, abandoned with the loss of the ship.

The men of the *Français* expedition reached France aboard a commercial ship on June 6, 1905. A venture that had begun with an unclear purpose had returned with significant results. Charcot and his staff had made comprehensive scientific observations; carried out hydrographic surveys of 550 miles of coast; and charted and explored almost 600 miles of new coast and islands, contributing many new names to the map. The scientific results eventually appeared in 18 volumes, all published at French government expense. Charcot's expedition narrative, *Le 'Français' au Pôle Sud,* appeared in 1906, the same year as the first scientific volume.

The *Belgica, Antarctic, Scotia,* and *Français* had all sailed south to a land where no other ships routinely traveled, a dangerous and risky business. But by the time the *Français* expedition ended, that world was changing. Men were already living on Laurie Island year-round, and the whalers were arriving.

Chapter 10

Whalers and Politics: 1904–1918

Permanent year-round human presence in the Antarctic began when William Speirs Bruce turned his South Orkneys meteorological base over to Argentina in early 1904. Another breed of long-term residents arrived the following summer. They were whalers, led by Carl Anton Larsen.

LARSEN AND THE BIRTH OF ANTARCTIC WHALING

Larsen had seen large whale populations in the Antarctic on his two *Jason* voyages to the Peninsula Region in the early 1890s and as captain of Nordenskjöld's *Antarctic*. He had also seen South Georgia and concluded that it offered excellent sites for a whaling base. Confident of his ability to overcome the technological problems that had defeated him and the Dundee whalers, he was determined to pursue matters.

Larsen was more than a simple, hopeful whaler when he arrived in Buenos Aires in late 1903 as a member of the rescued Nordenskjöld party. All of Nordenskjöld's men were celebrities, and in early December, a group of prominent Buenos Aires businessmen organized a banquet in their honor. After Nordenskjöld spoke, Larsen had his turn. He began by thanking the Argentines for the rescue. He then asked the audience why they were ignoring the whales right at their front door. There were thousands, he said, huge whales, just waiting to be taken.

The Argentine business community responded enthusiastically with an offer of capital to establish a whaling station on South Georgia. Larsen then returned to Norway in hopes of interesting his own countrymen. The Norwegians, however, proved to be a much tougher sell, and Larsen accepted the Argentine offer. The first Antarctic whaling company, Cia. Argentina de Pesca (or simply Pesca), was registered in Buenos Aires on February 29, 1904. The money came from Argentina, the personnel and equipment from Norway, and Larsen was the enterprise's profit-sharing manager.

Even before Pesca was founded, someone else had already fired the first shot in modern Southern Hemisphere whaling. At the close of December 1903, Adolf Amandus Andresen, a Norwegian who had settled in Punta Arenas, had harpooned a humpback whale in the Strait of Magellan. Andresen, too, had noticed the large local whale population, but since he was not a whaler by profession, he had gone home to Norway to learn the business. He then returned to Punta Arenas and mounted a whale gun on a tugboat to make his historic kill. Andresen's next efforts would be more professional, and like Larsen, he would become a major figure in the Antarctic whaling story.

As for Larsen, he was hard at work in Norway, recruiting veteran whalers and assembling equipment for the new company. The most critical element was a new steam-powered whale catcher named the *Fortuna*. Larsen had her specially built for Antarctic whaling. The key to taking the whales that had eluded him on the *Jason* voyages, she was larger, faster, and stronger than the ships used in the Arctic.

Larsen reached his pre-selected site on South Georgia on November 16, 1904. It was a small cove in Cumberland Bay that J. Gunnar Andersson of Nordenskjöld's expedition had seen in 1902. The Swede had named it Grytviken (Pot Cove) because of the many rusting try-pots that nineteenth-century elephant sealers had left scattered about—mute evidence that this had been a favored spot for an earlier generation of hunters. Now the whalers would use it for more than a temporary anchorage. Larsen's ships had sailed from Norway with a factory plant, materiel and stores, and two prefabricated houses. They had also brought along, disassembled, the house that Larsen had called home for nearly twenty years. It would serve as the manager's house and administrative office.

The *Fortuna* harpooned her first whale, a humpback, on November 27. She began whaling full time in mid-December, when the station was ready to process the catch. Larsen's company took 183 whales during its first year. There were so many whales right in Cumberland Bay that the *Fortuna* did not even need to leave the island.

Carl Anton Larsen and Grytviken, the South Georgia whaling station he founded in 1904–05 — **Left:** *Larsen, widely regarded as the father of Antarctic whaling, had fully recovered from his ordeal on Paulet Island when he founded the first Antarctic shore whaling station. (Nordenskjöld et al.,* Antarctic: Två År Bland Sydpolens Isar, *1904).* **Right:** *Grytviken, photographed by the* Deutschland *expedition in 1911 (see Chapter 12). (Filcher,* Zum Sechsten Erdteil, *1922)*

Grytviken whaling station workers atop a blue whale carcass in 1911 — *In the early years of South Georgia whaling, four-fifths of the whales taken were humpbacks, but the whalers were also taking blue whales—the largest whale species—from the beginning, including eleven in the first summer at Grytviken. The biggest blue whale ever killed by the whalers was brought into Grytviken in 1912. She was 112 feet long and weighted in excess of 170 tons. (Photo from Filchner,* Zum Sechsten Erdteil, *1922)*

Larsen was a whaler, but as he had often demonstrated, he was also a man of broad interests. He thus readily agreed when the Stockholm Museum of Natural History asked him to include Erik Sörling, a biologist, in his team. Sörling remained on the island until September 1905 preparing specimens from whales, seals, and birds. He and Larsen also began meteorological observations. These would continue without interruption until Britain and Argentina fought on South Georgia during the 1982 Falklands/Malvinas War.

Leases and Politics: The British Take Control

Early seeds of what would later become conflict over South Georgia had been planted by the whalers' arrival in November 1904. Larsen, managing an Argentine company, had established his station on the island without seeking anyone's permission. In the meantime, the British, who claimed the island, had somehow missed the fact that whalers had taken up residence. Indeed, a few months after Larsen arrived, they awarded South Georgia to another commercial venture, a group of farmers led by Ernest Swinhoe, an Englishman who had settled in southern Chile.

Larsen was already well established at Grytviken when Swinhoe's group formed their South Georgia Exploration Company in March 1905, chartered a small ship, the *Consort*, and sailed to Stanley in the Falklands in July to pick up their lease.* The lease granted them mining and grazing rights to all of South Georgia at a cost of £1 per year—a clear indication of what the British thought the island was worth. The would-be colonists next set out for their new home. Because their site needs were similar to those of the whalers, they headed right in when they saw Cumberland Bay.

Despite his surprise at seeing buildings already there, Swinhoe went ahead and landed, and then turned the *Consort* over to her captain for sealing around the island. On August 22, lease in hand, Swinhoe called at the whaling station. His meeting with the resident manager, Lauritz Larsen, Carl Anton's brother, was rather frosty, some say. But welcome or not, Swinhoe's party stayed where they were, living in tents near the whaling station. In late September, Swinhoe presented Larsen with a letter protesting Pesca's presence. By then, however, he was already concluding that South Georgia was much better suited for a whaling base than for a farming colony.

When the *Consort* returned in late November, Swinhoe reloaded his expedition and sailed for the Falklands. There he handed a report to the Falklands governor describing the situation at South Georgia. The British, however, had already learned that whalers were on the island as a result of reports that cargoes of whale oil had arrived in Buenos Aires. There had also been fallout from Swinhoe's contact with Lauritz Larsen. The Norwegian consul (who was also a Pesca director) and an Argentine government official had visited the British ambassador in Buenos Aires and produced Swinhoe's protest letter. Perhaps disingenuously, they claimed that the Pesca directors had not realized that Britain had claimed South Georgia. Now that they knew, they said, the company was prepared to pay for a lease.

The British clearly had to resolve this situation. Rumors were rampant at Stanley during December 1905 that the Admiralty would send a ship to South Georgia, and that the only whaling lease would go to Swinhoe's group. One part of this was true. On February 1, 1906, Captain Michael Hodges sailed H.M.S. *Sappho* into Cumberland Bay and spent several days meeting with Carl Anton Larsen. Hodges' report indicates that Larsen was cordial. Some whalers, however, later claimed that Hodges had been much less friendly. According to them, the British captain ordered Larsen to lower the Argentine and Norwegian flags. When Larsen refused, Hodges informed him that the *Sappho* would shoot them down in 30 minutes. Larsen gave in and lowered his colors. Even if this dramatic story is untrue, the *Sappho*'s visit did fulfill its intended purpose—a forceful statement that South Georgia was a British possession. Britain then granted Pesca a lease making the whalers' occupation legal in British eyes.

---

* Until the mid to late twentieth century, the capital town of the Falklands was commonly referred to as "Port Stanley." Today, however, it is simply called Stanley, and I have used this modern usage throughout.

Hodges' visit added an Anglo note to the site of Pesca's operation. Before the *Sappho* left the island on February 5, her cutter surveyed Cumberland Bay. The resultant chart renamed the small cove the whalers had called Grytviken. It became King Edward Cove, and the spit of land where the meteorology station had been established was re-christened King Edward Point. The title Grytviken, however, remained as a name for the whaling station.

In October, Britain tightened its control on the South Georgia whalers with an ordinance making it illegal to hunt from the island without a license or lease. She also imposed royalties on each whale caught.

After Pesca's, the next lease went to Swinhoe, at a cost of £250 a year. In 1908, having failed to raise the necessary capital to begin whaling, he sold the lease for £1,500 and dropped out of the picture. Others, however, had no difficulty finding capital for Antarctic whaling once Larsen had demonstrated that the right equipment made it feasible. The six years leading up to World War I saw the British grant seven more whaling leases for South Georgia, as well as leases in the South Shetlands and South Orkneys. During South Georgia's heyday, the whalers operated six major shore whaling stations on the island, all on the northeast coast.

The advent of Antarctic whaling had major political repercussions. Before the whalers arrived, the British government had paid little attention to South Georgia, indeed had even questioned whether the island was worth having. The whalers' success removed all doubts. On July 21, 1908, Britain became the first country in the world to officially assert sovereignty over territory in the Antarctic regions. The government formalized the claim in a "Letters Patent" that consolidated selected British claims and discoveries from James Cook onward. Specifically, the action designated everything within a pie-shaped sector of the Antarctic regions between 20° W and 80° W as the Falkland Islands Dependencies. This included South Georgia, the South Sandwich Islands, the South Orkneys, the South Shetlands, the Antarctic Peninsula, and Coats Land.[1] Many years later, both Argentina and Chile would stake their own competing claims to much of this area. In 1908, however, neither country objected to Britain's action. Nor did they protest when the British issued a second, clarifying, Letters Patent in 1917.

The natural next step was to establish an on-site government. In 1909, the Falklands governor appointed James Innes Wilson as resident magistrate for South Georgia. His job was to administer whaling and sealing licenses, serve as postmaster, and do anything else a magistrate might need to do in a small, isolated community. Wilson reached Grytviken at the end of November and moved into a house, rudely furnished with tables and chairs made from old packing cases, that had been purchased from the whalers. In the late 1920s, a whaler related a colorful story about Wilson's arrival:

> When the first magistrate came down from Stanley . . . he brought a Gladstone bag full of clothes and two hundred quid of Falkland Island [sic] stamps to start the post-office racket. He let his bags down over the side on a heaving line into the boat to go ashore, but he was so drunk he paid out too quick and the lot went in the water. When they'd finished shouting and roaring and pulling them out, everything was so wet he took it all back on board. He got a fireman to dry his clothes . . . , but the sheets of stamps he laid out sticky side up in his cabin all over the deck and his bunk. He had a booze-up . . . that night and staggered off to turn in about two. He pulled his clothes off in the dark and rolled into his bunk naked and lay there snoring all night like a sea elephant in a mud bath. . . . When this crazy bastard woke up next morning he was plastered from head to foot, and the steward had to peel the stamps off him with hot water and a sponge. The first mails he sent away had the stamps pasted on because all the gum was gone lubricating the sweaty hide of his Britannic majesty's proconsul. . . . [2]

True or not, the story suggests much about the whalers' feelings regarding the arrival of a magistrate.

That first official mail left South Georgia on December 23, 1909. Wilson also conducted a census during his first month on the island. At the end of December 1909, there were 720 people there, including himself, three women, and one child. Just over 80% were Norwegian.

Life on South Georgia

The South Georgia shore stations were more than just whale-processing plants. Grytviken, for example, grew to include a cinema, jail, church, soccer pitch, and cemetery. A few senior personnel even brought their wives and children along. And there were births. The first was a girl born at Grytviken on October 8, 1913. Her father, Fridthjof Jacobsen, the under-manager of the whaling station, was Carl Anton Larsen's nephew. He had arrived a year earlier with his wife and their three children. The proud parents named their fourth child Solveig Gunbjörb, and the magistrate issued her "South Georgia Birth Certificate No. 1."[3]

Most whalers left at the end of the summer, but a few remained year-round to keep the stations going and prepare for the next summer season. The winter was much more relaxed than the frantic summers, but even during summer, the whalers occasionally found time for fun. Early on, some of the more adventurous among them probed the edges of South Georgia's interior. Although no reports exist of anyone going deep into the heart of the island's forbidding mountains, at least some at Prince Olav Harbor spent their rare days off skiing across the narrow northern part of the island to King Haakon Bay and back.[4]

In November 1911, the Larsen brothers brought a living slice of Scandinavia to South Georgia—reindeer. They set ten free in Ocean Harbor, hoping these animals

South Georgia reindeer near the Stromness whaling station, 1995. *(Author photo)*

would multiply into a herd that could provide both sport and a different kind of meat for the table. The Larsens released a second group in Stromness Bay in 1912. The descendants of this second group all died in an avalanche in 1918, but in 1925, other whalers introduced seven replacements. The reindeer from the 1911 and 1925 groups adapted so well that several thousand of their descendants now populate the island. They exist, however, in two entirely separate herds, because glaciers and mountains between them prevent their joining up.

The Grytviken church arrived in 1913, a lovely little building imported from Norway by Carl Anton Larsen. The first pastor, Kristen Løken, consecrated it on Christmas Eve. The very next day, Løken presided over the first christening in the new church, that of Solveig Gunbjörb Jacobsen, the baby who had been born in October. Løken soon left, perhaps because he found that, as he wrote, "Christian life unfortunately does not wax strong among the whalers."[5] His successor lasted little longer, and after he departed in 1916, almost ten years passed before the next ordained pastor arrived. In the meantime, the building found its greatest use as a cinema. The whalers packed it on movie nights, at least twice a week during the high season, but after a dedicated cinema building opened in 1929, the church stood deserted much of the time.

The end of 1913 also saw the first South Georgia jail created. It was initially built to house a disruptive

The Grytviken church — *In 2009, restored and shining with fresh paint. (Author photo)*

Grytviken laborer. The magistrate and his official policeman, a man with so little to do that he usually served as a handyman, locked the prisoner in a storage shed and then constructed a cell around him. Once it was complete, the prisoner kicked a hole in the poorly built wall and ran up a nearby mountain. He surrendered after a cold night of freedom and promised to behave in the future. Although the British rebuilt the jail on a more permanent basis in 1914, there was little demand for it over the years. In 1931, a scientist working at Grytviken wrote that "it was chiefly when there was some work to be done, such as giving the quarters a new coat of paint, that the jail found a tenant."[6] But in later years it did find at least one use: in the 1950s, it served as a hostel for visitors to the island.

Reindeer, a church, a cinema, a jail, a wife in the parlor, curtains on the windows, and flowerpots on their sills. All were there, but little else was charming about the stations. They and their surroundings were filthy, and literally stank. This could be a shock to someone who had seen the island before 1904. Carl Skottsberg, who had spent that dreadful 1903 winter on Paulet Island with Larsen, visited his old captain at Grytviken in early 1909. He wrote, "I remembered a virgin Pot Harbor with luxuriant tussock-grass and roaring sea-elephants. . . . [Now] a strong smell of whale-oil mingles with the stink of the numerous carcasses on the shore where thousands of screaming gulls and cape-pigeons have an everlasting feast."[7]

Sealing Returns to South Georgia

Those were not only whalebones that Skottsberg saw among the carcasses. Some belonged to the great beasts whose roar he remembered, because elephant sealing on South Georgia had revived as a side effect of the whalers' arrival. Whalers were seeking oil from blubber, and that was just what elephant seals offered. It was thus inevitable that South Georgia whalers would begin to hunt them, just as their eighteenth- and nineteenth-century predecessors had.

Shortly after making its Falkland Islands Dependencies claim, Britain recognized the renewed sealing industry by imposing protective regulations. These prohibited sealing during the breeding season, November to February inclusive; forbade killing cows and pups; divided the coast of South Georgia into four divisions, allowing only three to be worked each season; and finally, proscribed killing fur seals, just in case anyone should find one. The regulations were so successful that South Georgia elephant sealing continued profitably until the 1964–65 summer, when the last whaling station on the island closed. Indeed, despite the 500,000 animals taken from 1910 to 1964, the elephant seal population actually thrived. There were as many, or more, in 1964 as there had been in 1910.

Pesca dominated South Georgia elephant sealing, but one significant independent sealer worked the island in 1912–13. That summer, a small American sailing ship arrived at South Georgia. The 384-ton *Daisy* and her captain, Benjamin

Dunham Cleveland, were both throwbacks to the whaling and sealing days of the nineteenth century—a sailing ship and a hard-bitten captain, determined to amass a cargo. There was something different, however, about this voyage. The American Museum of Natural History had paid Cleveland to take along a naturalist, 25-year-old Robert Cushman Murphy. Murphy was a new bridegroom who recorded his experiences in a diary and in letters for his bride. Decades later, Murphy, who had gone on to an illustrious career as an ornithologist, used his diary and letters to write a marvelous book, *Logbook for Grace,* which paints a vivid picture of South Georgia as he had found it.

Cleveland, for his part, was incensed by the sealing regulations presented to him when the *Daisy* reached South Georgia. Still, he went through the formalities and obtained the required license before beginning to hunt. Then, once out of sight of the magistrate, Cleveland simply ignored the new rules. The *Daisy*, the last sail-powered sealing ship to work South Georgia, left the island on March 15, 1913. An era dating back to the late 1700s had come to an end.

WHALING EXPANDS BEYOND SOUTH GEORGIA

By the time the *Daisy* sailed away, the modern Antarctic whaling industry had expanded well beyond South Georgia. When Larsen returned to Norway in mid-1905 to buy more equipment, he talked about the early success of his venture, and Christen Christensen, the ship owner who had sent Larsen on the *Jason* voyages, decided to give Antarctic whaling another try. Instead of building a shore station, however, he experimented with something new—a large mother ship that would serve as a floating whaling factory. Christensen purchased an old 2,400-ton vessel and converted her to his own design. He renamed her *Admiralen*.

The *Admiralen* left Norway in late October 1905 with two catcher vessels built along the lines of Larsen's *Fortuna*. Christensen's factory ship manager, Alexander Lange, began his whaling off the Falklands. He went on to the South Shetlands in mid-January and took 58 whales in a month. Then he headed home with his cargo. Although the *Admiralen* expedition was less profitable than his partners had hoped, Christensen felt that the knowledge gained had more than repaid the cost of this pioneering voyage. He was now convinced that there was money to be made in the Antarctic, and he began sending more and more factory ship fleets south.

Others were thinking along the same lines. In 1906–07, Adolf Amandus Andresen, the man who had shot that humpback whale in the Strait of Magellan in 1903, fitted out his own large ship, the *Gobernador Bories,* as a floating factory to be based at Deception Island. Andresen's wife, Wilhelmine, along with a parrot and an Angora cat, accompanied him for the summer.[8] Mrs. Andresen was the first recorded woman to reach the South Shetlands.

Three floating factories, including the *Admiralen,* joined Andresen's *Gobernador*

Bories at Deception Island in 1907–08. All anchored just inside Deception Island's Port Foster, at a small cove that soon became known as Whalers Bay. The whalers did look for other safe harbors in the region, in particular farther south along the Peninsula. Over the years, they used Port Lockroy, Foyn Harbor, Paradise Harbor, and Neko Harbor, among others. None of these, however, was as good as Deception Island for a summer's work. Port Foster thus became the center of Peninsula Region whaling beyond South Georgia.

Although the floating factories usually remained in their safe harbors, their catchers ranged all over the region. Before long, the whalers were revising the published charts, adding detail and corrections and contributing many new names to the map. They also left other reminders of their presence—rusting chains, mooring posts, and the remains of shipwrecks. And then there were the whalebones.... Today they still litter many shores in the region.

The Antarctic whaling industry was expanding furiously. The 183 whales processed by Larsen's Grytviken shore station in 1904–05 exploded to more than 1,000 in 1906–07, the first season that more than one whaling fleet operated in the South Shetlands. Four summers later, the South Georgia whaling stations and the fourteen floating factories in the Peninsula Region, most of them working in the South Shetlands and along the Antarctic Peninsula, took in excess of 10,000 whales. In recognition of the expansion of whaling beyond South Georgia, the Falkland Islands Dependencies government added a resident magistrate at Deception Island in 1911–12.

Whaling was a dangerous occupation, and many men and ships suffered accidents, ranging from minor to fatal, over the years. On January 22, 1908, Nokard Davidsen, a factory ship captain, fell overboard from a catcher vessel and drowned. His grave on a gentle slope at Deception Island's Whalers Bay was the first in what would become a full-fledged cemetery, complete with impressive stone monuments, sent from home by grieving families.

Just as the sealers had, the whalers looked out for one another, lending aid when one of them ran into trouble. Once sure that human life was safe, however, most were eager to claim a wrecked ship's cargo. The 1908–09 season saw a dramatic example.[9]

The *Telefon*, a supply ship for Christensen's whaling fleet, sailed to the South Shetlands in late 1908 with coal and other cargo for the fleet's factory ships, which were anchored at Deception Island along with Andresen's *Gobernador Bories*. On December 26, 1908, the *Telefon* hit an uncharted reef, now named Telefon Rocks, at the entrance to King George Island's Admiralty Bay. As water surged in through the gashed hull, the captain ordered everyone to abandon ship. That included his lone passenger, Mrs. Olava Paulsen, the wife of the *Admiralen*'s captain, who was aboard en route to join her husband for the summer.[10] The twenty men and one

The Deception Island whalers' cemetery, in 1961 — *From 1908 to 1931, 35 men, including Nokard Davidsen, were buried here, making this the largest concentration of graves in the Antarctic regions outside South Georgia. The fourteen-foot high concrete obelisk on the far left was the first monument in the cemetery, raised by the whalers in Davidsen's memory. It was completed in time for a memorial service that occurred less than three weeks after Davidsen's death. (Photo by J. B. Killingbeck, 1961. Reproduced courtesy of British Antarctic Survey Archives Service. Ref. no. AD6/19/3/C/B6b. Copyright: Natural Environment Research Council)*

woman, she wrapped in a tablecloth for warmth, shivered for hours in lifeboats before a whale catcher rescued them. (It was such an ordeal for Mrs. Paulsen that her husband refused to let her come south the next summer. Nonetheless, the Paulsens would later name their oldest son Arne Shetland.[11])

When the catcher delivered the *Telefon* party to Deception Island the next day, word of an abandoned ship with a valuable cargo launched a race to reach her and claim salvage. Andresen, aboard one of the *Gobernador Bories* catchers, won, and towed the wreck back to Deception Island. There he beached his prize at the north end of Port Foster in a bay later named Telefon Bay. Repairing the ship was impossible at that point, but Andresen had plans. Early in November 1909, he returned to Deception Island with a captain and crew of six specifically for the *Telefon*, as well as a specially trained diver and multiple bags of cement that would be used to patch the hull. Andresen's plan succeeded. In February 1910, the repaired *Telefon* sailed to Punta Arenas carrying 2,400 barrels of oil.

As time went on, improved charts showed more and more of the dangerous reefs and rocks littering the waters of the Peninsula Region. But sailing in these waters remained—and remains today—a dangerous business. Wireless, introduced to the Peninsula Region in 1912–13, did make it possible to call for help from a distance. The first wireless transmission from the area took place on January 25, 1913, when the Deception Island magistrate advised the Falklands governor in Stanley about the loss of a factory ship.

Whaling went right on booming after World War I broke out in August 1914, because the combatants used whale oil by-products in the manufacture of nitroglycerin explosives. The British, however, soon began pressuring the whalers, nearly all of them Norwegians operating under Falkland Islands Dependencies licenses, to stop selling to Germany. As a result, Britain gradually became the industry's only customer, and product prices temporarily plummeted. That, plus war-related demand for ships, led to a substantial drop in the number of Antarctic whaling fleets during the latter years of the war. Even so, a significant complement of whalers remained in the south throughout the conflict, not only in the summer, but also on South Georgia during the winter.

The coming of the whalers would prove to be a significant event for Peninsula Region exploration. Not only did the new hunters substantially improve the area maps, but they also supported the scientific explorers. Whalers provided information about local waters and shores, bases to sail from, and crucial supplies. Their ongoing presence in the south also offered the reassurance that rescue and relief were real possibilities. This was a huge shift from the grim circumstances faced by the *Belgica*, *Antarctic*, *Scotia*, and *Français* expeditions. From 1904 on, there was always a significant human presence in the Peninsula Region in the summer. People were there in winter as well, at least on South Georgia and at Argentina's tiny base on Laurie Island. Every Heroic Age expedition that visited this part of the Antarctic after Charcot's *Français* would receive significant help from the whalers. For the last one, that help would be crucial—almost certainly the difference between life and death.

Chapter 11

Charcot's Return with the *Pourquoi-Pas?*: 1908–1910

From 1907 to 1917, eight more scientific/exploring expeditions sailed to Antarctica. Three of the eight worked in the Antarctic Peninsula Region. Jean-Baptiste Charcot, back in the south again, led the first.

The Frenchman's objectives had been very vague in 1903, but in 1908, he knew exactly what he wanted to do—return to the area he had begun exploring with the *Français*. French delight with Charcot's first Antarctic expedition made raising funds for his second relatively easy. The public contributed generously, and the government made a substantial grant, lent Charcot three naval officers, and furnished numerous scientific instruments. Approximately half the budget went to pay for a new custom-built ship. Charcot christened her *Pourquoi-Pas?* (*Why Not?*) after several of his earlier ships. She was, in fact, the fourth with the name. Like the *Français*, she was a sailing vessel with an auxiliary steam engine. At 800 tons, in contrast to the *Français*'s 245, the *Pourquoi-Pas?* was a far more substantial vessel. And not only was her 450-horsepower engine more powerful, it was new and reliable.

Charcot had no difficulty assembling a talented scientific staff, but all except Ernest Gourdon, the geologist, were newcomers. Sadly for Charcot, the rest of his former close companions had unbreakable obligations elsewhere. As for the crew, eight men from the *Français* signed up. Charcot chose thirteen more from over 200 applicants.

The *Français* experience had taught Charcot a great deal, and he outfitted his new ship and expedition accordingly. Some of the returning crew members probably felt one change to be a distinct improvement: this time, Charcot brought a large supply of wine. He also planned to use innovative technology, the most important being an eight-horsepower motorboat specially modified for duty in the ice. He also took several experimental motor sledges. And for the *Pourquoi-Pas?* herself, he installed an eight-horsepower generator to provide electric lights. Originally intended as a luxury to be used only twice a week, the equipment worked so well that Charcot would use it constantly throughout the expedition.

The decision to try motor sledges grew out of joint trials Charcot had conducted with Britain's Robert Falcon Scott, who was planning his own, second, expedition to the Ross Sea. But neither Charcot nor Scott was the first to take a motor vehicle to Antarctica. In February 1908, months before Charcot left home with the *Pourquoi-Pas?*, Ernest Shackleton had landed a modified automobile onto the ice of McMurdo Sound, off Ross Island. The vehicle functioned poorly at best. (When Charcot loaded his own motorized vehicles aboard the *Pourquoi-Pas?*, he was unaware of Shackleton's disappointment.)

Charcot sailed from France at the end of July 1908. One special passenger, his new wife, Marguerite, was aboard. They had been married less than a year, and he wanted her with him for as long as possible before the expedition separated them. The *Pourquoi-Pas?* reached Buenos Aires in late October. Although Charcot's *Français* experience led him to expect a warm welcome, Argentina's reception exceeded his most optimistic hopes. Not only did the government vote him unlimited credit for the expedition's needs, but it placed the *Pourquoi-Pas?* in dry dock and supplied all the necessary materials to complete preparing her for the ice.

On November 23, the *Pourquoi-Pas?* departed Buenos Aires for Punta Arenas. There Mme. Charcot left the expedition, her physical presence replaced by a second steward. His addition brought the ship's complement to 30 men plus a cat, three kittens, and two pet dogs, the dogs another gift from the Argentine government.

Charcot sailed away from Punta Arenas and Marguerite on December 16, 1908. When he reached the South Shetlands, he headed immediately for Deception Island. Five years earlier, with no reason to stop there other than to leave a message, Charcot had bypassed the island to save time. This time, Deception Island had more to offer. The whalers were there. What Charcot saw when he entered Port Foster amazed him. It looked "almost uncanny, . . . like some busy Norwegian port."[1] Three factory ships, including Andresen's *Gobernador Bories,* lay in the harbor, and the 200 people in residence were more human beings in this one place than had visited the Peninsula Region with the *Belgica, Antarctic, Scotia,* and *Français* combined.

The whalers were happy to see Charcot, in part because he had brought mail for them. But beyond that, and to Charcot's delight, they wanted to thank him for the charts he had made on his first expedition, and to give him their corrections. The whalers also had material help for him. The mail Charcot delivered included a letter from Andresen's company instructing Andresen to provide the *Pourquoi-Pas?* with 30 tons of coal.

A medical emergency offered Charcot a way to repay the whalers. One of Andresen's workers had smashed his hand in an accident, and none of the whaling ships had a doctor. After carefully examining him, Jacques Liouville, the *Pourquoi-Pas?* physician, concluded that the man's hand would have to be amputated. Liouville returned the next day, accompanied by Gourdon to administer chloroform, and performed the operation. Had Charcot not arrived at Deception Island when he did, Andresen would have sent the injured whaler to Punta Arenas. Liouville told Charcot that the man could have died of gangrene en route.

Charcot spent two days at Deception Island picking up the coal and preparing reports on his work so far to send home with the whalers. While there, several of his crewmen located the *Uruguay*'s 1905 cairn containing the message that Galíndez had left for Charcot of the *Français*. Charcot of the *Pourquoi-Pas?* found reading it a moving experience. On December 24, Andresen delivered the coal to the

The *Pourquoi-Pas?* (on left) being coaled by a whale catcher (on right) at Port Foster, Deception Island. *(From Charcot, trans. Walsh, The Voyage of the 'Why Not?' . . . , 1911)*

Pourquoi-Pas?, by then anchored at the far end of Port Foster to escape the noxious smells of the whaling operation. Charcot's men celebrated Christmas Eve with presents from home. As on the *Français*, there was a cardboard Christmas tree gaily decorated with toys and candles. This one was a gift from Mme. Gourdon, who had heard about the first expedition's tree from her husband.

The *Pourquoi-Pas?* left Deception Island on Christmas Day. Just before Charcot departed, the whalers presented him with a very personal Christmas gift. Andresen, his wife, and several other captains used their only holiday of the season to ski the several miles from their base at Whalers Bay to where the *Pourquoi-Pas?* was anchored to say good-bye. They had other presents as well, promises for future help. Andresen told Charcot that in February he would visit Port Lockroy—the sheltered harbor on Wiencke Island that the Frenchman had discovered in early 1904—to collect any mail that Charcot wanted to leave. Further, he or another whaler would make a point of calling at Port Lockroy and, ice permitting, Booth Island in January 1910 to look for news. And finally, there would be coal waiting if Charcot returned to Deception Island the next summer.

Charcot made his first stop the next day, at Port Lockroy. His returns here and to Port Charcot, where he anchored on the 29th, were emotional experiences. As memories overwhelmed him and his fellow returnees, Charcot, who had a real sense of déjà vu, felt that "those of us who took part in the earlier expedition might well think ourselves four years younger!"[2]

As things turned out, the *Pourquoi-Pas?* was to stay at Port Charcot a bit longer than Charcot intended. Two days after he arrived, wind-driven swells filled the bay with ice and trapped the ship.

Although the ice held the *Pourquoi-Pas?* prisoner, the motorboat could escape. Charcot and three companions slipped out between the floes on New Year's Day 1909 and motored south through the Lemaire Channel to Petermann Island. There they found a harbor that seemed to offer a better anchorage for the *Pourquoi-Pas?* than Port Charcot did. Charcot named it Port Circumcision in honor of the date, the Catholic feast day celebrating Jesus's circumcision. By 10 p.m., the boat party was safely back at Booth Island. Two days later, the ice in Port Charcot opened just enough for the *Pourquoi-Pas?* to move to Port Circumcision.

January 4 was the first nice day in nearly a week. Charcot, Gourdon, and René Godfroy, one of the naval officers on loan to the expedition, seized the chance to make a reconnaissance trip in the motorboat. Since they expected to be back by evening, they took only the clothes they were wearing and enough food for one meal. They reached the nearby Berthelot Islands easily, landed in a few places, looked about a bit, and learned more about the ice to the south. Then, ready to head back, they sat down and ate nearly all their food, "a luxurious repast, which we . . . [were] destined soon to regret."[3]

The men reached the outer edge of the Berthelot Islands group only to find that ice floes now filled the open water they had enjoyed just a few hours earlier. Gone too was their lovely weather, replaced by snow and icy rain. The ensuing hours, and then days, were ghastly, as the hungry and freezing men fought the ice. After three days with virtually no progress, the boat's motor, which had been stopping and restarting, quit for good. The men tried to row, but it was impossible. Charcot finally decided to abandon the boat. As he saw it, their only hope was to travel over the ice to find some location from which they could signal Port Circumcision.

The three men were about to start hiking when they suddenly heard the *Pourquoi-Pas?*'s siren. They scrambled up a rock and yelled, and a call immediately answered them. Maurice Bongrain, the officer whom Charcot had left in charge, was as relieved to see the boat party as the three men were to see the ship. Those at the *Pourquoi-Pas?* had become increasingly concerned when Charcot had not returned, and had searched on foot and ski for the missing men for several days. Finally, Bongrain had decided to take the ship out. As the *Pourquoi-Pas?* slowly approached, Charcot's party returned to the motorboat to eat their last scraps of food, finishing just as the ship neared. Pride now took over: Charcot raised a flag, and with a desperate effort, Godfroy forced the motor back to life. When it stopped almost at once, the adventurers finished the last few yards with a paddle.

On January 8, just before the reunited party regained Petermann Island, the *Pourquoi-Pas?* ran aground with a jarring screech, and jagged bits of her hull bobbed to the surface. Unfortunately, it was already high tide. With little hope of the ship floating free on her own, the men worked frantically for hours to lighten the bow by shifting the heaviest gear to the stern. Then, at the next high tide, Charcot ran the engine full out, both ahead and astern. Suddenly, with another dreadful grinding noise, the ship was off and afloat. They had, Charcot wrote, "literally torn the *Pourquoi-Pas?* from off her fatal rock."[4]

Charcot left Petermann Island on January 12, to begin exploring farther South. The men knew they might be sailing in a dangerously damaged ship, but there seems to have been an unspoken agreement to ignore that thought. Two days later, the *Pourquoi-Pas?* neared the *Français* expedition's 1905 Loubet Land discovery. It was a beautiful day, and Charcot could see what appeared to be a large bay. He named it Matha Bay, after the *Français*'s second-in-command. Thanks to the excellent visibility, however, it was clear that Loubet Land was really part of Biscoe's Adelaide Island, and Charcot transferred President Loubet's name to the mainland coast east of Adelaide Island. He remained in the area only long enough for a preliminary survey and a brief detour to investigate whether Adelaide Island was genuinely an island. Although the view from the crow's nest suggested that it might in fact be a mainland peninsula, Charcot was unsure. He left the area with the question unanswered.

The *Pourquoi-Pas?* on the rocks — *The ship hit at high tide. As the tide fell, her bow, tilted up because of where she had hit, rose well out of the water while the aft deck became submerged. It was weight at the bow that was pinning her down, so Charcot's men worked to lighten the load there, moving heavy anchors and chains off the* Pourquoi-Pas? *to the rocks, emptying water casks, launching the boats with everything else heavy they could not shift to the rear. When they finally had her free, the exhausted men had to reverse the job, returning everything to where it had been. (Photo from Charcot,* Le Pourquoi-Pas? dans L'Antarctique, *1910)*

The *Pourquoi-Pas?* crossed the Antarctic Circle during the night, and rounded the southern tip of Adelaide Island early on the 15th. Before her lay a huge bay bordered on the east by a newly discovered mainland coast. Charcot soon moored his ship to the ice beside a tall rocky island that he named Jenny, after Mme. Bongrain. A few hours later, he and four of the staff, ironically not including Bongrain, landed and climbed to the 1,600-foot summit. The view was spectacular. Best of all, Charcot could see openings in the ice to the south. Distant fog, however, hid the land that he knew was there—Alexander I Land.

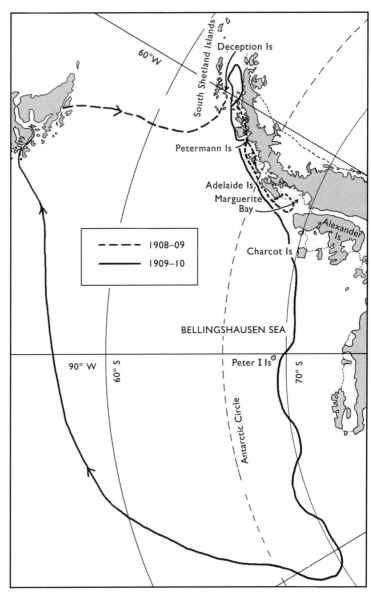

The track of the *Pourquoi-Pas?* — In 1908–09, Charcot sailed from Punta Arenas to Deception Island, then south, eventually reaching Marguerite Bay. From there, he returned north to Petermann Island, where he spent the winter of 1909. Following winter, he first sailed north to Deception Island. In January 1910, Charcot returned south, sailing past Marguerite Bay and into the Bellingshausen Sea. There he headed west until waning coal supplies and concern for his damaged ship forced him to end his exploratory work and leave Antarctica.

Charcot then headed southwest, out of the bay that he named Marguerite for his wife. By midnight, he saw mountains to the south glowing in the summer sun. It was Alexander I Land. The ice defeated Charcot's attempts to approach closer to either Alexander I Land or Marguerite Bay's southern coast the next day, and it continued to thwart nearly every move he attempted during the next two weeks. He did succeed in pushing through the pack to within a few miles of Alexander I Land, but his lone full success resulted from a three-day sledge trip that Bongrain made to investigate the status of Adelaide Island. Bongrain returned with a definitive answer: it was indeed an island. Another day, while moored off Jenny Island, Charcot tried using a motor sledge. Making it function was such a struggle that he

CHARCOT'S RETURN: 1908–1910

abandoned the effort. From this point on, the motor sledges gathered dust on the ship, a sad contrast to the early success of the motorboat.

Charcot longed to winter in Marguerite Bay, but the ice made it impossible. Years later, other parties *would* find a winter home here, but more than once, they experienced serious problems leaving the following summer. One group would have to be evacuated by air, two would spend an extra, unplanned, winter there, and others would escape only when modern icebreakers came to the rescue.

Charcot found the ice no more welcoming in Matha Bay, his second choice. And so, on February 3, he re-entered Port Circumcision, on Petermann Island. It was here that he would winter, a disappointing backup choice, because he had hoped to be so much farther south. Instead, he was less than 10 miles south of Port Charcot.

As soon as he had anchored the *Pourquoi-Pas?*, Charcot had the crew stretch cables across the harbor to block ice from entering, just as he had done at Port Charcot. Then everyone set to work converting the ship into a winter home and constructing a small scientific village on shore. Wires from the *Pourquoi-Pas?* fed electricity to the shore huts, a great luxury compared to the balky lanterns Charcot's men had battled in the winter of 1904.

Organizing the winter base took nearly a month, partly because the entire month of February served up only four agreeable days. On one of these, Charcot traveled to Port Charcot to leave a message about his location. He found walking about his old base a surreal experience. It was a place where the "impression of a persistence of life at our old winter quarters is so strong that . . . I am very frequently obliged to make an effort to convince myself that we are really all alone in the Antarctic."[5] Booth Island might be full of ghosts, but there were also very real and useful items there. Charcot returned several times to raid his 1904 winter base.

Shrove Tuesday, February 23, at Port Circumcision — *The staff declared a unilateral holiday from the work of establishing the Petermann Island base and dressed up to celebrate Mardi Gras. Charcot wrote, there was a "general masquerade, very merry, though simple. . . . The mess steward turns out in a most extraordinary garb, and the cook is disguised as the chef in a big hotel. . . . The crew are content with turning up their trouser-legs and displaying superb red under-clothing, which, with their blue knitted vests and sealers' boots and caps, makes a lovely uniform."[6] (Photo from Charcot, trans. Walsh,* The Voyage of the "Why Not" . . . , *1911)*

The *Pourquoi-Pas?*, specifically designed to house the expedition through an Antarctic winter, provided a comfortable home. She was adequately heated; electric light was available throughout; the cook was competent; there was that ample supply of wine; and the men had sufficient water to wash in, a rarity on polar expeditions of the day. More remarkably, they had clean clothes each week, because the *Pourquoi-Pas?* had a washing machine. The crew fed ice into it every Friday night. After the ice melted overnight, they heated the water with a seal blubber fire and washed the clothes.

As he had on the *Français,* Charcot made a point of keeping everyone well occupied through the winter days, weeks, and months. The *Pourquoi-Pas?*'s position at Port Circumcision, however, was unexpectedly precarious. Capsizing icebergs outside the entrance created huge swells that shoved blocks of ice into the little cove, ice that crashed frighteningly into the already damaged ship. Then, just when ice froze about the ship in mid-April and Charcot thought his vessel safe at last, a wild storm destroyed the protective buffer. This weather pattern plagued Port Circumcision for months. Gales would shatter the ice; a few days of good weather would allow it to refreeze; and then a new storm would roar in. The *Pourquoi-Pas?* was never safe for long in the winter of 1909.

Health problems were also a concern in mid-winter. Charcot himself had a persistent cough and was short of breath. Godfroy, the naval officer who had been stranded with Charcot in the Berthelot Islands back in January, was similarly afflicted, and both men suffered swollen legs. The cause was a mystery, especially since everyone else seemed to be fine. Charcot finally fell back on "polar myocarditis" as

The ***Pourquoi-Pas?*** **in winter quarters at Petermann Island** — *Just to the south of the Lemaire Channel. The southern entrance to this famous strait can be seen in the distance immediately above the ship. Booth Island, on the north coast of which Charcot wintered in 1904, is in the center left of the photo, above and to the left of the ship. (From Charcot, trans. Walsh,* **The Voyage of the 'Why Not?'. . . ,** *1911)*

a diagnosis. Even he was unsure what that might be.

During the latter days of winter, time started to drag, especially for Charcot. Perhaps because he was ill, he no longer felt the same excitement about this second expedition, and he was not nearly as close to the staff as he had been on the *Français*. As for the rest of the men, they were getting along well for the most part. Still, there were times when the lack of privacy weighed heavily on everyone.

All the men thus welcomed spring's arrival, with its greater opportunities for outdoor activity. Charcot and Godfroy were feeling much better, and Charcot eagerly began organizing his sledging program. Back in March, several men had taken the motorboat to the mainland and probed the fringe of the interior. One group had even traveled a few miles inland. Now it was time to follow up. This would be real pioneering work, the first effort to explore the Peninsula itself. A few days after preparations began, however, both Charcot and Godfroy suffered relapses. Dispirited, Charcot surrendered leadership of the first, depot-laying, trip to Gourdon, the geologist and *Français* veteran.

The main sledging began on September 18, again led by Gourdon. Two weeks later, the six-man party returned with a report of mixed success. Their attempt to climb to the Peninsula summit had failed when they reached a cul-de-sac where avalanches had littered the mountainsides with huge blocks of ice. But there were successes as well—valuable glaciological, topographical, and meteorological observations. Gourdon's team had been gone fifteen days, climbed to nearly 3,000 feet, and traveled about 15 miles overland. A modest distance, but all of it hard-won through difficult country.

Charcot abandoned his sledging program after Gourdon's failure to penetrate far inland. Instead, he would focus on exploration via the *Pourquoi-Pas?* He would leave Petermann Island in early November, timing that would allow time to visit Deception Island and collect the coal Andresen had promised him. Then he would spend a few weeks working in the South Shetlands before heading south to continue the previous summer's work. Escaping Port Circumcision that soon, however, was uncertain. The ice that Charcot had so longed for during the winter had finally arrived, and now stretched as far as he could see.

In the meantime, Louis Gain, the expedition biologist, reveled in the return of the penguins and resumed the bird studies he had begun in February. He was particularly delighted to see so many arriving Adélies sporting rings he had put on them the previous fall. Now he knew that these birds returned to the same rookery. When the first penguin eggs appeared, Charcot declared part of the rookery out of bounds to everyone except Gain. The biologist took full advantage and produced the first serious study of Peninsula Region penguins.

The Port Circumcision ice held the *Pourquoi-Pas?* captive until November 25. Charcot reached Deception Island two days later, after a tense time weaving his way past icebergs and through heavy pack ice in the Gerlache Strait. But would the whalers have arrived this early?

Charcot had his answer when he sighted a catcher. It was from Andresen's fleet, the best possible news. Andresen and his wife were equally delighted to see Charcot when he boarded the *Gobernador Bories*. And when he asked about the coal, Andresen's answer was an immediate yes. Charcot could have 100 tons. It was, the whaler explained, coal from the *Telefon*, wrecked just a day after Charcot had departed the previous summer. But the whalers had more than coal to offer. Mrs. Andresen gave the men fresh potatoes, apples, and oranges. She also had their mail and news of the outside world: that Ernest Shackleton had almost reached the South Pole in January 1909, but he and his three companions had turned back when, only 112 miles (97 nautical miles) short of their goal, they found themselves out of time and dangerously short of food; that Robert Peary and Frederick Cook both claimed to have captured the North Pole; and that a Frenchman, Louis Blériot, had flown a small monoplane across the English Channel. Perhaps most satisfying was learning that the whaler whose hand Liouville had amputated the previous summer had recovered completely.

Although Charcot knew that the *Pourquoi-Pas?* had suffered damage when she grounded the previous summer, he had no idea how serious it might be. Now he tried to find out, but it was difficult to learn much without a dry dock. Andresen solved the problem, offering to have the diver he had brought along to work on the *Telefon* take a look. Andresen's diver carefully examined the *Pourquoi-Pas?*'s hull on December 8. Before he went down, Charcot asked him to tell no one but Charcot himself if he found anything seriously wrong. Thus, when the diver came up, he said that there was only a minor problem. The diver's private report to Charcot, however, was entirely different. He concluded by saying, "You cannot, you must not navigate in such a state in the midst of ice. . . . Mere ordinary navigation is already dangerous, and the slightest shock might send you to the bottom."[7] Charcot listened, politely thanked both the diver and Andresen, and then made his plans as if nothing were amiss. He did admit some of the truth to the officers and staff but kept the crew in the dark.

Charcot was determined to continue, but he was not yet ready to begin his exploration program for the summer. It was too early, he thought, because there would still be substantial sea ice farther south. Instead, he would wait until the end of December to allow the pack time to break up. Meanwhile, he would survey and carry out natural history and geology studies in the South Shetlands. He was eager to start, but the weather was so frightful that he remained at Deception Island for two more weeks. Charcot finally left the island on December 23 and then spent most of Christmas week at King George Island. When the *Pourquoi-Pas?* sailed away on December 30, her men left Mme. Gourdon's cardboard Christmas tree standing proudly on shore. The next day, Charcot was back at Deception Island for a final visit with the whalers. Then he was ready to move on, but the weather was bad, so it was back to waiting.

The weather finally improved a bit, and Charcot left for the southern campaign on January 6, 1910. As she sailed away from Deception Island, the *Pourquoi-Pas?*'s bunkers were full. But below the waterline, she was as dangerously damaged as ever.

Charcot headed straight for the vicinity of Alexander I Land. January 11 saw the first exciting event of the new season. That morning, Charcot, who was in the crow's nest, glimpsed something through brief breaks in the clouds. But what was it? He kept silent until he could be sure. When the sky cleared at noon, the mountains of Alexander I Land shone brightly in the distance. Charcot, however, thought he had spotted something else, closer. He said nothing, but to everyone's surprise he altered course and headed east. After lunch, he climbed again to the crow's nest. There was indeed land out there, "a new land . . . , a land which belongs to us! . . . these two words, which I repeat to myself under my breath, a *New Land* [Charcot's emphasis]!"[8] Charcot's observations placed his discovery at 70° S, 77° W. Eventually, he named it Charcot Land (today, Charcot Island) at the urging of his men and others, a name he always insisted was in honor of his father, not himself. Blithely ignoring the damaged state of the *Pourquoi-Pas?*'s hull, Charcot pushed into the pack toward his discovery. But the ice was too dense, and after a few hours he abandoned the effort. By then, mist had swallowed the land. Neither Charcot nor anyone else would see it again until 1929.

Excited by the prospect of new discoveries, Charcot continued south and west. After crossing the 70th parallel, a new southern record at his longitude, he pressed on through snow squalls, following the pack ice edge, mostly to the west, occasionally a bit north, then west again.

On the afternoon of the 14th, Charcot spotted Peter I Island, unseen since Bellingshausen had discovered it in 1821. Unfortunately, fog and snow made it impossible to add anything to the Russian's cursory description, and the weather soon grew worse. The next few hours of howling gales and blinding snow were terrifying. Charcot wrote, ". . . icebergs and ice-blocks stud the boiling sea on every side. . . . The high icebergs, whose walls our yards seem to touch, tower over us and the smaller ones dance in front of the ship. . . . the hours fly on and our mad course through the unknown continues. . . ." Somehow, he made it through the densest part of the iceberg thicket, more through luck than anything else. Then, "All of a sudden . . . the black gulf turns brilliant and golden, dazzling with light . . . giving the impression of an entry into paradise after leaving hell. . . ."[9]

Four days after escaping the ice at Peter I Island, the *Pourquoi-Pas?* reached the vicinity where ice had stopped the Wilkes Expedition's *Peacock* in 1839. Charcot, however, found the sea more open. And in any case, he had an engine to force his way through ice that would have stopped the Americans. Eventually, he pushed his battered ship through the floes to 107° W, the approximate longitude where James Cook had attained his record latitude of 71° 10' S in 1774. Dense pack brought the *Pourquoi-Pas?*'s southward thrust there to an end at 70° 30' S. Charcot thought it might have been possible to reach Cook's latitude, but it would have cost him more time and coal than he was willing to invest. Perhaps he could have, perhaps not. In the years to come, other ships would try. All would fail until 50 years later, when two U.S. icebreakers finally succeeded.

Pressing on west into iceberg-filled waters that no other ship had penetrated, Charcot once again crossed the 70th parallel going south before the ice stopped him. At 118° 50' W, where a sounding found bottom at just over 3,000 feet, he concluded that land must be near, land that was part of a coast extending all the way from Marguerite Bay to Edward VII Land in the Ross Sea. He was wrong about being close to land: the Antarctic Continent was still more than 200 miles to the south. He was right, however, about the continuous coastline, an elusive stretch of land that would have to wait decades before human eyes would see it. But Charcot *had* sailed about two-thirds of the way along it at a very high latitude. The weather had been brutal, the ice a constant challenge, and he had done it in a gravely damaged ship. It was a remarkable achievement.

Charcot ended his exploratory voyage on January 22. He had gone as far as he could with his battered ship, tired men, and dwindling coal.

The *Pourquoi-Pas?* limped into Punta Arenas on February 11. The excited townspeople greeted the expedition enthusiastically, and congratulatory telegrams poured in from all over the world. Then it was on to Montevideo to repair the *Pourquoi-Pas?* before completing the voyage home. A dry-dock inspection in Montevideo confirmed Andresen's diver's warning about the *Pourquoi-Pas?*'s hull. The worst was a 50-foot-long hole on the port side that cut completely through the outer planking in several places. A few inches more would have been fatal. Nonetheless, the *Pourquoi-Pas?* was fundamentally sound, and after a few weeks of repairs, Charcot sailed her home.

Charcot reached France in early June 1910. His book about the expedition, *Le Pourquoi-Pas? dans l'Antarctique (The Voyage of the "Why Not?" in the Antarctic)* was in the hands of an eager French public before the end of the year. Analyzing and publishing the scientific work took much longer. The first of 28 volumes appeared in 1911, the last in 1921. There was a great deal to report. The expedition had surveyed 1,250 miles of Antarctic coasts; produced numerous new charts and maps; and carried out a comprehensive scientific program. With the exception of Alexander I Land, Charcot had discovered and been the first to chart everything known about the west side of the Antarctic Peninsula Region to the south and east of Adelaide Island. That would remain the case for nearly twenty years. Charcot, who had begun his Antarctic career in such an uncertain fashion, had indeed done France proud.

After 1910, Charcot and his ship shifted their polar work to the Arctic. In September 1936, a wild storm smashed the *Pourquoi-Pas?* onto the rocks of Iceland's south coast. The sole survivor said that when he last saw Charcot, the leader of the *Français* and the *Pourquoi-Pas?* Antarctic expeditions was standing on the bridge, gently releasing his pet seagull from its cage.

CHAPTER 12

Filchner's Battles in the Weddell Sea: 1911–1912

William Speirs Bruce was just as eager to return to Antarctica as Charcot had been. In April 1908, he published a bold plan for an expedition featuring a sledge trip across the Antarctic Continent, from the Weddell Sea coast to McMurdo Sound in the Ross Sea. But by the time Bruce began actively promoting his idea, in 1910, someone else was already considering it. This man came from Germany.

Wilhelm Filchner, a German army staff officer, was 32 years old in 1909 when he decided to lead an expedition to Antarctica. Filchner, who had never been to either of the polar regions, seems to have turned his attention south simply because he wanted a new challenge, and Antarctica looked like a good place to find it. Indeed, his only previous exploring experience had been on several lengthy trips to Russia and central Asia. There he had had a number of adventures, including encounters with brigands, wild animals, and local warlords. None of this, however, had really prepared him for organizing and leading a major polar expedition.

Filchner's original plan copied liberally from Bruce's 1908 proposal, and his army superiors enthusiastically embraced it. In contrast, Kaiser Wilhelm definitely did not. He had been so bitterly disappointed with the results of Erich von Drygalski's 1901–03 *Gauss* expedition to East Antarctica that he opposed any new German Antarctic venture.[1] Since that meant no government funds would be available, Filchner had to scale back his plans. He eliminated the crossing attempt and focused his program on the Weddell Sea side of the continent. He would establish a wintering station there as a base for scientific work and sledging trips into the interior.

Early on, Filchner organized the "German Antarctic Expedition Association" and turned all his administrative tasks and authority over to its people. He took this step so he could devote his own time to the technical aspects of the expedition. Then he began preparing. He found his ship in Norway, a strong whaler with an excellent 300-horsepower auxiliary steam engine.[2] After buying her, he renamed her *Deutschland*. Selecting a captain was not nearly so straightforward. Drygalski recommended Richard Vahsel, who had been his second officer on the *Gauss* expedition. Filchner had misgivings, especially when he heard Vahsel widely disparaging the *Deutschland*'s previous captain, a Norwegian who was Filchner's personal choice for the job. His concerns deepened when Hans Ruser, Drygalski's captain, warned him strongly against taking Vahsel. But Filchner lacked the final say in the matter. The German navy insisted on a German captain, and the Expedition Association, bowing to the navy's wishes, appointed Vahsel captain. Vahsel then appointed the rest of the ship's officers. Filchner would soon find himself suffering the consequences.

Wilhelm Filchner at the time of the expedition. *(From Filchner,* Zum Sechsten Erdteil, *1922)*

The *Deutschland* left Germany at the beginning of May 1911. Vahsel and four others aboard had already been to Antarctica with Drygalski, and the ice pilot and several crew members were Norwegians with significant Arctic ice experience. As for the scientific staff, the geographer, Heinrich Seelheim, and the meteorologist, Erich Przybyllok, had been to Spitsbergen with Filchner on a practice polar region trip in 1910. Another staff member, the Austrian mountain climber and naturalist Felix König, had spent time in Greenland learning how to handle the dozens of sledge dogs now aboard the *Deutschland*. (Filchner also planned to use Manchurian ponies to pull his sledges, taking a cue from Shackleton, who had used their Siberian cousins on his 1908–09 South Pole attempt. The ponies were traveling to Buenos Aires separately.)

Just before the expedition sailed, the German navy offered Filchner the honor of flying the naval flag. When he, an army officer innocent of naval matters, accepted, the *Deutschland* became subject to naval regulations. Only later did Filchner realize that this made the captain master aboard ship. He had unwittingly yielded much of his authority to Vahsel.

Filchner remained behind when the *Deutschland* left home. He would join the expedition in Buenos Aires after tidying up loose ends at home. In the meantime, he appointed his friend Seelheim interim leader aboard the *Deutschland*. It was a happy voyage at first, but problems emerged as the ship crossed the equator. Officers, staff, and men began bickering, and, most seriously, Seelheim and Vahsel were soon at loggerheads. Relations between the two men deteriorated so seriously that Vahsel wired Filchner an ultimatum: he refused to continue the voyage with

Seelheim; one of them must go.

Vahsel had been able to send his telegram directly from the ship because Filchner had outfitted the *Deutschland* with wireless equipment. Despite the fact that the best equipment available had an operational range of only 400 miles, Filchner hoped that this new technology would allow him to communicate with civilization once he sailed into the ice. (An Australian, Douglas Mawson, who was taking an expedition to East Antarctica, was also attempting to introduce wireless to the Antarctic that same summer. In February 1913, Mawson would achieve the first two-way wireless exchange between the Antarctic Continent and the outside world, using a relay station on sub-Antarctic Macquarie Island.)

The *Deutschland* captain's ultimatum at first appeared to put Filchner—who received it at sea en route to Buenos Aires—in a difficult position. He would have to sacrifice Seelheim or find a new captain on short notice. Filchner, however, had already met a man he thought would do very nicely as a replacement for Vahsel. He was Alfred Kling, a young officer on the merchant ship transporting Filchner to join the *Deutschland*. Filchner was ready to replace Vahsel with Kling and keep Seelheim when the geographer took the decision out of Filchner's hands. Seelheim resigned and departed the expedition at the *Deutschland*'s first port of call in South America.

The *Deutschland* reached Buenos Aires on September 7, 1911. Filchner and Kling, whom Filchner still wanted to take along as a ship's officer, had arrived ten days earlier. The Argentines treated Filchner very well, but his relationship with Vahsel was another matter. Vahsel, the Antarctic veteran, questioned the polar neophyte Filchner's preparations and did all he could to belittle the expedition leader. Vahsel also flatly refused to accept Kling as a new first officer. Already distressed over losing Seelheim, Filchner stubbornly insisted on retaining Kling and appointed him to the scientific staff. Kling's first job would be to accompany the expedition's sledge-pulling ponies, currently housed in the Buenos Aires zoo, to South Georgia on a whaling ship.

Filchner and Vahsel had not even reached the ice and their relationship was already in serious trouble. But without authority to dismiss Vahsel, Filchner's only options were either to quit his own expedition or to make the best of a bad situation. Filchner soldiered on, he said, because the expedition was his baby and he did not want to abandon it.

The *Deutschland* sailed from Buenos Aires on October 4, bound for South Georgia. Filchner planned a slow passage so that he could do oceanographic work en route, but twelve days out of port, Ludwig Kohl, one of the two expedition doctors, suffered a severe attack of appendicitis. When Kohl experienced another attack the next day, Wilhelm Goeldel, the second doctor, immediately performed an emergency appendectomy. Filchner then headed straight for South Georgia.

When the *Deutschland* reached South Georgia four days later, Carl Anton Larsen welcomed Filchner warmly. The two men had met through Nordenskjöld in Norway, and the whaler had given Filchner valuable advice. Larsen was even more helpful at South Georgia, including taking the recuperating Kohl into his home.

Filchner's team remained at South Georgia for a month and a half engaged in scientific work, a bit of inland island exploration, surveying and charting the South Georgia coast, and completing preparations for the impending expedition. The *Deutschland* also made a short exploratory cruise to the South Sandwich Islands during these weeks. Nearly everyone went along—even Kohl, who insisted on participating despite his recent operation. The ship battled her way south through four days of storms to reach the islands. Once there, fierce snow squalls followed one after another. After impossibly heavy surf defeated the single landing attempt, on Zavodovski Island, Filchner headed back to South Georgia. Not only had he concluded that he could spend his time more profitably there, but Kohl had had a relapse. Vahsel, too, was ill, complaining of rheumatism and saying he needed time to rest.

The Germans resumed their work on South Georgia, and all was going well when tragedy struck. On the evening of November 26, Walter Slossarczyk, the *Deutschland*'s third officer, rowed out into Cumberland Bay. When the man failed to return, Filchner mounted a search, but no one found any sign of him. Three days later, an inbound whale catcher brought his empty dinghy back. With no body to bury, Filchner's men erected a simple memorial cross on a hillside above the Grytviken cemetery. The cross also marked the death of the expedition's wireless hopes, because Slossarczyk had been the only man on the *Deutschland* who was trained for wireless work.

Kling and the ponies arrived on December 3. Eight days later, on December 11, 1911, the *Deutschland* left South Georgia for the Weddell Sea. Aboard were 33 men, eight ponies and several dozen dogs, a cat, two steers, two pigs, and a small flock of sheep donated by the whalers. (Kohl was not among the men; his relapse on the South Sandwich Islands voyage had made it clear to all that he was too ill to continue.)

Filchner's chosen course relied heavily on information from Bruce's *Scotia* expedition, as well as on advice from Bruce himself. The plan was to sail toward Coats Land and then follow the coast as far south as possible before landing the wintering party. No one knew how far that might be, but Filchner hoped that it would be well beyond the 74° S that Bruce had reached.

Only three days out of Grytviken, the *Deutschland* met outliers of the pack ice, and from then on, the ice ruled her progress. It was stop and go, and stop, and go.... There were a few days when she broke into enough open water to advance 100 miles or more south in 24 hours. Then the ice would close up again and she would be back to creeping forward, or even finding herself retreating. The slow progress heightened tensions among the men, now split into factions. There were

Filchner and his three staunch supporters—Kling, König, and Przybyllok—on one side, and Vahsel and most of the officers and scientific staff on the other. The crewmen were caught in the middle.

The *Deutschland* finally crossed the Antarctic Circle on January 11, 1912, a full month after leaving South Georgia. Two days later, an oceanographic sounding at 70° 02' S, 27° 26' W, discovered what would be one of the most important scientific results of the expedition, an ocean depth in excess of 15,000 feet.[3] Land was clearly a long way off. And so Filchner pressed on, battling his way through the ice. He passed Weddell's 1823 record south on January 29. The next day, the *Deutschland* reached 76° 40' S, 31° 32' W, and at last Filchner had something to show for his effort: he could see land, an ice-covered slope rising gently to about 2,000 feet. He named it Prinzregent Luitpold Land after his most important expedition patron. He knew that Bruce had seen no landing sites on Coats Land. Would Luitpold Land, which he was sure was a continuation of Coats Land, be any more welcoming? A frustrated Filchner had no way to know initially, because intermittent fog blocked his view.

The skies finally cleared early the next morning, and the icecap glistened magically in the sun. It was magnificent. Even better, Filchner spotted a possible landing place. It was in a small bay, 3 to 4 miles wide and about 7 miles long, formed by the junction of Luitpold Land on the east and, another great discovery, a massive ice shelf to the west. (Today we know it as the Filchner Ice Shelf.) The date was January 31, 1912, and, at 77° 44' S, the *Deutschland* had reached as far south as a ship could in the Weddell Sea. Perhaps in hopes of improving his relationship with his captain, Filchner named the promising bay Vahsel.

Filchner had Vahsel moor the *Deutschland* to the edge of the fast ice in the captain's namesake bay, and then landed a scout team to look for a base site. Their report was so discouraging that Filchner decided to search along the ice shelf for something better. Only about 60 miles to the west, however, the *Deutschland* met dense pack. Filchner turned back to try again at Vahsel Bay.

Once back at Vahsel Bay, a new reconnaissance team reported that moving everything onto the icecap or the top of the ice shelf would be difficult. Undaunted, Filchner was determined to find a way. He sent two more groups out the next day to investigate further. König's team located a route onto the icecap and reported that the surface looked like a good place for the base. In contrast, the geologist, Fritz Heim, recommended using one of the tabular icebergs seemingly cemented to the ice shelf on the west side of the bay. Filchner preferred König's idea. Although more difficult logistically, he felt it was safer, especially if he established the base several miles inland.

Filchner was ready to move ahead. Not so, Vahsel. Although the tension between the two leaders had been unpleasant, even vicious at times, it had not seri-

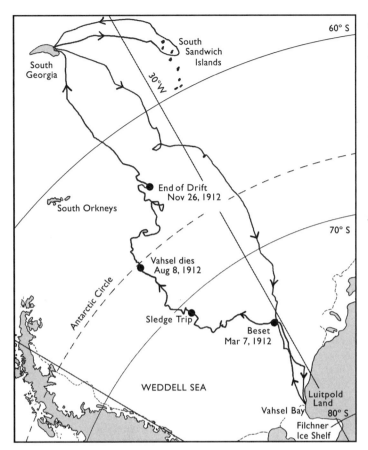

Deutschland voyage track — *Most of Filchner's predecessors in the Weddell Sea had difficulty with ice. That certainly included d'Urville, Ross, and Bruce. The one exception was Weddell, who had reached 74° 15' S in 1823, where he found himself in open water, with nothing but three tabular icebergs in sight. The location where Weddell achieved his then record latitude was less than 50 miles to the west of where the* Deutschland *became beset in an impenetrable ice-covered sea, trapped into a nearly nine-month-long drift.*

ously affected the expedition's program to this point. Now it did. Openly challenging Filchner's decision, Vahsel adamantly refused to furnish the manpower needed to establish the base on the icecap. Vahsel equally disliked Filchner's second choice, the ice shelf, arguing that mooring the ship alongside for unloading would be too dangerous. By the time Filchner had reluctantly bowed to all of Vahsel's objections, the only location left was the one Heim had recommended. Filchner, uncomfortable with this option, decided to try again to find a site on the ice shelf farther west. If he was no more successful this time, he would return to Vahsel Bay. The *Deutschland* sailed away for the second time on February 3. She reached a few miles farther west along the ice shelf than she had two days earlier, but nowhere did Filchner see a place to land. And so, two days later, he was back. Vahsel Bay it would have to be.

Vahsel Bay, yes, but the question remained, where? The ice topography had changed so much in just three days that Filchner needed new reconnaissance. That should have warned him about the instability of the location, but somehow he, and everyone else, evidently missed the implications. And so Filchner considered several sites before choosing one that he thought lay on the ice shelf. A snow ramp led up to the site from the adjacent fast ice, which meant that the ship would not have lie alongside the ice cliff for unloading. That made the place acceptable to Vahsel. Before unloading began, however, Filchner realized that his chosen location was actually an iceberg. That made him uneasy, but Vahsel and Heim assured him that it was safe, because it was firmly attached to the ice shelf, and even if it were to break out eventually, they asserted, it would be the last piece of ice to go. Filchner, still concerned, asked Vahsel to consult with Paul Bjørvik, the Norwegian ice pilot. Vahsel replied that he had already done so: Bjørvik had had no reservations. Only later did Filchner learn that Vahsel had not talked to Bjørvik at all, and that the Norwegian's real opinion was "Only a stupid kid would think of building a station on an iceberg."[4]

And so Filchner set to work establishing his base on what he named the Station Iceberg (Stationeisberg). His intent was that this would be an interim base, to serve only until he could shift his camp onto the icecap proper during the winter. The first indication of trouble came just two days later, when a gale shattered the fast ice in Vahsel Bay. By evening, two large icebergs were drifting out of the bay. Despite this second implicit warning, the major unloading proceeded, beginning on the 9th with lumber for the hut, followed by the dogs and ponies. That afternoon, rising seas forced the ship to move out to sea. Before the *Deutschland* left, Filchner hurriedly landed provisions and camping gear for the construction crew, which remained behind on the ice. Two ponies broke loose in the rush, and one of them, Maggi, fell into a crevasse.

Hauling Maggi the pony out of a crevasse on the Station Iceberg. *(From Filchner,* Zum Sechsten Erdteil, *1922)*

When the *Deutschland* returnd two days later, the building crew greeted Filchner with good news. Not only had they finished the hut foundation, but Maggi, the crevasse pony, was still alive. On Filchner's orders, König, whom Filchner had left in charge, had taken several shots at her to prevent her suffering. He thought he had killed her, but when he returned a few hours later to check, he discovered a very live pony standing at the bottom of the crevasse. König tossed her some hay from the unloaded stores and then left Maggi to await rescue until the ship returned with enough manpower to pull her out.

On February 13, the men held a party to celebrate the near-completion of the base hut. Four days later, Filchner wrote in his diary, ". . . we [have] reached a [positive] turning point in the history of the landing."[5] He had written one day too early.

The winter station building on February 13 — *Although the building not yet closed in, the men held a party to celebrate getting the roof framing up. One of the bitches contributed her own blessing, giving birth to a litter of eight puppies that day inside the unfinished hut. (From Filchner,* Zum Sechsten Erdteil, *1922)*

Early on February 18, a "cannonade . . . as if hundreds of heavy guns . . . [were being] fired at once" assaulted the men's ears.[6] The Station Iceberg had broken away and was on the move. Filchner hurriedly launched a boat to rescue the men and as much of the base materiel as possible. By evening, nearly everything was back aboard the *Deutschland*. One dog remained behind, an escapee who evaded all attempts at recapture. (Fog defeated Filchner's attempt to shoot her, to save her from death by starvation.) The small inlet Filchner had named Vahsel Bay no longer existed. Instead, there was now a much larger bay, created when over 350 square miles of icebergs had broken free from the Filchner Ice Shelf.[7] The Station Iceberg had been the first to go, not the last as Vahsel and Heim had so confidently predicted. It now glided out into the Weddell Sea, "as if leading a parade."[8] The men could hear the solitary dog howling as it passed.

Because the men had managed to salvage almost everything, Filchner decided to treat the near disaster as a setback, rather than an end to his plans to build a winter base. He would simply find a new location in the now-enlarged Vahsel Bay. Time permitted, because Vahsel had agreed to have the *Deutschland* remain for the winter rather than sail back north as originally planned. The captain did, however, want to delay establishing the base until the ship was safely frozen in. Filchner could accept that, because Vahsel was at last willing to provide the labor needed to establish the station on the icecap.

Then, at the beginning of March, Vahsel changed his mind. Having the ship winter in Vahsel Bay, he told Filchner, was too dangerous. They should head for the open sea as soon as possible. Filchner pleaded with him to stay, but Vahsel was master of the ship. They would go.

The *Deutschland* quit Vahsel Bay on March 4. Three days later, about 300 miles to the north, she met dense pack ice. Then the ice closed about her. Like the *Belgica* before her, she was now a prisoner of the pack, embarked on a helpless drift.

As soon as he realized that they were indeed trapped, Filchner established a winter routine. There were the usual efforts to celebrate anything and everything; scientific programs to pursue; the daily ship's work; the dogs and ponies to be cared for; recreation on the ice.... The men had adequate clothing, plenty of food (including fresh meat from penguins and seals), and a cook who knew how to make their meals appetizing. In short, many of the problems that had plagued the *Belgica* expedition would not be issues for the men of the *Deutschland*.

Filchner's men, however, faced their own, unique, challenge. They had split into two opposing camps, camps that had become even more polarized after the weeks at Vahsel Bay. Vahsel made a point of deriding Filchner in front of everyone, even blaming him for the events at the Station Iceberg. Since Filchner refused to defend himself publicly, many among the crew were unsure whom to believe. It was a poisonous atmosphere, one that could only exacerbate the difficulties of drifting through an Antarctic winter night.

The ponies and dogs, especially the puppies, offered badly needed diversion from the rank bitterness of warring human factions. They were neutrals in the human conflict, and many of the men looked to them for a companionship uncomplicated by personal rivalries. On April 1, Filchner moved the dogs off the ship, and the ponies joined the canines on the ice a few days later. The ponies soon found regular employment pulling the water-collecting sledges; the dogs would have to wait for work until someone took a sledging trip. Employed or not, all the animals had to be cared for, and that could be a challenge. Stasi, a pony who one day "ate a copy of the *Hamburger Fremdenblatt* as well as a complete cigar box, with lid, and thereafter gnawed on a bone..."[9] was a particular trial. The men also had to cope with interspecies problems when, at the beginning of May, the dogs developed a taste for pony meat after being given flesh from one that Filchner had had shot. But the ponies had their own answer. A few expertly aimed kicks and the dogs and ponies were friends again. Filchner might well have longed for so simple a solution to the men's differences.

The dogs finally found a job in June, when the *Deutschland*'s drift carried her to within 40 miles or so of where Benjamin Morrell had reported finding New South Greenland in 1823. Filchner, seeing an opportunity to do some sledging, as well as to escape life aboard the *Deutschland*, decided to investigate.

It was June 23, in the heart of the Antarctic winter night, when Filchner, König, and Kling set off for a trip over the pack ice. They had two sledges, sixteen dogs, and provisions for three weeks. The ice soon proved a difficult and confounding sledging surface, a jumbled maze of ridges and floes that forced the men into constant zigzag detours. Kling's position fix on the 26th indicated that they were very close to Morrell's reported land. The next morning, the men traveled a few

On left, emperor penguins on the ice of the Weddell Sea, the beset *Deutschland* in the background. On right, Alfred Kling riding one of the expedition ponies on the ice during the beset winter. *(Both from Filchner,* Zum Sechsten Erdteil, *1922)*

more miles before stopping at a lead to take a sounding. Although Filchner lacked the equipment to measure the full depth, it was clear that the sea there was at least 3,000 feet deep. No land in sight, combined with water this deep, was proof as far as he was concerned: New South Greenland did not exist, at least not where Morrell had reported it. With the purpose of the sledge trip achieved, it was time to return to the *Deutschland*, wherever she was.

The three men used their outbound tracks as a guide for the return trip, sometimes losing them but always, eventually, re-finding them, even after several wild chases when the dogs took off after seals. On the morning of the 29th, Kling optimistically climbed a small ice hillock to look for the *Deutschland*. To his surprise, there she was, about 10 miles away. They set out toward her, only to be stopped by a gaping lead across their route when they were about 4 miles from the ship. While they were searching for a way across, they heard shouts from the other side. They had been seen, and some of the crew had come to help them. The lead, however, yawned too wide to cross, and another night on the ice ensued. Their experience was a virtual replay of what had happened to the winter sledging party from the *Belgica*. When the lead was still impossibly wide the next morning, Filchner sent the men who had again arrived from the ship back for a boat. That solved the problem, and all were safely aboard the *Deutschland* by mid-afternoon on June 30. The longest sledge trip of the expedition had ended.

The *Deutschland* emerged from a two-month polar night in mid-July. By then, the imprisoning ice had carried her northward past 69° S, into a region where the pack was in turmoil. Here, winds and currents were smashing floes against one another and anything else in their way. It was pure luck that the *Deutschland* had become solidly encased in a single large, protecting floe. Filchner probably expressed everyone's feelings when he wrote in his diary, "If only the ship floe survives until we have passed the zone of most severe ice pressures!"[10] Fortunately, it did.

The *Deutschland*'s captain did not fare as well. Vahsel had complained of rheumatism and other ailments several times early in the expedition, but by June, he was showing signs of really serious illness—coughing, faintness, and shortness of breath. Vahsel's condition deteriorated in July and early August, and he died on August 8. For the record, Dr. Goeldel announced the cause as heart and kidney failure. It was a polite, but incomplete, diagnosis. In private, Goeldel confided to Filchner that Vahsel had probably died from complications of syphilis. Filchner was not entirely surprised. He had heard rumors in Germany that Vahsel might have this venereal disease. Vahsel, however, had stoutly denied it, and Filchner had naively taken him at his word. That proved to be a serious mistake. Late-stage syphilis often results in brain damage and mental illness, including paranoia and irrationality, the precise behaviors that Vahsel had repeatedly displayed on the expedition.

Richard Vahsel, Captain of the *Deutschland* — *Taken when he was second officer of the 1901–03* Gauss *expedition, before syphilis destroyed his health. Filchner did not include a picture of his nemesis Vahsel in* Zum Sechsten Erdteil, *his account of the* Deutschland *expedition. It is highly likely that this was a deliberate omission. In fact, the only photos of named expedition members illustrating* Zum Sechsten Erdteil *are a single group picture of the crew (without officers), each man identified by name, plus individual pictures of Filchner's small group of loyal supporters — two of Kling and one each of Przybyllok and König. (Photo from Drygalski,* Zum Continent Des Eisigen Südens . . . , *1904)*

Two days after Vahsel died, his shipmates draped his body with the imperial German flag and lowered him through the ice. The captain's death saddened everyone, even Filchner to some degree. Still, his death encouraged Filchner to hope that the expedition's interpersonal tensions, especially those affecting him personally, might ease. Naval regulations, however, meant that Wilhelm Lorenz, Vahsel's hand-picked first officer, automatically became captain. Filchner quickly realized that his own position was little improved, because the dead captain's antagonism had thoroughly infected Lorenz, the rest of the officers, and most of the staff. In fact, the same Goeldel who had confided the truth about Vahsel's death to Filchner told his own friends a different story. To them he said that Vahsel had attributed his illness to problems with Filchner.

The *Deutschland* continued drifting more or less north for months. To the men, it must have felt as if the Weddell Sea ice were toying with them as the drift carried them north, but then, now and again, looped back to the south. . . . But they *were* creeping basically north, and summer was approaching. On November 26, the pack finally loosened enough for the *Deutschland* to move under her own power. She was at 63° 36' S, 36° 34' W, and her nearly nine-month drift was over. On December 16, after three weeks of working her way slowly through the ice, the *Deutschland* reached the northern edge of the pack. She anchored at Grytviken three days later.

Uproar broke out between Filchner's and Lorenz's partisans as the *Deutschland* docked, but there was no question whose side Larsen and the local authorities were on. That Filchner was the expedition leader was all they needed to know. Although the British magistrate came aboard once the ship docked and offered to arrest Filchner's opponents, Filchner, embarrassed to have non-Germans witness his sordid difficulties, asked him not to. He did accept Larsen's help in separating the hostile parties. He then remained aboard ship with his own supporters while Lorenz's group was housed on shore.

Filchner felt that the near riot he had just endured was sufficient justification to appoint Kling captain. Together, they took the *Deutschland* to Buenos Aires, leaving the opposition behind at South Georgia. Filchner then sailed to Europe on a commercial ship to raise money to continue his work in the 1913–14 season. As for Lorenz and company, they reached Buenos Aires aboard a whaling ship, then sailed home on a passenger liner and began a vitriolic anti-Filchner campaign. Not only did they accuse Filchner of cowardice, of sabotaging the landing at Vahsel Bay, and of deliberately denying the expedition great discoveries, they portrayed Vahsel as a martyr and expedition hero. Filchner's supporters urged him to respond with charges of his own, but he refused, even though the damage to his reputation had made it impossible for him to raise money to continue the expedition. By the time he realized this, Filchner no longer cared very much. As he put it to König, "My urge for the Antarctic was satisfied. . . ."[11]

Wilhelm Filchner sold the *Deutschland* to Felix König for use in an Austrian expedition to the Weddell Sea planned for 1914–15. World War I terminated the project, because Austria drafted all of König's men into the armed forces and commandeered the *Deutschland* for the navy's use. This strong ship, which had survived months of besetment in the Weddell Sea, sank in the Adriatic before the war ended—a victim of guns, not ice.

One member of Filchner's expedition did make it back to the Antarctic regions. He was Ludwig Kohl, the doctor left behind at South Georgia to recover from his shipboard appendectomy. Kohl later married Carl Anton Larsen's daughter Margit and added her surname to his. In the early 1920s, he accompanied his father-in-law to the Ross Sea. And in September 1928, Ludwig and Margit Kohl-Larsen returned to South Georgia and spent eight months exploring the island. Their efforts included a two-week sledge trip into the island's mountainous interior to study the glaciers. Kohl-Larsen christened one the König Glacier in honor of his friend from the *Deutschland*.[12]

Kohl-Larsen may have honored the *Deutschland* expedition, but as far as the German public was concerned, it had been a failure. Filchner himself did little to improve its reputation. His official expedition narrative, *Zum Sechsten Erdteil (To the Sixth Continent)*, took ten years to appear, and when it finally did, in 1922, it presented a highly sanitized version of events. Years later, Filchner wrote a second work that he called "Festellungen" ("Exposé"). This at last told his version of the full story; the account was published in 1985, 28 years after his death.

Although Filchner had failed to meet his original objectives, his expedition did have several very significant achievements. He had both a major land discovery and an ice shelf discovery to his credit, had measured a 15,000-foot depth in the Weddell Sea, and had achieved an ultimate farthest south in the Weddell Sea. Even his beset winter had been productive. Not only had he disproved the existence of Morrell's New South Greenland, but he had also collected important meteorological data, and the course of the beset drift provided valuable information on currents in the Weddell Sea. The Luitpold Coast, the Filchner Ice Shelf, and Vahsel Bay—the *Deutschland* expedition left a firm imprint on the Antarctic map. Even so, the expedition faded into obscurity over the years.

Obscure yes, but a few people remembered. In 1956, Filchner received a letter from England that read, "It was on Sunday, 29th January, this year that I reached Vahsel Bay and continued to the west along the ice shelf edge for 20 miles. There I established our base.... I close by assuring you that throughout the planning of this expedition the name FILCHNER [Fuchs's emphasis] has been continually in our minds...."[13] The letter was signed by British explorer Vivian Fuchs, organizer of the Commonwealth Trans-Antarctic Expedition. Two years later, in 1957–58, Fuchs, starting from the base he established in January 1956, would lead the first overland crossing of the Antarctic Continent.

CHAPTER 13

Endurance, Shackleton's Triumphant Failure: 1914–1916

Vivian Fuchs, who had written so thoughtfully to Filchner in 1956, was not the first to attempt crossing the Antarctic Continent, nor was he the first to take note of the *Deutschland* expedition's results. More than 40 years earlier, the Anglo-Irish Antarctic veteran, Sir Ernest Shackleton, had led the way.

Three years after Shackleton's barely failed attempt to reach the South Pole, which the whalers had told Charcot about, two other men succeeded: Norway's Roald Amundsen, who had been on the *Belgica* with de Gerlache, and Britain's Robert Falcon Scott, who had earlier led the 1901–04 British *Discovery* expedition to the Ross Sea. Both spent the winter of 1911 on the southern coast of the Ross Sea, several hundred miles apart. They then set out for the South Pole in the spring. Amundsen's five-man team arrived first, on December 14, 1911. All in his party returned safely. Scott's group of five reached the pole five weeks later, only to find the Norwegian flag already flying there. All died on the return trip.

With the South Pole conquered, Shackleton, eager to return to Antarctica, felt that "there remained but one great main object of Antarctic journeyings [sic]—the crossing of the South Polar continent from sea to sea."[1] He was familiar with Bruce's plan, and when he announced his "Imperial Trans-Antarctic Expedition," the gist of his program closely followed the *Scotia* leader's proposal. Shackleton's plan was to set out across the Antarctic Continent from a base established as far south as he could reach on the Weddell Sea coast, probably at Vahsel Bay, so recently discovered by Filchner. A support party out of McMurdo Sound in the Ross Sea would lay supply depots for him to pick up for the closing stages of the trip. Separate ships would transport and support the two groups of men—the *Endurance* to the Weddell Sea and the *Aurora* to the Ross Sea. In effect, Shackleton was planning two separate, but closely linked, major expeditions. What follows is the story of the *Endurance* and the Weddell Sea party.

The *Endurance* was a new steam sailer, built in Norway in 1912 at the instigation of de Gerlache to serve as an Arctic cruise ship or for polar expeditions. She had yet to do either when Shackleton bought her and renamed her for his family motto,

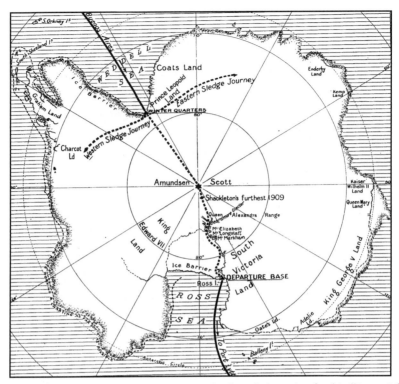

Shackleton's proposed trans-Antarctic route and other sledge trips for his "Imperial Trans-Antarctic Expedition" — *The crossing team and scientific personnel would be landed on the Weddell Sea side of the Antarctic Continent, somewhere in the vicinity of Vahsel Bay. The crossing team, led by Shackleton, would set off from there while others who had landed would make exploratory/scientific sledging trips to the east and west. In the meantime, another team that had been landed at Ross Island in McMurdo Sound, on the Ross Sea side of the continent, would lay supply caches for the crossing team to pick up after they had passed the South Pole. (Illustration from Shackleton,* The Imperial Trans-Antarctic Expedition, Prospectus, *1913)*

Fortitudine Vincimus (By Endurance We Conquer).

Shackleton's plan excited the British public, and nearly 5,000 men (and three "sporty girls") applied to join him. Very few made the cut, because he already had a number of Antarctic veterans in mind. The most important was Frank Wild, a man with even more southern experience than Shackleton himself. Wild, who was devoted to the man he called The Boss, had been on three previous Antarctic expeditions, including Shackleton's 1908–09 South Pole attempt. His most recent foray south had been from 1911 to 1913 with Douglas Mawson in East Antarctica. Shackleton also took along Mawson's photographer, Frank Hurley, and Tom Crean, who had been on both of Scott's expeditions. The *Endurance*'s captain was a New Zealander, Frank Worsley. Although he lacked ice experience, he was a superb navigator and small boat expert—skills that would both prove critical.

War clouds darkened the European horizon as Shackleton completed his expedition preparations in mid-1914. On August 3, when he was almost ready to sail, the newspapers announced that the British government had issued a Naval Mobilization Order. Shackleton immediately cabled the Admiralty offering his ship and men for the war effort. The response was terse, a one-word telegram from Winston Churchill, then First Lord of the Admiralty. It read "Proceed."[2] King George V sent for Shackleton the next day and assured him that the expedition would be sailing with his approval, despite the gravity of the military situation.

And so the *Endurance* sailed from England on August 8, 1914, to fight what Shackleton called the "White Warfare of the South."[3] He himself remained behind to deal with last-minute details before following on a commercial vessel to join the *Endurance* in Buenos Aires.

Three days after the *Endurance* left Buenos Aires on October 27, her men found a stowaway. A nineteen-year-old Welshman named Perce Blackborrow, he had slipped aboard in the Argentine capital with the help of two crewmen. Shackleton was officially furious. Frank Wild, who was present at the interview, reported that he shouted, "Do you know that on these expeditions we often get very hungry, and if there is a stowaway available he is the first to be eaten?" Wild went on, "Shackleton was . . . fairly heavily built, and the boy looked him over and said, 'They'd get a lot more meat off you, sir!' The Boss turned away to hide a grin and told me to turn the lad over to the bos'un, but added, 'Introduce him to the cook first.'"[4] Blackborrow quickly became a welcome member of the crew, and Shackleton, privately delighted with the young man's spunk, was happy to have him.

When Shackleton reached South Georgia on November 5, the whalers at Grytviken greeted him with bad news. The ice this year was unusually heavy in the direction of the Weddell Sea. Accordingly, he decided to stay longer than originally planned, to allow the ice time to break up. The first weeks of waiting passed reasonably quickly, the scientists working on the island while Shackleton traveled about, courtesy of the whalers, and met a number of the station managers. But at the beginning of December, Shackleton grew impatient and decided to leave despite the whalers' warnings that the ice was still dense to the south.

It was December 5, 1914, when the *Endurance* departed from South Georgia with 28 men, dozens of sledge dogs, and provisions for two years. She met her first serious ice only two days later. It was an ominous development, but Shackleton's only alternatives were to push on or turn back and try later. He chose to continue, and for the first 100 miles, the pack was fairly open. Then the ice closed up and progress slowed to a crawl, much as it had for Filchner.

The *Endurance* finally found open water on January 9 at about 70° S, 17° W. She reached 72° S the next day, and Shackleton saw Coats Land. He passed Bruce's farthest south on the 12th and discovered new land; it was the missing link between Coats Land and Filchner's Luitpold Land. Like his two predecessors along this coast, Shackleton named his discovery after his most generous patron. The land he saw thus became the Caird Coast, after James Caird, who had contributed £24,000 to the expedition. Five days later, Shackleton sighted Luitpold Land. He had confirmed the existence of a continuous stretch of land along the eastern side of the Weddell Sea.

On January 19, the *Endurance* reached 76° 34' S, 31° 30' W. There the ice closed about her. Shackleton was unconcerned: he had experienced this sort of thing before on this voyage. Perhaps a day or two would pass and he could continue on to Vahsel Bay. A frustrating delay, but no more than that. But this time, things were different: the *Endurance* had sailed her last miles under Shackleton's control.

A month would pass before Shackleton accepted the fact that he was beset, caught for the winter. In early February, long leads opened in the pack just beyond the ship, and he thought he might break out. But the *Endurance*'s engine was not strong enough. A few days later, a new effort, this time to cut through the ice with saws, proved equally futile. Shackleton finally surrendered on February 24 and converted the *Endurance* to a beset winter base. Only two days earlier, the ship had drifted to 77° S, 35' W, the farthest south and the closest to Vahsel Bay she would reach.

Shackleton's first step was to move the dogs off the ship, onto the ice where the men constructed what they called dogloos. The dogs scorned them except in the most severe weather. With more room available on deck for cargo, Shackleton cleared space in the hold and had the men convert it to a living area for the officers and staff. The occupants soon christened their new home The Ritz.

Life aboard the drifting *Endurance* was relatively comfortable, and spirits remained high for the most part even after the winter night began on May 1. Shackleton, though himself depressed over the probable demise of his crossing hopes, worked hard to maintain his men's morale. That was his chief concern in the early months after the *Endurance* became beset. After all, the *Belgica* and the *Deutschland* had both survived their beset drifts. But Antarctic pack ice is fickle. Sometimes simply a stubborn jailer, at other times it can be a nightmare of devastating pressure that can crush the hull of even the strongest ship. Nordenskjöld's men, who had watched the *Antarctic* sink beneath the ice, could certainly testify to that.

The Weddell Sea currents initially pushed the *Endurance* along a zigzag route that took her as much to the west as to the north. By July 8, the ship was much farther west than anyone had been in the Weddell Sea at such a high latitude. Although that was good for geographical science, it was hazardous for the *Endurance*. She was moving into an area where winds and currents drive the ice against the east coast of the Antarctic Peninsula, accentuating the potential for crushing pressure on a ship. On July 22, four days before the sun returned to end the Antarctic night, ice smashed into the *Endurance*'s sides. A week and a half later, the floe surrounding and protecting the ship disintegrated in a fierce gale. Shackleton wrote, "The effects of the pressure around us were awe inspiring. Mighty blocks of ice, gripped between meeting floes, rose slowly till they jumped like cherry stones squeezed between thumb and finger. The pressure of millions of tons of moving ice was crushing and smashing inexorably."[5] The men had hustled the dogs back onto the ship just in time. The dogloos were in ruins, but somehow the damage to the *Endurance* herself was relatively minor. The blow to Shackleton's confidence in her safety, however, was severe.

Several more times through early October, the Weddell Sea ice attacked the *Endurance*. The ship suffered some damage—buckled stanchions, stoved-in planks, and a smashed rudder—but nothing was irreparable. The drift had turned north, and the *Endurance* was now above the 70th parallel; temperatures were rising; and there were more and more seals and penguins about on the ice. Spring and freedom seemed to be on their way when the officers and scientists moved out of The Ritz and back into their regular cabins on October 12.

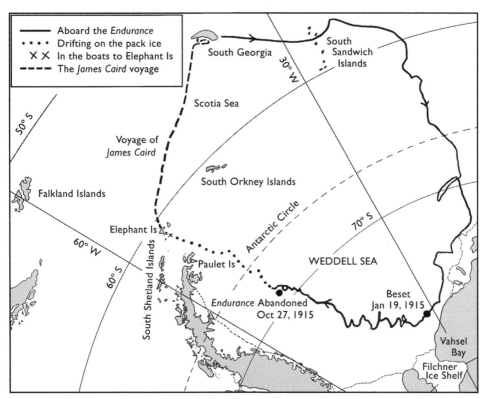

Track of *Endurance* expedition — *Like Filchner, Shackleton had to fight his way south into the Weddell Sea, battling often dense pack ice. But the* Endurance *never reached Vahsel Bay, instead becoming beset in late January. Her drift, entered into significantly farther south than the* Deutschland's, *placed her in the control of currents that carried her much farther to the west, closer to the Antarctic Peninsula. There the danger of crushing pressure from the pack ice was greater than in the central Weddell Sea where the* Deutschland *had spent her trapped months. Either ship could have been lost, but it was the* Endurance *that faced the greater danger, and it was she that lost the battle to the ice.*

Five days later, the ice began what would be its final onslaught. The attack started slowly, the first day more a feint than anything really serious. On the 18th, however, the floe on the port side cracked. As huge slabs of ice shot up beside her, the *Endurance* heeled over until she was listing 30 degrees. The men had to move about the ship like mountain goats until evening, when the pressure eased off and the ship righted herself. The pack churned about them for days. Then, on October 23, tons of seething ice bashed a hole in the hull. The men worked all night at the pumps while Harry McNeish, the carpenter, cobbled together a dam to block the massive leak.

"The End" — *Shackleton's title for this photo of the ruins of the* Endurance, *watched by a dog team. (Royal Geographical Society)*

Renewed pressure assaulted the *Endurance* three days later, and water poured in past McNeish's dam. That night, eight emperor penguins suddenly appeared on the ice and "proceeded to utter weird cries that sounded like a dirge for the ship."[6] The next morning, October 27, 1915, a desolate Shackleton gave the order to abandon ship.

Twenty-eight men, nearly 50 dogs, and a single cat stood on a large ice floe in the Weddell Sea at 69° 5' S, 51° 30' W. They were more than 300 miles from the nearest known shelter, Nordenskjöld's hut at Snow Hill Island, and even farther distant from the emergency supply depot that Captain Irízar had left on Paulet Island when he rescued the Swedish expedition in 1903. No one in the outside world had any idea of their plight. Survive or die. It was up to them. Shackleton, however, seemed to be at his best when faced with seemingly insurmountable odds. Once the men had established a temporary camp on the ice, he gathered everyone together and told them his plans. They would sledge hundreds of miles across the pack to Paulet Island. They could do it, he said. His words, wrote Worsley, the ship's captain, "had an immediate effect: our spirits rose, and we . . . [took] a more cheerful view of a situation that, actually, had not one element in it to warrant the alteration. . . . We knew that if mortal man could lead us to safety, Shackleton was that man."[7]

Shackleton immediately began organizing the sledge trip. He limited each man to two pounds of personal gear in addition to his clothing and sleeping bag. They would have to abandon everything else. Shackleton set the example, pointedly throwing down his gold cigarette case, his gold watch, coins, and, most dramatically, a Bible that Queen Alexandra had given him. He kept only the Bible's flyleaf, the 23rd psalm, and a page with verses from the book of Job that began "Out of whose womb came the ice?" Shackleton made an exception for a banjo that belonged to the meteorologist, Leonard Hussey. Its music, he said, would be good for morale. He failed to see the same value in some of the animals. As far as Shackleton was concerned, the youngest puppies were simply extra mouths to feed. He ordered them shot. The carpenter's cat also had to go, he explained, to save him from being killed more brutally by the remaining dogs.

On October 30, Shackleton led the way out into the pack. Fifteen men dragged heavy sledges laden with the *James Caird* and the *Dudley Docker,* two of the *Endurance*'s ship's boats. The rest accompanied dog teams pulling smaller sledges. The dogs negotiated the pack reasonably well, but moving the boats over the tortuously scrambled ice was excruciatingly difficult. As a result, two days of exhausting work netted a bare 2-mile advance. But abandoning the boats was not an option: they were essential should the men come to open water before reaching land. Shackleton had already left the *Stancomb Wills,* the smallest of their boats, and he was unwilling to cut the safety margin further.

"The first attempt to reach the land three hundred and forty-six miles away" — *Shackleton's title for this painting by expedition artist George Marsten, showing the men of the* Endurance *man-hauling one of the boats over jumbled pressure ridges in the pack ice. The wreck of the* Endurance *is behind in the distance. (Marsten painting, from Shackleton,* South, *1919)*

Shackleton admitted defeat on November 1. His new plan was to camp on the ice where they were and wait until sledging conditions improved or the ice opened up sufficiently to use the boats. In the meantime, the northerly drift should bring them closer to Paulet Island. He sent the men back to the wreck of the *Endurance* to retrieve all the food they could find, as well as anything else useful. During the next several weeks, the increasingly elaborate collection of tents and other structures the men now called Ocean Camp became well stocked with salvaged items. Among these were some of Frank Hurley's photographic plates. Without consulting Shackleton, the photographer had returned to the *Endurance*, dived into the icy water below deck, and retrieved hundreds of them. When Shackleton learned of this and objected to the weight of the heavy plates, Hurley argued that the photographs would be of enormous value when they reached home. Shackleton reluctantly agreed that Hurley could save the best of the lot, and the two men then selected over 100 of them. They smashed the rest to avoid the temptation to preserve any more. Shackleton also allowed Hurley to keep his exposed movie film, a small pocket camera, and three rolls of unexposed film.[8]

The ruins of the *Endurance* sank below the ice on November 21. The men saw her go with a mixture of relief that her agonies were finally over and a sense of desolation that their home of so many months was gone. Morale plummeted in the weeks after the *Endurance*'s hulk disappeared. Living in their canvas tents on the ice was miserable; their northward drift was slower than hoped; and only Charles Green, the cook, had a regular job. By December 20, they had gained about 100 miles since establishing Ocean Camp, and Shackleton, increasingly concerned about his men's

Frank Hurley (left) and Sir Ernest Shackleton (right) in front of a tent at Patience Camp — *The two men flank the camp's communal cook stove, crafted from empty oil drums salvaged from the wreck of the* Endurance *and fueled with seal blubber and penguin skins. In the upper right background, a man stands in front of one of the ship's boats that would become crucial to their survival in just a few months. (Royal Geographical Society)*

mental state, decided to give sledging another try.

Shackleton's men set off on December 23, once more dragging the heavy *James Caird* and *Dudley Docker*. (They had retrieved the *Stancomb Wills*, but again abandoned her to lighten the load.) Even with the help of an advance party hacking a path through the pressure ridges, it took the entire man-hauling group to pull one boat at a time. When seven days of backbreaking work gained only 7 ½ miles, Shackleton conceded for a second time that sledging was hopeless. They would camp anew, once more pinning their hopes on the northward drift. On December 29, Shackleton established a new camp on the largest floe he could find.

With so much left behind, the new location, which Shackleton christened Patience Camp, was a good deal more Spartan than Ocean Camp had been. Food and fuel were also a growing problem, because the seals and penguins they depended on had become scarce. In mid-January, Shackleton made the wrenching decision to kill most of the dogs. The end of his sledging hopes had made them superfluous, and he could no longer afford to keep them simply for sentimental reasons. He ordered all but two small teams shot. The men understood the necessity, but still, many felt like murderers, killing dear friends.

The pieces of the pack were moving north at slightly different rates, and as the weeks went by, Ocean Camp's floe drifted closer to the new camp. At the beginning of February, Alexander Macklin, one of the two doctors, and Hurley took the surviving dog teams to see if they could salvage anything from the old camp. Although Ocean Camp was a shambles, the two men located a lot of food and a few books. They brought back as much as they could. Shackleton sent another group to collect more the next day. The food was important, but, Shackleton now realized, the most valuable thing Ocean Camp had to offer was the twice-abandoned *Stancomb Wills*. They retrieved her just in time. A day later, a third foraging team had to turn back because new leads between the floes had placed Ocean Camp out of reach.

Even after salvaging all they could from Ocean Camp, the men of the *Endurance* were on the thin edge of starvation much of their time at Patience Camp. By the end of March, Shackleton realized they had too little food for both themselves and the remaining dogs. It was a sad day when they shot their last animal companions. The men, however, were less sentimental this time. Dogs who had been food consumers now became food themselves. But Shackleton's decision to kill the dogs stemmed from more than food considerations. Pieces of the Patience Camp floe had been breaking off for weeks, and the pack about them was now liberally laced with open water. It looked as if they might have to launch the boats very soon: there would be no room for dogs.

Patience Camp had drifted past the latitude of Snow Hill Island in early March. At mid-month, the men were 60 miles east of Paulet Island. But 60 miles or 600, the island was unattainable. There was too much ice to reach it by boat, and not enough to sledge there. The peaks of Joinville Island appeared a few days later. The first land the men had seen in a year, it was no more reachable than Paulet Island. They could only pray that the ice would open in time for them to launch the boats before they passed the little land that remained between them and the Drake Passage.

The tense wait came to an end on April 9. Shortly before noon, the Patience Camp floe broke right across the middle. It was a bittersweet moment. Shackleton wrote, "The floe had become our home. . . . Now our home was being shattered under our feet, and we had a sense of loss and incompleteness hard to describe."[9] They all gathered on the largest remaining bit of that home. Then, after so many months moving at the whim of wind and current, they launched the boats, at last taking their fate into their own hands.

There were three boats. The 22-foot-long *James Caird*, the largest of the trio, was the most seaworthy. The slightly smaller *Dudley Docker* was also fairly sturdy. The weak link was the *Stancomb Wills*, the boat Shackleton had twice left behind and rescued at the last moment from Ocean Camp. When fully loaded, her sides rose only 17 inches above the water. Shackleton put thirteen men in the *James Caird* under his own command, ten in the *Dudley Docker* with Worsley, the *Endurance*'s captain, and five in the *Stancomb Wills* under Hubert Hudson, the navigating officer.

The night after leaving Patience Camp nearly ended in disaster. That evening, Shackleton tied his tiny flotilla to a small floe and the men set up camp on the ice. During the night, a swell cracked their campsite in two right under one of the tents, and a man inside plunged into the icy water. His frantic tentmates hauled him out just before the pieces of the floe slammed back together. Then the ice opened again and Shackleton, who had rushed to help, found himself on the wrong side of a widening gap along with the *James Caird* and two other men. After all three men pushed the *James Caird* across to the rest of the camp, Shackleton's two companions jumped across. By the time it was The Boss's turn, however, the gap had grown impossibly wide. Fortunately, Wild saw what was happening and launched the *Stancomb Wills* even before Shackleton called for help.

Shackleton's men rowed and sailed the boats north for the next two days, sometimes fighting their way through bits of pack, at other times battling stormy seas that forced them to tie up at floating ice to wait for calmer water. Despite the experience of their first night, they camped on the ice the second night. All of the men, too bone-tired to think about the danger, slept soundly. They spent the following night in the boats, fending off ice floes that bumped and ground against the small boats. This time, because of the cold and the threat of the ice, no one slept.

The men were so focused on battling the ice that they failed to consider the consequences when they left the pack behind, around noon on April 13. Their first thought was simple delight because "At last we were free from the ice, in water that our boats could navigate."[10] Navigate, yes, but not drink. Their euphoria upon reaching open water withered when they realized that no one had thought to take on a supply of ice for drinking water. And it soon became clear that the pack ice had provided protection from rough water. Now the heavily laden boats were in constant danger of being swamped.

At dawn the next day, the men saw Elephant Island, one of the most easterly islands in the South Shetland chain, shining in the sun 30 miles away. It was a gorgeous windless day . . . for sightseeing. The already exhausted men turned to the oars and rowed as if their lives depended on it. They did. It was evening before the winds at last cooperated and Shackleton could use the sails. The *James Caird* had the *Stancomb Wills* in tow. That automatically kept those two boats together, but Shackleton soon lost sight of the *Dudley Docker*. It was a stormy night, and some of the men—exhausted, thirsty, and frozen almost senseless—were in such poor condition that Shackleton feared they might not survive until morning. Somehow they all did. He could only pray that the *Dudley Docker* and her men had survived as well.

On April 15, the seventh day since leaving Patience Camp, Shackleton at last found a landing place, at Cape Valentine on Elephant Island's far eastern end. To his immense relief, the *Dudley Docker* turned up shortly after. Soon all 28 men were together on the beach. Hurley wrote,

> We were a pitiful sight; the greater number terribly frostbitten and half delirious. Some staggered aimlessly about and flung themselves down on the beach, hugging the very rocks, and trickling the pebbles through their hands as though they were nuggets of gold. It is hard to describe the joy we felt . . . [to] feel land under our feet—land that would not split and disintegrate![11]

Elephant Island was more than just solid land. It offered water, food, and fuel. The famished men killed an elephant seal shortly after reaching shore, and Green, the cook, worked for hours as his "blubber-stove flared and sputtered fiercely . . . [cooking] not one meal, but many meals, which merged into a day-long bout of eating."[12] This was the first time they had eaten their fill in months. Shackleton, however, quickly realized that Cape Valentine's stony beach, a place that was clearly regularly inundated by storm-driven tides, was a precarious refuge at best. The day after landing, he sent Wild in the *Stancomb Wills* to search out a safer location. Wild found a rocky spit that looked a bit better, 7 miles to the west along the north coast of the island.

Shackleton moved his men to the place he named Cape Wild (today called Point Wild) on April 17. Their new home welcomed them with a fierce blizzard that continued into the next day. But the place, only an acre or two of rock and shingle, was pretty bleak in any weather.

The men of the *Endurance* had reached the safety of land, but it was a very tenuous safety. Many of them were extremely weak, and food was a serious concern, especially with winter approaching. Given everything, Shackleton felt that waiting on Elephant Island, in hopes of rescue by the whalers the following summer, could be a death sentence for some of his men. He thus came to a desperate decision. He would sail the *James Caird*, the best of their boats, to civilization to summon a rescue ship. Stanley in the Falklands, about 550 miles away to the north, was the closest inhabited place, but the prevailing winds blew in the wrong direction for a small sailing boat attempting to head north. The best alternative was South Georgia, nearly 800 miles distant to the northeast. The winds would be right, and Shackleton knew that he would find sympathetic whalers in residence.

Shackleton was proposing to sail hundreds of miles at the beginning of winter over one of the roughest seas in the world. In a 22-foot whaleboat. Despite the dangers, many of the men wanted to go. Shackleton chose Frank Worsley for his navigation skills and small boat knowledge, Tom Crean because he was strong and reliable, and Harry McNeish because his carpentry skills could prove vital. Two seamen—Timothy McCarthy and John Vincent—completed the party. Shackleton turned down Frank Wild's pleas to be included because he needed the man at Elephant Island to hold things together there.

The men prepared the *James Caird* as best they could with their very limited means. They used sledge runners, box-covers, and canvas to construct a deck. The artist's oil paints provided caulking material to make the boat more watertight, and Elephant Island supplied the ballast, 2,000 pounds of rocks.

Shackleton's desperate voyage in the *James Caird* began on April 24, 1916. As he and his five companions sailed away, they looked back at the 22 men left behind, "a pathetic little group on the beach, with the grim heights of the island behind them and the sea seething at their feet, but they waved to us and gave three hearty cheers. There was hope in their hearts and they trusted us to bring the help they needed."[13]

Icy spray soaked the men from the beginning to the end of the trip, and life under the makeshift deck was little better. Sleep was a luxury they seldom enjoyed. And cooking in the cramped quarters was a great struggle as the *James Caird* rolled and pitched in the often violent seas. But the effort was worth it. Hot drinks and meals were the highlights of their miserable days.

The first of many gales hit the third night out. On the fifth night, when they were nearly 300 miles from Elephant Island, Shackleton hove the *James Caird* to for the night so they could all get some rest. When dawn broke, the six men below deck sensed that something was wrong. They crawled out to see why the boat seemed to be wallowing and met a terrifying sight. The *James Caird* was encased in ice, on the verge of sinking from the added weight. Shackleton immediately tossed two of the by-now sodden sleeping bags and all of the extra oars overboard. Then the six men set frantically to work to hack off the ice.

SHACKLETON'S TRIUMPHANT FAILURE: 1914–1916

Ice they could remove, storms they could sail through, but none of that mattered if they strayed off course. Worsley had obtained a sun sight just before leaving Elephant Island. After that, however, overcast skies forced him to navigate by dead reckoning, a treacherously risky business for a boat heading for a single island in a vast ocean. Finally, on May 1, the sun reappeared, and Worsley was able to check their position. They were right on course, roughly midway between Elephant Island and South Georgia. It was an astounding feat of navigation.

The voyage nearly ended prematurely on the night of May 5–6. About midnight, Shackleton saw what he first thought was a line of clear sky on the horizon. Then, to his horror, he

> realized that what I had seen was not a rift in the clouds but the white crest of an enormous wave. . . . It was a mighty upheaval of the ocean, a thing quite apart from the big white-capped seas that had been our timeless enemies for many days. . . . We felt our boat lifted and flung forward like a cork in a breaking surf. We were in a seething chaos of tortured water. . . .[14]

Somehow the *James Caird* survived, but it took the men hours to bale her out. Another blow hit the next day, May 7. They had loaded two casks of fresh water and 250 pounds of ice aboard on leaving Elephant Island. With the ice long gone and the first cask now empty, they opened the second one. To their shock, the water it held was so brackish that they were confronted by a repeat of the hellish thirst they had

"In sight of our goal: nearing South Georgia" — *Shackleton's title for this painting, "drawn from material supplied by the Boat Party," of the* James Caird *fighting her way through violent seas toward the glacier-covered mountains of South Georgia. (from Shackleton,* South, *1919)*

experienced on the way to Elephant Island. At least the end of the voyage seemed near. Worsley calculated that they had just 70 miles to go.

The *James Caird* did reach South Georgia on May 8, but daylight was fading as she neared the island's southwest coast. Because their only map, a copy of Filchner's rough chart of the coast, offered little detail, Shackleton decided that landing in the dark was too dangerous. They would haul off and wait for morning.

These six men, wracked by thirst, spent an awful night. Then things grew worse. At 5 a.m., a violent storm hit. For hours, Worsley fought the gales for control of the *Caird* while his companions baled frantically. The storm, driving the small boat fatally toward the rocky cliffs of Annenkov Island, was winning by late afternoon. Worsley was just beginning to think about how no one would ever know they had reached South Georgia when the wind suddenly shifted and then dropped. The storm, however, had lasted all day, and it was again too dark to land safely. They had to endure another brutally parched night at sea.

On May 10, sixteen days after leaving Elephant Island, the *James Caird* slipped into a small welcoming cove in South Georgia's King Haakon Bay. Shackleton, supported by Worsley's navigation and boat-handling skills, had completed a voyage that would become a legend in the lore of the sea.

The men of the *James Caird* spent the next five days recovering. But the journey was not over: they had landed on South Georgia's uninhabited southwest coast. All the whaling stations were on the other side of the island. Shackleton had three choices. He could try to sail the *James Caird* around. That one was out because the battered boat was no longer up to it. He could winter where he was and use the time to repair the *Caird* or wait for the whalers to show up in the spring. Shackleton rejected that option because he was determined to return to his men on Elephant Island immediately. Finally, he could cross South Georgia's unexplored, glacier-covered mountains to reach a whaling station. It was the only acceptable choice.

Traversing South Georgia was as daunting a prospect as the voyage in the *James Caird* had been. It was true that whalers had been crossing the narrow northern neck of the island for years, on summer recreational trips between the Prince Olav Harbor whaling station at Possession Bay on the northeast coast and King Haakon Bay on the southwest shore, where the *James Caird* had landed. But that was not the route that Shackleton proposed to take. The nearest whaling station open during the winter was at Stromness Bay. And that was more than 20 miles away as the crow flies, along a diagonal over the entirely unexplored mountains in the middle of the island. This was the path Shackleton would have to take. In early winter.

Shackleton sailed the *James Caird* to the head of King Haakon Bay on May 15. There they hauled the boat ashore, overturned her and converted her into a crude hut with a foundation of rocks and tussock grass. It was such a crude affair that Shackleton named their new home Pegotty Camp in honor of the proprietor of a boat-made-house in Charles Dickens's novel *David Copperfield*. Shackleton then used the next several days to prepare for the overland trip. He had already decided who would go—himself, Crean, and Worsley. Shackleton and Crean had mountaineering experience from earlier Antarctic expeditions, and Worsley had done some climbing in the New Zealand Alps. The other three men, two of whom had not yet recovered from the *James Caird* voyage, would remain behind, to be picked up in due course.

It was 3 a.m. on May 19 when Shackleton, Worsley, and Crean set out, three debilitated but determined men. Their equipment consisted of provisions for three days, a primus stove, a few matches, about 90 feet of rope, McNeish's carpenter's adze (to serve as an ice axe), a chronometer, two compasses, a set of binoculars, a small piece of Filchner's chart of South Georgia, and the clothes on their backs.[15] That was all they had—that plus a full moon, clear skies, and desperation.

At dawn the first morning, they found themselves high atop a ridge looking down on what appeared to be a huge frozen lake. Shackleton decided to descend to it, since it seemed to be along their route. On the way down, the fog lifted enough to reveal that the supposed lake was in fact Possession Bay. Not only that, it was well north of where they wanted to be. And so they climbed back to the ridge and changed course to the east. This would be only the first of many false starts. Crossing South Georgia turned out to be a matter of negotiating their way around and across crevasses; retreating from ridge tops that ended in precipices; slipping and sliding on the ice; and never really knowing just where they were or what lay ahead.

Shackleton's small party took risks that better-equipped mountaineers would have refused to consider. They simply had no time to search out the safest routes. At one point, they found themselves high in the mountains atop a razorback ridge. It was late at night, foggy, and turning colder by the moment. Unless they descended quickly, they would freeze. Shackleton began by carefully cutting steps in the ice as the three men inched their way down, but it was too slow. In a leap of faith, Worsley wrote, "We each coiled our share of the rope until it made a pad on which we could sit. . . . Shackleton sat on the large step he had carved, and I sat behind him, straddled my legs round him and clasped him round the neck. Crean did the same with me. . . . Then Shackleton kicked off. We seemed to shoot into space. . . ."[16] The wild ride ended in a soft landing—in a snow bank. They had hurtled down more than 1,000 feet in a couple of minutes.

Shackleton, Worsley, and Crean reached the east side of South Georgia early on their second day. While taking a short break for breakfast, they heard a steam whistle from a whaling station. That simple sound meant so much: that the station was near and that the whalers were in residence. In 1965, a well-equipped party retraced Shackleton's route. Their leader wrote, ". . . nearly half a century later, we stood in the same [place]. . . . We heard no whistle, but, at that moment, we could still experience an atmosphere charged with powerful emotion. It had taken Shack-

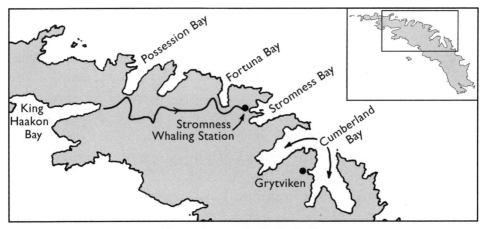

Shackleton's route across South Georgia.

leton just 24 hours to reach this point. We had taken 12 days...."[17]

Although the air easily carried the voice of the whistle to them, the three men still had to struggle to cover the final miles to the source of that wonderful sound. At last, they topped a ridge and saw a ship entering Stromness Bay and tiny men moving about a distant station. Shackleton's party paused, shook hands, and then went on. The last challenge was descending through a 25- to 30-foot waterfall. Of reaching the bottom, Shackleton wrote,

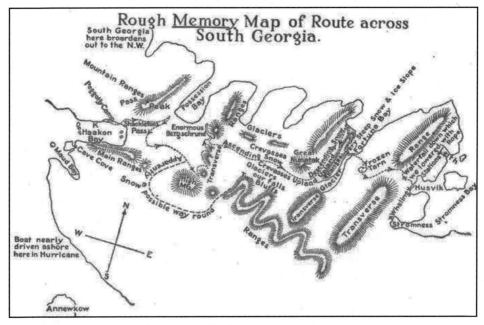

Frank Worsley's "Rough Memory Map" of the route that he, Shackleton, and Crean took across South Georgia — *Note that Worsley shows, incorrectly, that they completed the crossing at Husvik in Stromness Bay, rather than Stromness where they in fact ended. This error results from the fact that he has mistakenly interchanged the locations of these two whaling stations on his map. (From Shackleton,* South, *1919)*

> ... we had flung down the adze from the top of the fall and also the log-book and the cooker wrapped in one of our blouses. That was all, except our wet clothes, that we brought out of the Antarctic. . . . That was all of tangible things; but in memories we were rich. We had pierced the veneer of outside things. We had 'suffered, starved, and triumphed, groveled down yet grasped at glory, grown bigger in the bigness of the whole.' We had seen God in his splendours, heard the text that Nature renders. We had reached the naked soul of man.[18]

The three men walked into the Stromness whaling station at 3 p.m. on May 20, only 36 hours after leaving King Haakon Bay—30 miles away by the route they had taken. No one recognized these scarecrows, filthy and covered in blubber soot, with long hair and tangled beards. Two small girls saw them and fled, as did the first adult they saw.[19] Another man, however, did not, and despite his doubts, took them to the station manager's house. There he went inside to tell Thoralf Sørlle, the station manager, about the three strange men who claimed to have crossed the island.

Sørlle, curious, came to the door. Shackleton, who had met Sørlle when the *Endurance* was at South Georgia in 1914, wrote that the following conversation ensued: "'Well?' 'Don't you know me?' I said. 'I know your voice,' he replied doubtfully. . . . 'My name is Shackleton,' I said. Immediately he put out his hand and said, 'Come in. Come in.'"[20]

Sørlle could not do enough for them. He invited them into his home, fed them, and provided hot baths and clean clothes to replace their rags.[21] That evening, he sent a catcher with Worsley aboard to King Haakon Bay to pick up the three men waiting there. The whalers also retrieved the *James Caird*, treating her with reverence as they secured her on deck.[22]

Shackleton and Crean slept that night in soft beds at Sørlle's house, safe from a blizzard that had begun soon after they reached Stromness. It was a storm that would have killed them had it hit while they were crossing South Georgia. They had been miraculously lucky. Shackleton, ordinarily a man relatively indifferent to religion, later wrote,

> I have no doubt that Providence guided us. . . . I know that during that long and racking march of thirty-six hours over the unnamed mountains and glaciers of South Georgia it seemed to me often that we were four, not three. I said nothing to my companions on the point, but afterwards Worsley said to me, 'Boss, I had a curious feeling on the march that there was another person with us.' Crean confessed to the same idea.[23]

The next morning, Sørlle told them about the war and other world events. But he also had news of a personal concern: Shackleton's Ross Sea Party was in trouble. A storm had torn their ship, the *Aurora*, from her moorings in McMurdo Sound, and she had been trapped in her own months-long drift. She had eventually escaped the ice, but ten men had been left behind in the south. Although that was all

Sørlle knew, Shackleton now realized he had *two* marooned parties to worry about. His first concern, however, had to be the men on Elephant Island. His one comfort when he thought about them was that Frank Wild was there. Still, Shackleton had seen enough of Elephant Island to realize that it was a brutally inhospitable place.

For their part, the 22 men waiting at Elephant Island grew to know Point Wild intimately, and the best thing they could say of the place was that it was solid land. Frequent blizzards and violent winds swept over their tiny spit, and the air was constantly cold and damp. Since shelter from the elements was crucial, Wild immediately set the men to work building a home. What they created was a combination structure—foundation walls from local stone, topped with the *Stancomb Wills* and *Dudley Docker* upturned and laid side-by-side for upper walls and a roof. It was cramped, cold, dark, and fetid inside, a miasma of blubber smoke from their only source of heat. Food was a constant problem. They had a few remaining boxes of provisions salvaged from the *Endurance*, but the great bulk of their menu had to come from the meager resources of Elephant Island. Since most of the seals and penguins had already decamped for the winter, they depended on taking whatever game turned up, when and if it did. It would be just barely enough.

Most of the men were amazingly healthy, given their circumstances, even though Alexander Macklin and James McIlroy, the two doctors, had warned Shackleton that some of them would not last more than a month. That was one reason Shackleton

Cutaway view of the inside of the hut on Elephant Island — *Composite Marsten drawing and Hurley photograph. The hut's inside measurements: 18 feet long, 9 feet wide, 5 feet high at the highest point. The expedition physicist, Reginald James, who was in the Elephant Island group, wrote a song he called "Antarctic Architecture" with a chorus that went "My name is Frankie Wild-o and my hut's on Elephant Isle // The wall's without a single brick and the roof's without a tile // Yet nevertheless, you must confess by many and many a mile // It's the most palatial dwelling-place you'll find on Elephant Isle."*[24] *(Royal Geographical Society)*

had left when he did, as well as why he was desperate to return as quickly as possible. Remarkably, several men who had looked very shaky recovered quite quickly. Others took longer, but gradually came around. And then there was Perce Blackborrow. The former stowaway's left foot, frostbitten on the boat trip to Elephant Island, developed gangrene, and it was clear by mid-June that he would die unless something was done. On the 15th, the doctors cleared the hut of everyone but themselves, Wild, Hurley, Blackborrow, and two other invalids. Hurley was there to build a fire so they could vaporize their tiny stock of chloroform. He stuffed the stove with blubber and penguin skins and raised the hut temperature to about 80° F. Then Macklin administered the chloroform, and McIlroy operated with Wild's assistance. A few minutes later, all five of Blackborrow's gangrenous toes were gone.

The months dragged on. June became July and then August, and still the men waited. Had the *James Caird* reached South Georgia? The men were sure she had. It was inconceivable that Shackleton would fail. But where was he? When would a rescue ship show up?

Shackleton, in fact, was doing everything he could to get back. He made his first try only three days after reaching Stromness. The whalers lent him a catcher called the *Southern Sky*, and he headed for Elephant Island with a volunteer crew. When ice stopped the ship about 70 miles from the island, Shackleton asked the ship's captain to take him to the Falklands.

In the Falklands, Shackleton cabled the press with an account of his experiences. He also asked the British government for help. The government was understandably more concerned with fighting World War I than rescuing stranded explorers, but it did offer a ship if Shackleton were willing to wait a few months. He was not. He thus contacted the Uruguayan government at the British government's suggestion. Uruguay, a noncombatant, was happy to provide a ship immediately. The *Instituto de Pesca,* a battered vessel with dubious engines, collected Shackleton, Worsley, and Crean in the Falklands in mid-June, and then set sail for Elephant Island. Ice blocked the way 20 miles from shore. Shackleton, too anxious to wait for Uruguay's generous offer to prepare the *Instituto* for a second try, sailed to Punta Arenas to look for another ship.

The British community in Punta Arenas rallied to Shackleton's support. Within three days, they collected enough money for him to charter the *Emma*, a small sailing yacht. In addition, the Chilean government provided a steamer, the *Yelcho*, to tow the *Emma* part way. Shackleton left for Elephant Island on July 12. This time the ice stopped him about 100 miles out. Then it was back to the Falklands, where Shackleton learned that the ship the British government was sending should arrive about the middle of September. Again unwilling to wait, Shackleton backtracked to Punta Arenas and asked the Chileans if they would lend him the *Yelcho* again. They agreed.

The *Yelcho* left Punta Arenas on August 25 with a Chilean crew under the command of Luis Pardo. Shackleton, Worsley, and Crean were along as passengers. The *Yelcho* reached Elephant Island on the 30th, more than four months after the *James Caird* had left for South Georgia. At last, the ice was open. When Worsley spotted the camp shortly before noon, he saw men running out of the hut, yelling at the sight of the ship. Wild wrote, "[We] were in the hut at lunch. I was serving out the stew made from seal's back-bones, when the yell 'Sail Ho' brought us all tumbling pell mell from the hut; that stew is still there."[25]

Hurley used his last three film exposures shooting the scene as Shackleton and Crean approached shore in a small boat. Shackleton's first words were a question. Was everyone alive? The answer was yes. Shackleton, with crucial support from Worsley on the *James Caird* and from Wild on Elephant Island, had achieved a near-miracle. Every single man who had left South Georgia on the *Endurance* had survived.

A wild welcome greeted the *Yelcho* when she reached Punta Arenas on September 3, 1916. After weeks of jubilant celebrations in Chile, Shackleton sailed to New Zealand to join the Ross Sea Party rescue. Many of the rest of the men enlisted in the British army or navy. All had survived the wreck of the *Endurance*, but several would lose their lives to World War I.

Three men from Shackleton's Imperial Trans-Antarctic Expedition did not make it home. They were members of the Ross Sea Party, which had arrived in McMurdo Sound aboard the *Aurora* in January 1915 to establish supply depots for Shackleton's crossing team. After placing some preliminary caches along the anticipated route, they were ready to winter on the ship. Then, as Sørlle had told Shackleton, a gale tore the *Aurora* from her anchors and sent her drifting north in the Ross Sea with eighteen men aboard as well as most of the expedition's supplies and equipment. The ten who were marooned spent the winter at Robert Falcon Scott's Cape Evans hut. Unaware that Shackleton was in his own struggle for survival and would never come, they heroically established the rest of the depots the next summer, but three died in the effort. In the meantime, the *Aurora* broke free of the ice and eventually reached New Zealand. In January 1917, with Shackleton aboard, she returned to Cape Evans and rescued the seven survivors.[26]

World War I delayed the publication of *South*, Shackleton's book about the expedition, but it was an instant best-seller when it appeared in 1919. The public also flocked to showings of Hurley's movie account. This production was a creative combination of the still photos and movie film he had salvaged from the wreck, with additional wildlife footage shot at South Georgia in 1917 on a return visit made expressly for the purpose.

The rescue of Shackleton's Ross Sea Party ended the Heroic Age throughout the Antarctic. Fifteen exploring expeditions had sailed for the Antarctic from 1897 to 1914, seven of them to the Peninsula Region. All these seven had come from

Europe, but Argentina had also been active, generously providing support to every one of them, as well as taking over and maintaining William Speirs Bruce's South Orkneys base. And Chile and Uruguay had participated by supplying ships to rescue Shackleton's men. More than 200 men had taken part in the Peninsula Region expeditions, 191 of them spending one or more winters. In less than two decades, they had increased mankind's knowledge of the area more than had been learned in all the previous years since de la Roché first saw South Georgia in 1675.

When the Heroic Age dawned, the map of Antarctica's Peninsula Region showed the South Shetlands, South Orkneys, South Sandwich Islands, and South Georgia at least semi-accurately. Both sides of the Antarctic Peninsula, however, were depicted as vague groups of speculative coasts, islands, and possible mainland down to about 68° S. Beyond that, all was a question mark. As for the Weddell Sea, it was charted as a deep theoretical embayment in a tentative continent somewhere far to the south. Maps published after 1916 were strikingly different. The new maps showed the west coast of the Antarctic Peninsula and its adjacent archipelagos far more completely and fairly accurately—in concept if not in detail—as far south as Alexander I Land. The Peninsula's east coast (the Weddell Sea's west coast) was less defined and less well mapped, but portions of it were more or less correct down to 68° S. To the east, the Weddell Sea now had a definite east coast (Coats Land, the Caird Coast, and Luitpold Land), plus a southern boundary at a known latitude (the Filchner Ice Shelf). Inland, however, the maps of the Peninsula Region were still almost completely blank. Although Heroic Age explorers elsewhere in the Antarctic had made significant trips into the continent's interior, even to the South Pole itself, the Peninsula Region expeditions had concentrated their efforts along the coasts. Later expeditions would have to color in the interior detail for this part of the Antarctic Continent.

Chapter 14

The Decade Following World War I: 1919–1927

The carnage of World War I engrossed the world's attention for four years, and polar exploration was one of the victims. At least two planned expeditions to the Antarctic were canceled, leaving Shackleton the only major explorer to head south after the war began. Following the rescue of his men on Elephant Island and his marooned Ross Sea Party, the entire Antarctic was almost deserted by humanity. Almost, but not quite: Argentina continued to man her tiny base in the South Orkneys, and a few whalers remained to hunt the waters of the Peninsula Region. When the Armistice silenced the guns of war in November 1918, the whalers quickly returned in large numbers. Concurrently, there were many who called for a resumption of exploratory activity. Although nearly a decade would in fact pass before the world responded with a major program, two small expeditions did sail south at the outset of the 1920s. Both were Peninsula Region efforts.

TWO MEN ALONE IN THE ANTARCTIC

In late 1919, John Lachlan Cope, a 27-year-old survivor of Shackleton's Ross Sea Party, announced a grandiose plan for an Antarctic expedition. Among other ideas, he proposed using surplus British Royal Flying Corps (RFC) airplanes to fly in a relay to the South Pole. The proposal, however, was ill-defined and hugely expensive, and even though the RFC was prepared to donate the planes, Cope could not raise the necessary funds. He then developed a more modest project, one concentrating on the aircraft aspect of his original idea. At this point, George Hubert Wilkins, a 32-year-old Australian, entered the picture. Wilkins had significant Arctic experience and had worked with the *Endurance* veteran Frank Hurley as a photographer during the latter days of World War I. His time with Hurley had infected Wilkins with the Antarctic bug, and Cope's project excited him. Wilkins, an early aviator who had flown with the Royal Australian Flying Corps during the war, recognized that it offered him the chance to be the first to fly an airplane in the far south.[1] Cope, however, created such a financial and organizational tangle that he had to scale his plans back once again. This time, he dropped the aircraft plans altogether and announced that he would simply take four men to the east coast of the Antarctic Peninsula. Wilkins, upset by the cancellation of the aviation plans, almost quit. He decided to continue because he wanted to see Antarctica.

Cope's four-man team included himself and Wilkins, both with polar experience, and Thomas Bagshawe and Maxime Lester. The latter two signed on for the adventure, Bagshawe as geologist and Lester as surveyor. There was more than a bit of résumé puffery here. Bagshawe was only 19, with no university training in geology or anything else relevant, and the 24-year-old Lester's only quasi-pertinent experience was as second mate on a tramp steamer.

Cope and Wilkins met in Montevideo on the way south. There Wilkins learned that Cope was broke, and that the whalers, on whom he depended for transportation, had refused to take him and his equipment any farther. Although Wilkins stepped in and persuaded the whalers to help, as originally promised, this new evidence of Cope's incompetence made him once again question his own participation. In the meantime, Lester and Bagshawe were on their way to Deception Island with other whalers. Cope and Wilkins arrived a full month later with eight supposed sledge dogs picked up in the Falklands.

Cope planned to have the whalers deliver the party to Hope Bay, where three of Nordenskjöld's men had spent the winter of 1903. Men and dogs would then continue in their own small boat to Snow Hill Island. Once there, they would winter in Nordenskjöld's hut. Ice conditions near Hope Bay, however, made that impossible, and Cope thus moved on to yet another plan. He would begin on the west side of the Peninsula and trek overland to the Weddell Sea coast. This was one more exceedingly ambitious idea, given that no one had ever penetrated farther than a few miles inland from either side of the Peninsula.

On January 11, 1921, Cope's team left Deception Island with the whalers. The next day, they reached the north side of Paradise Harbor, a sound on the west coast of the Peninsula along the east side of the Gerlache Strait. There, the whalers helped the four men unload all their expedition equipment and stores onto a tiny island separated from the mainland at high tide. An abandoned whaleship waterboat beached on shore inspired Cope's name for the islet—Waterboat Point. More important, the waterboat became the nucleus for a makeshift hut that would serve to shelter the men until they could cross the Peninsula and occupy Nordenskjöld's old camp. Provision cases provided material for several walls. Then the men nailed sacks around the sides of the waterboat itself and hung quilts on the inside walls for insulation—a décor, Bagshawe later wrote, that "resembled my idea of the appearance of a padded cell in a lunatic asylum."[2]

The four men spent a week creating the base at Waterboat Point. They then set off in the boat to find a place to start the trip across the Peninsula. Cope spent more than a month commuting from Waterboat Point as he pursued his search. Nothing

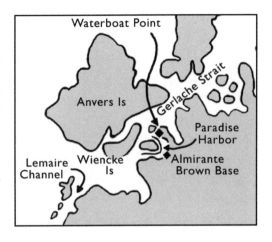

Bagshawe and Lester's Waterboat Point hut location — *The hut was near the northern opening of Paradise Harbor, an arm of the southern Gerlache Strait. In early 1951, Chile established Presidente Gabriel González Videla base on Waterboat Point. This operated year-round until the mid-1960s, after which it functioned in summers only for several more years. The same summer that Chile opened its Videla base, Argentina expanded a nearby refuge into a year-round base, Almirante Brown. This station, too, would function year-round for many years.*

availed, and in late February he decided to retreat to Montevideo to regroup. He would return the following summer. Disgusted with Cope and the whole expedition, Wilkins said he had had enough. He was going home. In contrast, Lester and Bagshawe impulsively decided to stay and winter at Waterboat Point. Cope promised to pick them up in February 1922.

Cope, Wilkins, and Lester took the boat to find a whaling ship willing to transport the two experienced, departing, members of the expedition north. Bagshawe remained behind to take care of the dogs. His solitary wait ended a week later when a whale catcher arrived with his three companions. Cope and Wilkins then scrambled about, hurriedly collecting their belongings, and sailed away within the hour. It was the last that Lester and Bagshawe would see of either man in the Antarctic.

It was now March 4, 1921, and winter was on its way. What Lester and Bagshawe were about to do smacked of genuine lunacy. They were two men alone, two men from vastly different backgrounds. Would they be able to get along with each other through an Antarctic winter? And even if they could, what could they accomplish? Not only were they untrained in anything relevant, but Cope had left them little in the way of supplies or equipment. The day after Cope and Wilkins left, another whale catcher showed up. The two factory ship captains aboard had come to talk the would-be winterers out of staying. When the two young men stubbornly refused to change their minds, one of the whalers, a Captain Anderson, promised to pick them up himself next season if Cope failed to return. The whalers left shaking their heads, and Lester and Bagshawe were once more alone.

As for Wilkins and Cope, they were heading north for civilization. When they reached Montevideo, Wilkins went on to the United States, at that point determined to buy planes and return to Antarctica on his own the next year. Cope told Wilkins that he was going to go to England, get a ship, and then reorganize the expedition. What he actually did was go to the Falklands, where he hung about until the British authorities, Wilkins wrote acidly, "got him a job peeling potatoes on a Scottish steamer" and sent him home.³

For Lester and Bagshawe, now alone in the makeshift hut at Waterboat Point, the weather grew increasingly nasty as winter approached. All the while, the two neophyte explorers were discovering more and more problems with their equipment and stores. Some were trivial, such as the fact that they had only one fork between them, an implement from Bagshawe's personal picnic set. More important was their almost total lack of medical supplies. And Cope had left them such limited food that they had to subsist mostly on seal and penguin. Further, the stove, created from an old oil drum, was a temperamental beast that seemed to go out deliberately when they needed it most. Lester and Bagshawe were never comfortable at Waterboat Point, their living conditions wavering between not truly bad, at best, to nearly unlivable, at worst. Despite it all, they tried to lead as civilized a life as possible, even using a tablecloth for meals and faithfully cleaning house every Saturday.

Their scientific program had to focus on work on and about their tiny island, because the motley collection of mutts Cope had gathered in the Falklands was hopeless as sledge dogs. The two men did set out on one brief man-hauling sledge trip late in August, but they returned at the end of the day, sensing that it was too dangerous to do more than that with only two men. Besides which, what would they have done with the dogs had they been away longer? The animals were useless for any practical purpose, a great deal of work to care for, and a check to the men's ability to explore. Even so, Lester and Bagshawe were glad to have these fellow creatures because of the companionship they provided. They thus felt a deep sense of loss after a dog drowned at the beginning of April. When a second died in mid-August, Bagshawe wrote, "This tragedy brought the thought of death very close to us. We did not speak of it, but the realization of how much we depended upon each other was foremost in our minds. . . . The thought of being left in the Antarctic alone, wondering whether rescuers would ever come was a nightmare on which we dared not dwell."⁴

On a beautiful summer day, Paradise Harbor deserves its name. Then it is a place illuminated by glowing sunrises and sunsets; the domain of sparkling blue waters, accented with icebergs of fantastic shape and color, that reflect rugged mountains, ice cliffs, and chaotic glaciers spilling off the Peninsula mainland; and a home to playing seals, penguins, and whales. In early spring 1921, however, the skies were leaden and dull, the weather unsettled, and rising temperatures rotted Lester and Bagshawe's stored meat. They threw it away and turned to killing seals and penguins as they needed them. Although there was no problem replacing the meat, the impact on their morale was another matter. Rotting food; erratic but usually nasty weather; an island home so tiny that life there was "very much like being in a small prison camp . . .";⁵ a cramped, leaky, and often frigid hut. Near the end of September, Bagshawe wrote in his diary, "The man who called this spot Paradise

Two Men in the Antarctic — Left: *Bagshawe in his winter furs.* **Right:** *Lester in front of the hut shortly after the whalers arrived to pick them up. Bagshawe captioned a photo of himself in the same pose, "Bagshawe looks happy too."*[6] *(Both from Bagshawe,* Two Men in the Antarctic, *1938. Reprinted with the permission of Cambridge University Press)*

. . . should have the honor of living here!"[7]

More than 60 years later, the base leader at Argentina's Almirante Brown base, located in Paradise Harbor a few miles from Waterboat Point, apparently felt even more strongly that it was an honor he questioned. In April, 1984, unable to face the idea of spending the winter there, he deliberately set fire to the base to force the government to evacuate him. An American ship responded to the base's SOS call the same day and quickly picked up the seven men of the wintering party, whose Paradise Harbor home was now "a black heap of burned timbers and twisted corrugated metal."[8] The rescued men were eventually taken to another Argentine base and flown home. Argentina, after rebuilding Almirante Brown, converted it to a summer-only operation.

Although early departure was not an option for Lester and Bagshawe, full-blown spring eventually arrived, bringing with it something that changed their entire outlook—the return of the penguins. Waterboat Point was home to a gentoo rookery of around 12,000 birds, and a nearby islet boasted nesting grounds of both gentoo and chinstrap penguins.[9] The men developed a real affection for these birds and spent long hours studying them. Lester and Bagshawe, neither of whom had formal naturalist training, would produce a surprisingly worthwhile, and historic, study of gentoo and chinstrap penguins.

Spring also turned their thoughts to home, and on December 18 a whale catcher arrived. Captain Anderson, who had visited them in March, had kept his promise, and more. Because so many in England were worried about the two young men, he had made a special trip from Deception Island to pick them up. To Anderson's astonishment, Bagshawe and Lester preferred to stay a bit longer to complete their penguin work. Although the whaler found their wish difficult to comprehend, he agreed to return in a few weeks. In the meantime, he took the six surviving dogs to Deception Island and distributed them among the whaling ships as pets. Anderson returned on January 13, 1921, precisely a year and a day after Lester and Bagshawe had first reached Waterboat Point. Leaving proved to be emotionally difficult for the two men. Bagshawe wrote,

> [We] had grown strangely fond of the old hut and the inhospitable island.... It seemed like desertion to leave it to its loneliness.... We wondered if... [the] penguins would miss us. They had been our friends and we knew that we should miss them.... It had been an experience which, though I would not particularly care to repeat, yet I would not have missed for worlds.[10]

Lester published an article about the expedition in the *Geographical Journal* in 1923. Bagshawe, however, wrote nothing until 1939. Then he finally responded to the urging of his Antarctic peers and produced a book, *Two Men in the Antarctic*, and two articles about his experiences. In all three, he was rather dismissive of his and Lester's work. Bagshawe was wrong. Despite their youth and inexperience, these impulsive young men had done far more than survive. They had collected more data per man than any expedition before them. Today, the remains of their waterboat hut are recognized as a historic monument under the Antarctic Treaty.

Lester and Bagshawe were finished with the Antarctic. So, too, was Cope. He never came back. Wilkins, however, was far from done. He would, in fact, return eight times. The first occasion was during the same summer that Lester and Bagshawe departed, on an expedition led by a much more experienced returnee—Sir Ernest Shackleton.

The *Quest*: Shackleton's Last Expedition

In the years immediately following the *Endurance* expedition, Shackleton had occupied himself with war work for the British government. But he was an explorer at heart, and in 1920 he decided to tackle the Canadian Arctic. Discussions with the Canadian government over funding continued through 1920 and into 1921. In the meantime, Shackleton purchased an Arctic sealer that he renamed *Quest*. Then a new government took office in Canada and opted out of the project.

With the attraction of Canadian government money gone, Shackleton turned his attention to the part of the polar regions he knew best, the Antarctic. An old school friend, John Quiller Rowett, solved the money problem by offering to fund nearly the entire expedition. The plan he and Shackleton settled on was to use the *Quest* for a two-summer voyage to explore and chart sub-Antarctic islands; map 2,000 miles of the coast of East Antarctica; search for mineral deposits; and carry out oceanographic, biological, and meteorological studies. It was a rambling, ambitious plan that just grew and grew after Rowett opened his deep pockets. Rowett's money also allowed Shackleton to equip the ship with wireless and a small plane. Unfortunately, industrial unrest in England made it impossible for him to replace the ship's original geriatric engine.

Eight members of Shackleton's 21-man *Quest* party had been with him on the *Endurance*. These included Frank Wild as second-in-command and Frank Worsley as ship's captain; the two doctors, Alexander Macklin and James McIlroy; and Leonard Hussey, the young meteorologist whose banjo Shackleton had spared because of its morale value. Wilkins, who now hoped to fly with Shackleton, was among the newcomers. Shackleton also invited the British Boy Scouts to send an Eagle Scout. After a national competition drew 1,700 applicants, Shackleton selected two, seventeen-year-old James Marr and a second boy, Norman Mooney, who would leave the expedition early, at Madeira, because of severe seasickness.

The *Quest* left England on September 24, 1921. Although all the expedition members were aboard, they had only part of their equipment, because the *Quest* was too small to carry everything. Shackleton had solved that problem by sending the plane's floats and wings, as well as much of the cold weather gear, ahead to Cape Town. He soon realized, however, that inadequate cargo capacity was far from the *Quest*'s only shortcoming. The engine he had been unable to replace was both unreliable and underpowered, and the uncomfortably cramped ship leaked and corkscrewed with every wave. Even the seasoned sailors struggled with seasickness. Wilkins commented caustically that the agent who had bought the ship for Shackleton must have been "drunk and seeing double."[11]

Shackleton spent a week in Lisbon to repair the ship and a further month in Rio de Janeiro trying to fix the engine. Because of these delays, he decided to postpone the early part of his program and head first for South Georgia, and that created new problems, given that Shackleton had sent so much on to Cape Town. Although he thought he could find replacements for most things at South Georgia, substitutes for the missing airplane parts would definitely not be available there.

The *Quest* was not the only expedition participant in questionable health. Two days before he left Rio, Shackleton felt faint and sent for Dr. Macklin. But by the time Macklin arrived, The Boss said he had recovered and refused to be examined. The *Quest* sailed for South Georgia on December 19. It was a stormy voyage, and the engine boiler soon sprang a leak. Clearly, the *Quest* still had serious problems. And so did Shackleton. Macklin, who thought his leader looked tired and ill, tried to persuade

The *Quest* sailing past the Tower of London as she leaves England for the voyage — *This small vessel with an elderly and unreliable engine proved a most uncomfortable home for the men of the expedition. (Photo from Wild,* Shackleton's Last Voyage, *1923)*

him to let someone else take the bridge. Shackleton refused.

The expedition reached South Georgia on January 4, and Shackleton spent the day on deck with a pair of binoculars, happily reminiscing about the voyage of the *James Caird*. He wrote in his diary that night, "At last . . . we came to anchor at Grytviken. . . . A wonderful evening. . . . 'In the darkening twilight I saw a lone star hover // Gem like above the bay.'"[12]

Just after 2 a.m. on January 5, 1922, Macklin, who was on watch, heard a noise from Shackleton's cabin. When the doctor looked in, Shackleton said he could not sleep. Macklin responded that he had been doing too much. Shackleton joked, "You are always wanting me to give up something. What do you want me to give up now?"[13] Then, as the horrified Macklin watched, Shackleton suffered a massive heart attack. Macklin raced for help, but The Boss was dead before he could return with Dr. McIlroy. The two doctors sadly woke Frank Wild to give him the awful news.

Wild took over. In the morning, he delivered the news of Shackleton's death to the rest of the men and then announced that he was determined to carry on. His first priority, however, was taking care of Shackleton's body. Wild's initial thought was to bury his leader at South Georgia, but he was unsure about Lady Shackleton's wishes. With no way to ask because the *Quest*'s wireless had failed, he decided to send Shackleton's embalmed remains home. A whaler lovingly constructed a special coffin. Then Hussey, whom Wild had chosen to accompany Shackleton on his last voyage, set off with the coffin on a whaleship for the first leg of the journey, to Montevideo.

The *Quest*, for her part, left South Georgia on January 18, a sad ship with a new leader determined to honor Shackleton by making the expedition a success. Three days later, the *Quest*'s leaks intensified, forcing the men to spend hours at the pumps every day. In addition, Wild wrote, "The *Quest* rolled like a log and the seas in the waist rushed like a swollen flood from side to side, so that one rarely passed about the ship without a wetting."[14] All this, added to the lateness of the season and the questionable engine's heavy coal consumption, forced Wild to abandon his plan to map the coast of East Antarctica. Instead, after reaching the continent, he would turn west, sail across the northern Weddell Sea to the South Shetlands, and then return to South Georgia.

Wild entered the pack at 65° 07' S, 15° 21' E, on February 4. Despite the *Quest*'s problems, both veterans and newcomers seemed excited, or so Wild thought. He reached what would be the expedition's farthest south, 69° 18' S, on February 10. There, Wild, remembering the fate of the *Endurance*, a much stronger ship than the *Quest*, decided to retreat because new ice was forming on the open sea.

Shortly before he turned back, Wild realized that many of the men were unhappy with his leadership. They had signed on to sail with Shackleton, and with The Boss gone, they were far less enthusiastic about the voyage. Inevitably, they studied everything Wild did with a critical eye. This was especially true of those aboard who had not known him on the *Endurance* expedition. Wild, a superb second-in-command, was not a particularly charismatic leader on his own, nor did he now have Shackleton's impending return to inspire the men with. He did what he could, however, starting with airing his concerns and announcing that he would not tolerate the open carping that was undermining his authority. Wild believed his efforts helped. Perhaps, but some of the men thought that the true response was more a matter of a change in the level of open unhappiness than in the way people really felt.

Wild headed west across the Weddell Sea, toward Elephant Island, where he wanted to pick up elephant seal blubber to eke out the ship's dwindling coal. He

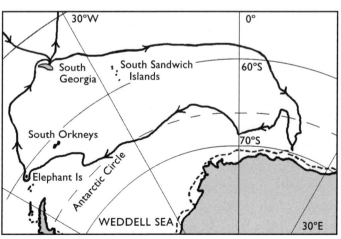

Track of the *Quest* — *Shackleton's original plan had been to survey much of the coast of East Antarctica. The Quest, however, was not up to the job, and Frank Wild, who took over after Shackleton died, wisely decided to curtail the program.*

also had another, less practical, reason—to revisit the place where he and his fellow Elephant Island veterans had spent four and a half miserable months in 1916. With pack ice littering the northern Weddell Sea, Wild resisted any temptation to explore, but despite his caution, the ice trapped the *Quest* on March 15. She was already at 63° 51' S, however, and the pack was carrying her north. Six days later, it loosened enough for the *Quest* to break free. Escaping the ice proved a mixed blessing, because the ship returned to her wild rolling and was soon covered in icy spray. Everyone suffered, even Frank Wild, that most veteran of Antarctic veterans. One man's diary recorded, "On the bridge [one day] Commander Wild remarked: 'The man who comes down here for the sake of experience is mad; the man who comes twice is beyond hope; while as for the man who comes five times (himself) ——' [sic]. Words failed him."15

Frank Wild, shortly after returning from the *Endurance* expedition — *Wild, Shackleton's devoted and loyal second-in-command on both the* Endurance *and* Quest *expeditions, was the most experienced Antarctic explorer of his day. He had first gone to Antarctica in 1901–04 with Robert Falcon Scott on the* Discovery *expedition. The* Quest *was his fifth—and last—Antarctic expedition, the only one on which he did not spend at least one winter on the ice. On his four previous forays south, he had spent a total of six winters in the Antarctic, including the two on the* Endurance *expedition. (Royal Geographical Society)*

The *Quest* reached Elephant Island on March 25. Wild first anchored at the corner opposite Point Wild. There, the men landed and took some elephant seals. Another landing the next day garnered more blubber. Several days later, Wild sailed around the island to Point Wild, only to find that the weather made a safe landing there impossible. Since there was no desperate need to reach Point Wild's shore, Wild gave up and headed back to South Georgia.

The winds en route were so strong that the ship made the voyage completely under sail. With the engine having loafed the entire way, there was no need for the blubber the men had collected at Elephant Island, and they tossed it all overboard when the *Quest* reached South Georgia on April 6. Hussey, to everyone's surprise, was there to greet them.

As Wild had directed, Hussey had taken Shackleton's body to Montevideo and cabled Lady Shackleton the news of her husband's death. She responded with her wish that he be buried at South Georgia. But before Hussey had returned Shackleton's body to the island, the people of Montevideo, including the Uruguayan president, had packed the city's English Church, adorned with wreaths from around the world, for a memorial service. Mourners had then loaded Shackleton's coffin onto a gun carriage for a trip through crowd-lined streets to a British naval vessel. Af-

The whalers' cemetery at Grytviken, where Sir Ernest Shackleton lies buried — *The cemetery, today enclosed by a white picket fence, is in the far lower right, with Shackleton's monument, the tallest gravestone in the cemetery, at the left rear. Frank Wild wrote of Shackleton's final resting place, "The graveyard is a simple little place. In it are already a few crosses, some of them very old, mute reminders of forgotten tragedies. . . . There are some newer crosses [as well]. . . . All of them [mark] the graves of strong men."* [16]

Behind the cemetery loom the mountains of South Georgia, and on the hillside above it are two memorial crosses. The lower one is for Walter Slossarczyk of the Deutschland *expedition, who died in Cumberland Bay in late November 1911. The higher is in memory of seventeen men who perished when a fishing boat sank off South Georgia in a storm in 1998. (Author photo, 2009)*

ter firing a seventeen-gun salute, the ship had transported Hussey and Shackleton's body back to South Georgia. There, on March 5, the whalers had conducted another service in the Grytviken church. Afterward, they buried Shackleton in the simple Grytviken graveyard, a lovely spot at the base of steep hills only 100 yards or so from the whaling station.

Early in May, a month after the *Quest* arrived at South Georgia, Shackleton's men built a memorial cairn honoring their leader on a prominent site above King Edward Point. They topped it with a cross and cemented a brass plaque into the stone. The plaque read "Sir Ernest Shackleton // Explorer // Died here, January 5th, 1922 // Erected by his comrades."[17] Only a wooden cross marked Shackleton's actual grave until early 1928. Then the Falklands governor visited South Georgia and formally dedicated the elaborate stone grave marker that still stands prominently in the cemetery. (The cairn also remains, although it lost its base when its location was shifted slightly at the beginning of World War II to make room for a gun emplacement.)

The *Quest* left South Georgia on May 8, 1922. On the way north, Wild paid week-long visits to Tristan da Cunha and Gough Island. Then it was on to Cape Town for repair and refit. The *Quest* expedition ended there, however, because Rowett was unwilling to continue funding the venture without Shackleton in command.

Frank Wild, faithful to the man he had accompanied on four Antarctic expeditions, had done his best with a questionable ship and reluctant companions. As a result, the voyage did boast some accomplishments. Her men had taken numerous deep-sea soundings in the Southern Ocean, and had done dredging and trawling for bottom sediment and sea life outside the pack ice. There was other good scientific work as well, especially the geological efforts. But one thing they had not done: fly an airplane in the Antarctic.

The *Quest* expedition was the last to consider Antarctic aviation, or to plan significant exploration anywhere in the far south, until the late 1920s. Unlike the quiet years in the nineteenth century, however, the Peninsula Region was far from deserted. Not only did whaling remain strong, but a program born of an effort to preserve and control the whaling industry also developed into the first ongoing Antarctic scientific effort.

The cross and cairn at King Edward Point erected by the men of the *Quest* in memory of Shackleton — Wild wrote, *"There were no expert masons amongst us, but the work when completed had a most pleasing appearance."*[18] *This photo, taken by Hubert Wilkins, is in color, with pale tints of pink and blue. It serves as the frontispiece for Wild,* Shackleton's Last Voyage, *1923.*

THE DISCOVERY INVESTIGATIONS

Groundwork for this scientific program began even before World War I ended. In 1917, the British Colonial Office set up a committee to consider ways to preserve the whaling industry through related scientific research. The resulting report, published in 1920, outlined a series of proposed studies into the natural history of whales. It also stressed the need for a complete hydrographic survey of Britain's Falkland Islands Dependencies claim. A second committee developed the final plan in 1923. Early the next year, the British government purchased Robert Falcon Scott's old ship *Discovery* and appointed an executive committee, now called the Discovery Committee, to oversee the recommended investigations. The Discovery Committee began its work in February 1925 with a five-man biological laboratory at Grytviken, on South Georgia. This lab would operate during each whaling season until 1930–31.

In late 1925, the *Discovery* sailed from England to begin the Discovery Committee's principal work. She reached South Georgia in late February 1926 after a voyage slowed by oceanographic studies en route. The scientists then worked at the island's whaling grounds for two months, but as things turned out, they learned more about the *Discovery* than about whales. The ship the Discovery Committee was named for was underpowered and rolled far too much for scientific work at sea. The *Discovery* accordingly spent the winter in Cape Town being altered, in hopes of solving her problems. In the meantime, the Discovery Committee acquired a second ship, the *William Scoresby*. She was designed to carry out whale marking—shooting tag darts into whales—and also was equipped for oceanographic work.

The two ships worked around South Georgia from early December 1926 to the end of January 1927. They then separated, the *Scoresby* to continue whale marking near South Georgia, the *Discovery* to sail to the South Shetlands and the Antarctic Peninsula for hydrographic work. The *Discovery* was the first vessel in these waters capable of receiving Greenwich time signals directly, and her men used the signals to check longitudinal positions on the maps. They soon realized that only one landmass, Deception Island, was accurately located on existing charts. When they were finished for the season, the *Discovery* team had completed a significant oceanographic voyage, including identifying the existence of the Antarctic Convergence (or Polar Front, its modern scientific name).

Another ship had confirmed the phenomenon virtually simultaneously. The German scientific vessel *Meteor* was working in the Antarctic at the same time that the *Discovery* made her 1925–27 voyages. One of the *Meteor*'s primary objectives, right from the outset, was to investigate conclusions a German meteorologist, Wilhelm Meinardus, had reached from studying data obtained on Drygalski's 1901–03 expedition. These data, Meinardus thought, pointed to an ecological boundary in the Southern Ocean, a place where water densities and temperatures changed systematically between sub-Antarctic and Antarctic waters. The *Meteor* scientists' Antarctic-wide observations corroborated Meinardus, and they christened the discovery the Meinardus Line—a name soon replaced by today's more familiar terms.

The *Meteor*'s Antarctic days ended when she sailed home in 1927. The *Discovery* would serve one more Antarctic expedition, but not with the Discovery Committee and not in the Peninsula Region. She was a good exploring vessel, but the second cruise had confirmed that she was completely inadequate as a scientific ship. The Discovery Committee replaced her with the *Discovery II,* a new ship specifically designed for the job.

Building the new ship was the first step in converting the Discovery Committee work into what became the Discovery Investigations. This ambitious program of

Southern Ocean scientific voyages was pursued every summer, and during several winters as well, through the end of the 1930s. The results appeared as a series of seminal papers on Antarctic hydrology, marine life, and related topics. This massive body of work, collectively called the *Discovery Reports,* eventually exceeded 14,000 pages. The Discovery Investigations failed, however, in their objective of altering whaling practices to preserve the stock of whales.

PELAGIC WHALING COMMENCES

One reason the whalers were able to ignore the Discovery Investigations' findings was that they expanded their hunting beyond the Peninsula Region at virtually the same time the scientific study began. The Falkland Islands Dependencies (FID) government had exerted at least some limited control over the industry because whalers using FID harbors or shore locations were subject to restrictive clauses in their license agreements. Inevitably, these restrictions plus the royalties the FID demanded encouraged whalers to look for whaling grounds elsewhere in the Antarctic. In the summer of 1923–24, the same Carl Anton Larsen who had pioneered South Georgia whaling took the first significant steps to move the industry beyond FID waters. His experimental whaling voyage to the Ross Sea found ample stocks of whales but had difficulty processing them without a safe harbor for his factory ship. That very problem was about to be solved by a technological breakthrough developed by Petter Sørlle, a Norwegian whale gunner. His breakthrough revolutionized the industry.

Sørlle's innovation stemmed from an idea that had come to him while working at the South Orkneys in the 1912–13 season. That summer, two factory ships arrived off the islands in mid-November and found them so blocked with ice that it was impossible to reach a safe harbor. While waiting for the ice to break out, the catchers began whaling along the ice edge. Sørlle, a gunner on one of the catchers, saw how difficult it was to process the whale carcasses when the factory ships were at sea. During those frustrating weeks, Sørlle thought of a solution. His idea was to insert a ramp, which he called a slipway, into the stern of a factory ship. The whalers could then use power-driven winches to haul whale carcasses directly on board for processing.

It took Sørlle ten years to work out the kinks in his idea. Two years after Norway granted him a patent on his device in 1923, engineers installed the first slipway on the *Lancing*, a Norwegian factory ship. When her crew pulled a massive Antarctic blue whale aboard in December 1925, it was clear that Sørlle's invention was a success. Full pelagic whaling—hunting and processing whales completely at sea, without need for a shore station or safe harbor—was now possible. That, in turn, freed the whalers to pursue their prey anywhere in the Southern Ocean, and the numbers of whales taken throughout the Antarctic exploded. In 1925–26, the summer the *Lancing* introduced the slipway, Antarctic whalers killed just over 14,000 whales, virtually all of them taken in the waters of the Peninsula Region. Five years later, the catch had nearly tripled, to more than 40,000. Pelagic whaling fleets, many of them operating in the Ross Sea and along the coast of East Antarctica, were responsible for over three-fourths of this catch. The Discovery Committee, originally formed to work in the FID, followed the whalers as they moved beyond the Peninsula Region, and, like them, expanded its efforts to the entire Antarctic.

As the 1920s ended, the explorers returned, and the whalers, now joined by the Discovery Committee, once again provided support. The principal change from the Heroic Age was that help was now available continent-wide, rather than only in the Peninsula Region. But one of the first new exploring expeditions would draw heavily on the Antarctic whalers' aid in the very locale where they had first offered their support. That place, where Andresen had supplied coal and so much other help to Charcot in 1909 and 1910, was Deception Island, in the South Shetlands.

Chapter 15

The First Aviators Arrive: 1928–1936

The pause in Antarctic exploration ended in the summer of 1928–29. Then, two historic expeditions arrived, bringing with them technology—the airplane and reliable radio—that would be as revolutionary for exploration as the slipway had been for whaling. The larger effort, led by America's Richard E. Byrd, worked exclusively in the Ross Sea Region. Byrd, who claimed to have flown over the North Pole in 1926, brought three airplanes, a motor vehicle, nearly 100 dogs, and 42 men to winter at the Bay of Whales in a base he christened Little America. In November 1929, he would make a round-trip flight over the South Pole from Little America. But even before Byrd sailed south from New Zealand in December 1928, another man had already flown the Antarctic skies. He was Hubert Wilkins, back in the Antarctic for the third time. And once again, he went to the Antarctic Peninsula Region.

Hubert Wilkins Flies at Last

Wilkins had been disappointed twice in his hopes of flying in Antarctica. After the second time, on Shackleton's *Quest* expedition, Wilkins turned his attention to the Arctic. There, in early 1928, he and his American pilot, Carl Ben Eielson, flew 2,500 miles from Point Barrow, Alaska, across the Arctic Ocean to Spitsbergen, about 400 miles to the north of Norway. This first heavier-than-air flight over the region earned Wilkins a knighthood and furnished him the fame he needed to raise money for an Antarctic expedition.[1] Most of his funds, including $25,000 from the Hearst Press, came from the United States. A Norwegian whaling company made a major in-kind contribution by transporting the expedition party to Deception Island and providing living accommodations aboard a factory ship while Wilkins was there.

Wilkins's goal was to fly across the Antarctic Continent from Deception Island to the Bay of Whales in the Ross Sea. But his planned route—2,000 miles long—was along a tangent, not over the South Pole, because he considered a polar route too ambitious for the aviation technology of the day. He also thought a nonstop flight would be overly risky. Instead, he intended to start with reconnaissance flights to locate a site for an advance base 500 to 600 miles south of Deception Island, perhaps in the vicinity of Alexander I Land, last seen by Charcot in 1910. He would use this base as a staging point for the major leg of his transcontinental flight.

The airplane Wilkins and Eielson had flown to Spitsbergen had performed splendidly, and Wilkins was confident it would do equally well in the Antarctic. It was an innovative design for the day, one he felt very fortunate to have found. While in San Francisco in late 1927, he had happened to look out a window and had seen a plane flying by that looked ideal for his upcoming Arctic flight. After an intense search of local airfields, he located the aircraft and learned that she was a prototype built by the Lockheed Company. He immediately visited the company in Los Angeles to consult with the plane's designer, Jack Northrop. Their conversation convinced Wilkins that this was indeed the right plane for his purposes, despite the fact that the prototype had never made a major flight. Shortly after his visit, that plane was lost en route in a race to Hawaii. Wilkins nonetheless remained confident that this was the craft for him, and had two built. One crashed before he could take possession. He named the second one *Los Angeles* and used it for his Spitsbergen flight.

Perhaps Wilkins, a somewhat superstitious man, felt that the luck that had led him to the plane had attached itself to the aircraft itself. He certainly needed a plane he could trust, because, just as he had in the Arctic, he was planning to fly over places where rescue might well be impossible. His Lockheed plane and its single 220-horsepower Wright J-5 Whirlwind engine—the same engine Charles Lindbergh had used for his historic 1927 trans-Atlantic flight—had already proven themselves, and in an era when aircraft design for long-range flights was still in its infancy, that was enough for Wilkins. When his backers insisted he take a second plane, he ordered a twin to the *Los Angeles* and named it the *San Francisco*. He equipped both aircraft with skis, wheels, and pontoons so they could operate from any surface. Ben Eielson, his Arctic pilot, was a part of Wilkins's plan from the beginning, and Joe Crosson, another American, joined the team to pilot the second plane. An aircraft mechanic and a radio operator completed Wilkins's team.

Wilkins's intended destination for his transcontinental flight, the Bay of Whales, was just the place where Richard Byrd would establish his wintering base. Although there was no doubt that Byrd's expedition was an American one, the nationality of Wilkins's was less clear. The question was politically important should Wilkins make any claims. He spoke of his expedition as American, because most of his money had come from the United States. At heart, however, he was Australian.

Politics was becoming an increasingly significant issue in the Antarctic at the time Wilkins and Byrd were making their plans. In 1923, fifteen years after she had claimed nearly all of the Peninsula Region as the Falkland Islands Dependencies, Great Britain claimed the Ross Sea Region on behalf of New Zealand. A year later, France asserted sovereignty over Adélie Land in East Antarctica. Then Argentina challenged Britain, implicitly putting forth a claim to the South Orkneys in 1925 and explicitly to South Georgia plus the South Sandwich Islands in 1927. It was thus almost inevitable that the British government would be interested in an expedition led by an Australian.

Wilkins reached Montevideo aboard a commercial ship in October 1928. There he transferred to the *Hektoria*, a whaling factory ship bound for Deception Island. On the way south, the *Hektoria* made a routine call at the Falklands. The visit, however, was not routine for Wilkins. Here, to his surprise, the governor presented him with a confidential memorandum from the British Foreign Office authorizing him to make claims for Britain if he discovered new land from the air. It would be an approach unprecedented in international law.

The *Hektoria* reached Deception Island on November 6. Four days later, the whalers helped Wilkins unload the *Los Angeles* into the water on her pontoons. But when Eielson taxied across the water in preparation for taking off to fly to shore, clouds of birds that had been gorging themselves on rotting whale carcasses flew with him, some right into the propeller. Wilkins surrendered and had the plane towed ashore.

Wilkins soon realized that he faced far more serious problems than suicidal birds. Instead of the snow-covered land and thick sea ice that he had expected to find, there was no snow where the land was flat enough for a runway for a ski-equipped plane, and the bay ice was too thin to safely serve as an airstrip for such a plane. His best alternative seemed to be a hillside with a short site for a wheeled-aircraft runway. At least he could use that for test flights.

On November 16, 1928, Wilkins, with Eielson at the controls, took off in the *Los Angeles*. They landed after only 20 minutes because of deteriorating weather, but this was the first Antarctic airplane flight, and Wilkins at once radioed their achievement to the world.[2] About a week later, Crosson took the *San Francisco* up briefly. Both airplanes flew again for several hours on November 26, this time in search of a better runway site. Unfortunately, heavy cloud cover made it impossible to see much of anything.

Wilkins became so frustrated that he decided to chance the bay ice. But when Eielson took off in the *Los Angeles* from shore and landed on the frozen bay, the plane broke through the thin ice. She would have sunk completely had her wings not rested on the edges of the hole the plane had created. Wilkins next tried using pontoons. When taking off with the full fuel tanks essential for long flights proved to be impossible, Wilkins reluctantly concluded that his only option was a land runway and wheels. And that doomed his plans for a staged crossing, because a wheeled aircraft could not take off from the snow farther south. Even a long round-trip flight would be dangerous, since a forced landing could strand the plane.

Despite the risks, Wilkins was determined to at least use his plane for exploration. To accomplish anything significant, however, he needed a full load of fuel, and

Carl Ben Eielson (left) and Sir Hubert Wilkins (right) in front of the *San Francisco* at Deception Island in 1928–29 — *Although Antarctic historians have credited Wilkins with having made the first Antarctic flight, it was pilot Ben Eielson who was at the controls of the plane for the historic first on November 16, 1928. Wilkins acknowledged this to at least one person. The evening of the flight, he wired Eielson's father, "Ben made first Antarctic flight today. Regards, Wilkins."[4] (Photo from reproduction in Burke,* Moments of Terror, *1994)*

the default to wheels still left him with problems. The runway he had been using was simply too short. He would have to build one. Wilkins wrote, "To make a runway . . . through the mounds of lava that resembled piles of large-sized coke, looked at first like an impossibility. . . ."[3] But the whalers pitched in, and with hands, buckets, shovels, rakes, and wheelbarrows, they and Wilkins's team cleared tons of rock to create a 2,500-foot runway. It was barely long enough for a fully loaded takeoff, and because of the terrain, achieving even that distance had forced Wilkins to accept two twenty-degree bends along the runway's length.

Wilkins and Eielson took off in the *San Francisco* on December 20 with fuel for 1,400 miles and 30 days of emergency rations. They began by flying down the west coast of the Antarctic Peninsula. Near the southern end of the Gerlache Strait, they climbed to 9,000 feet, crossed the Peninsula, and then flew down the east coast. It was a thrilling feeling, this exploring from the air. Eight years earlier, Wilkins had tried unsuccessfully to reach the top of the Peninsula with Cope, and it had taken them weeks to map just 40 miles along the coast. Now he was covering 40 miles in 20 minutes. Wilkins doled out names to prominent features as he flew south over this previously unseen landscape. The names honored his American sponsors, but simultaneously, he dropped documents making British claims.

At about 66° S, Wilkins spotted a glacier-filled valley ending at the Weddell Sea coast. It appeared to be one terminus of a strait cutting across the Peninsula. Still farther south, he thought he could see two more such channels. He named the southernmost of these Stefansson Strait, for the Canadian/American explorer he had accompanied to the Arctic in 1913–17. People had been speculating about possible passages through the Antarctic Peninsula for decades. Now, Wilkins thought, he had found them. The *San Francisco* continued south to 71° 20' S, 64° 15' W, before Wilkins directed Eielson to turn back. Ten hours after taking off, they were safely back at Deception Island after a round-trip flight of 1,200 miles, nearly all over previously unseen land. Wilkins lost no time radioing the world of his accomplishments and his exciting discovery that Graham Land was actually an archipelago.

Unfortunately for Wilkins, aerial mapping in the Antarctic is fraught with problems. He had been the first to try it, and like many who followed him, he had made mistakes. The channels he thought he had seen, soon to appear on all new maps of the region, did not exist.

Weather conditions at Deception Island precluded any further significant flying for several weeks. Finally, on January 10, Wilkins and Eielson flew 250 miles south along the west coast of the Peninsula in search of a more southerly base for the following summer. Clouds defeated the effort. Wilkins then stored his planes at the Deception Island whaling station and sailed north.

Unaware that the Hearst Press was willing to finance another season, Wilkins approached the Discovery Committee in London for support. He already had his planes, and the whalers were prepared to continue helping him. When the Discovery Committee agreed to put the *William Scoresby* at his disposal and granted him £10,000, Wilkins had no more need for American money. His second effort would be unequivocally British.

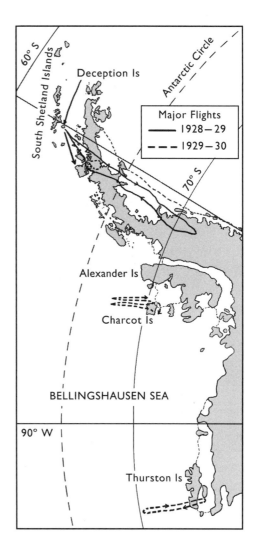

Wilkins's major flights — *Wilkins made a number of flights in the summer of 1928–29, all from Deception Island. These included his 20-minute flight on November 20, 1928, the first time an airplane had flown the Antarctic skies. But it was his December 20 flight that was truly historic, the first major feat of aerial exploration in the far south. The following summer, Wilkins used the* William Scoresby *to find new places to fly from, leading to several flights that took off from locales far beyond Deception Island.*

Wilkins also changed pilots, because Eielson and Crosson had accepted new jobs at home. He hired Al Cheesman, a Canadian bush pilot, and Parker Cramer, an American, as replacements.

Wilkins reached Deception Island on a whaling ship early in December 1929. The planes were in excellent shape, but that was the only good news. Once again, both the bay ice and the snow on shore were hopelessly inadequate for flying purposes. And distressing news of another sort soon arrived, a radio message that Ben Eielson's plane had been lost in the Arctic. Joe Crosson was leading the search. (Weeks later, Crosson would find Eielson's wrecked airplane on the Siberian coast of the Bering Strait. Searchers located Eielson's body a few days later.)

The *Scoresby* arrived shortly after Wilkins learned that Eielson was missing. On December 12, Wilkins loaded the *Los Angeles* aboard and set sail south down the west coast of the Peninsula, in search of smooth pack ice for a ski takeoff. Finding nothing suitable, he gave up at 67° 30' S and retreated to Port Lockroy's sheltered waters to use the pontoons. Wilkins made his first flight of the season on December 19. With Cheesman at the controls, he flew across the Peninsula to reexamine the previous summer's discoveries. On the return, Wilkins spotted an ice-covered bay just south of Port Lockroy that seemed to offer a possible runway. The next day, he unloaded the *Los Angeles* onto the bay ice. Things looked good until the plane's skis began to sink. Wilkins hastily reloaded and sailed south again to resume the search.

The hunt for smooth pack ice was no more successful this time, but Wilkins did find acceptable open water within the pack. On December 27, he and Cheesman took off on pontoons to fly toward Charcot Land, about 150 miles to the south. This flight over the pack ice was nerve-wracking right from the start, because, Wilkins wrote, "the water was thickly strewn with small fragments of ice that would have wrecked any seaplane attempting a landing."[5] They had begun their flight flying at 2,000 feet but soon had to descend to clear a bank of clouds. In less than an hour, the *Los Angeles* was skimming over the jumbled pack at only 500 feet, this near Charcot Land, where the cliffs were supposed to be 2,000 feet high. Not only that, Wilkins realized, but "a sudden turn might lose us enough altitude to put us in danger of running into icebergs. . . ."[6] Wilkins prudently decided that it was safer to turn back. The weather cleared two days later, and this time the *Los Angeles* reached Charcot Land. It was, Wilkins now realized, an island to the west of Alexander I Land, linked to that landmass by a neck of what others later named the Wilkins Ice Shelf. As Wilkins flew over, he dropped canisters containing documents claiming the land for Britain, something that Charcot had not done for France. (Charcot, in fact, had never claimed any of the land he discovered on either of his Antarctic expeditions.)

Wilkins then headed back north to refuel the *Scoresby*. The ship stopped at Port Lockroy en route, and Cramer, who had yet to fly, persuaded Wilkins to leave the ship and fly the rest of the way to Deception Island with him. The two men landed there on January 5, 1930, to find the whalers excited about an earthquake that had just shaken the island. Although there had been no major damage, the harbor bottom *had* sunk about 15 feet. Moreover, this was a second recent warning that Deception Island could be a dangerous place. Almost exactly five years earlier, the beach sands at Whalers Bay had started moving wildly. A deafening roar followed as a piece of the crater wall collapsed into the sea on the outside of the island. Then the waters of Whalers Bay began boiling. With their engine fires damped for the summer, the factory ships could not escape. All the whalers had been able to do was wait and hope. Luckily, things grew no worse then, or in 1930 for that matter. (Nearly 40 years later, other men at Deception Island would not be so fortunate.)

The *Scoresby* reached Deception Island a few hours after Wilkins and Cramer, and then sailed on to the Falklands to refuel. When she returned three weeks later, Wilkins sailed again in search of a takeoff point. This time he continued more than 1,000 miles to the southwest before stopping at an inlet in the edge of the pack at about 70° 10' S, 98° W. Wilkins made his final flight on February 1. He thought he might have spotted land to the south when he reached 73° S, but with a snowstorm swirling about the plane, he decided not to investigate. Just in case land was near, however, he did drop one final claiming canister.

After that, the *Scoresby* returned to Deception Island. Wilkins then left for the north on a whaling ship. Both planes were aboard, later to be sold to the Argentine government. Wilkins had at last achieved one goal. He had flown in the Antarctic. The continent, however, remained uncrossed by air.

Wilkins had been the first, risking his life every time he took off. He knew that rescue was unlikely should his plane crash. He knew that all earlier versions of his plane *had* crashed. And when he flew in 1929–30, he knew that Eielson had been lost in the Arctic. Despite it all, he had taken to the air, as had the others who had followed him into the Antarctic skies. In addition to Richard Byrd's flights from the Bay of Whales, two other exploring expeditions, both working along the coast of East Antarctica, used aircraft the same summer that Wilkins made his flight over Charcot Land. And one more expedition used a plane that summer. The whaling factory ship *Kosmos*, also working off East Antarctica, carried a tiny two-seater Gypsy Moth for whale spotting. On December 26, 1929, Leif Lier, the pilot, and Ingvold Schreiner, the ship's doctor, took off near the Balleny Islands. They never returned. Any of the other pioneer Antarctic aviators could have shared their fate.

THE DISCOVERY INVESTIGATIONS AND WHALING IN THE 1930s

After supporting Wilkins, the *William Scoresby* returned to her work with the Discovery Committee. The Committee's new ship, *Discovery II,* had begun her career the same summer the *Scoresby* was with Wilkins. The work of the Discovery Investigations, which ceased only with the outbreak of World War II in 1939, was challenging and productive throughout the decade. Not only did the scientists carry out studies of Antarctic marine life that laid the foundation for all such later work, they undertook oceanographic surveys of large parts of the Southern Ocean, developing the science of oceanography itself as they worked. And there were historic surveys of Antarctic lands. The first took place during the 1929–30 season, when the *Discovery II* made a routine cruise to the South Sandwich Islands and found the area entirely free of ice. Stanley Kemp, the scientific and expedition leader, determined to take advantage of this rare occurrence, abandoned his planned program and concentrated on what became the first in-depth survey of the area. A similar opportunity at the South Orkneys in 1932–33 yielded its own groundbreaking study. And during 1932, the men of the *Discovery II* achieved another historic first—a winter circumnavigation of the Antarctic regions, during which the scientists successfully plotted the approximate location of the Antarctic Convergence around the continent.

Sometimes the Discovery Investigation's scientists stumbled on surprises. One of the most exciting took place in the summer of 1933–34, when the *Discovery II* chanced upon a colony of 36 adult fur seals and pups on Bird Island, off South Georgia. This was a significant find, because no one had seen more than a solitary fur seal in the area for many years.[7] The men reported their discovery but kept the location secret to protect the animals.

As for the whalers who had supported Wilkins, they too continued their work, but they would operate from the Deception Island base for just one more year. The shore station there closed at the end of the 1930–31 season because pelagic whaling offered greater flexibility. Most of the South Georgia stations ceased operations around the same time. The Antarctic-wide whaling catch, however, was exploding. Recognizing this, the League of Nations drafted an International Convention in 1930 to control the industry. All the countries then seriously involved in Antarctic whaling signed it in 1931. Sadly, it had little effect. The one thing that did reduce whaling activity throughout the Antarctic, at least for a time, was the worldwide economic collapse of the early 1930s. Much of the industry, in fact, shut down in the 1931–32 season. Although Antarctic whaling bounced back the next summer, it was years before the industry fully recovered. When it did, the whalers shattered the previous record catch. In 1937–38, Antarctic whalers would take more than 46,000 whales, 83% of the world's total catch.

The Great Depression also hit the explorers hard. Some simply abandoned their hopes. Others managed by scaling back plans. Wilkins, who was as eager to return to Antarctica as many before him, found another way. In 1933–34, he signed on as manager for an expedition that had no money problems, because its leader could draw on his own large personal fortune.

THE FIRST AVIATORS ARRIVE: 1928–1936

LINCOLN ELLSWORTH FLIES ACROSS THE ANTARCTIC CONTINENT

The man who hired Wilkins was an American, Lincoln Ellsworth. Although he was already 53 years old when he first saw Antarctica in December 1933, he had been fascinated with the polar regions for years. He had undertaken his first personal polar expedition in 1925 when, with his father's backing, he had financed and accompanied Roald Amundsen's attempt to fly from Spitsbergen to the North Pole. Engine problems with one of their two planes had forced an emergency landing on the pack ice less than 200 miles short of their goal. The six-man party then spent almost a month on the ice before managing to take off in their one intact plane and fly back to civilization. During that month, Ellsworth's father died, leaving his son millions of dollars. Ellsworth and Amundsen returned to the Arctic the next year. Using a dirigible named the *Norge*, purchased from the Italian government largely with Ellsworth's money, the two men led a team that flew from Spitsbergen over the North Pole to Teller, Alaska. This was the first successful transpolar flight of any sort.

Four years later, in 1930, Wilkins visited Ellsworth, and the two men talked about Wilkins's Antarctic flights. Meeting again the following year, they made plans for an Ellsworth Antarctic expedition that would continue Wilkins's work. Ellsworth hired Wilkins as both technical adviser and expedition manager and Bernt Balchen, Richard Byrd's pilot for the 1929 South Pole flight, as his pilot. Together, the three men agreed that Ellsworth and Balchen would attempt a 3,400-mile flight from Byrd's Little America base in the Ross Sea to the head of the Weddell Sea and back.

Ellsworth had a plane specially designed and built for the expedition. The first aircraft from the newly formed Northrop Aircraft Corporation, founded by the man who had designed Wilkins's *Los Angeles*, Ellsworth's plane had a single 600-horsepower Pratt & Whitney engine and enlarged fuel tanks that gave her a maximum theoretical range of 7,000 miles. Ellsworth proudly named her *Polar Star*. He also purchased a Norwegian herring trawler as an expedition ship. After Wilkins supervised her remodeling, Ellsworth made the final change, re-christening the vessel the *Wyatt Earp*, in honor of the nineteenth-century American frontier sheriff who had been his boyhood hero. He wrote, "I don't suppose any vessel ever sailed before so filled with the presence of the figure whose name it bore...."[8] Not content with two biographies of Earp in the ship's library, Ellsworth hung his hero's personal cartridge belt in his cabin and wore a gold wedding band given to him by Earp's widow.[9]

Brimming with enthusiasm, Ellsworth and his team sailed to Little America in the summer of 1933–34. Several days after they arrived, Balchen made a successful test flight. The men then tied the plane down on the bay ice to be ready for the main event the next morning. But during the night, disaster struck: the ice broke, damaging the *Polar Star* too seriously for repair on the spot. A deeply disappointed

Ellsworth sailed back to New Zealand and left the *Wyatt Earp* there, in Dunedin, for the winter. He shipped the *Polar Star* home for repairs.

The accident gave Ellsworth time to reconsider his plans. Not only would he cut back to a one-way transcontinental flight, as Wilkins had originally planned, but he shifted his departure point to the opposite side of the continent. He would take off from Deception Island and fly from there to the Bay of Whales, where Richard Byrd, now back for his second expedition, would be in residence. Byrd's Little America base would afford Ellsworth and his pilot a comfortable place to wait while Wilkins sailed the *Wyatt Earp* around the continent to pick them up.

The *Wyatt Earp* reached Deception Island in a swirling snowstorm on October 14, 1934. The next morning, in slightly better weather, everyone went ashore and began scouting for a runway. Balchen finally found a snowfield he thought would do. And then the storms returned. After waiting several days for the snow to end, Ellsworth's team unloaded the *Polar Star* and dragged her up the steep beach. They then spent another week preparing the plane to fly.

On October 29, the day of the planned test flight, more snow squalls hit. Disappointed that he had to scrub the flight, Ellsworth thought he should at least run the plane's engine a bit before closing down for the night. The mechanic started the motor and the propeller began to turn. Then the motor abruptly jammed and a loud crack announced that a connecting rod had snapped. That would have been a minor problem if they had had a replacement. They had none. This time, however, Ellsworth had a solution. Wilkins ordered a replacement rod by radio, and then set off in the *Wyatt Earp* to pick it up in Punta Arenas, Chile.

Five men, including Ellsworth and Balchen, stayed behind and camped in one of the abandoned whaling station buildings while continuing flight preparations. They also watched helplessly as the snow runway they had selected melted. By the time the *Wyatt Earp* was back on November 16, the snowfield was unusable.

Ellsworth stayed at Deception Island only long enough to repair the engine. He then sailed south along the west coast of the Peninsula looking for a spot—somewhere, anywhere—he could use as a runway. With no better luck than Wilkins had had, he reversed course and headed around to the east coast. Ellsworth found what he was looking for at Nordenskjöld's Snow Hill Island. The men unloaded the plane there on December 2. In the meantime, Ellsworth visited Nordenskjöld's hut. The first to call there since 1903, he found a wealth of artifacts from the Swedish expedition. Ellsworth gathered a selection, which he later gave to New York City's American Museum of Natural History. Then it was back to his own expedition.

When things were nearly ready, Ellsworth established radio contact with Byrd to exchange weather information. Things were still hopeless at his own end, however—day after day of fog, gales, and blizzards. Ellsworth also had another problem. Balchen, who had been moody during the wait at Deception Island, now insisted that Ellsworth take a third man on the flight in case they had to clear runways at stopping points along the way. Ellsworth refused because he felt another man would add too much weight. He could only hope that Balchen would change his mind when the time came to fly.

December 18 was the first completely clear day the team had had since reaching Deception Island two months earlier. The men spent the morning digging the plane out of the snow. Then Balchen and Ellsworth made a test flight. Everything, including the overhauled engine, worked perfectly. Ellsworth was confident that he would take off to cross Antarctica the next morning. Instead, bad weather returned, making it impossible to fly that day, or the next, or the next.... When pack ice began to close about Snow Hill Island, Ellsworth decided he would leave if he was unable to take off by January 1. The weather relented one day before the deadline, and once again, the men dug out the plane. Balchen, however, persuaded Ellsworth to delay the flight until morning. By then, the fog was back.

On January 3, 1935, the crew began moving the fuel drums down from the flying field in preparation for departure. Then, seemingly miraculously, the skies cleared. Balchen agreed to make an immediate try, and an excited Ellsworth dictated a message for the press: "Flash—Balchen and I took off ... this evening, heading for the unknown. The great adventure so long awaited is at hand."[10]

The *Polar Star* was only an hour out from Snow Hill Island when Balchen reversed course. His response to Ellsworth's puzzled "Why?" was that there was too much bad weather ahead. On landing, Balchen told Wilkins "Ellsworth can commit suicide if he likes, but he can't take me with him."[11] He had decided that without a third man along, he would continue only if they had a 100% chance of a nonstop flight. Ellsworth was furious, because his plan had assumed that they would land if the weather turned bad. Neither man had accepted the other's position.

The aborted January 3 flight was Ellsworth's last that season. After fighting the weather for six days to bring the plane down to the ship, he and his men finally left Snow Hill Island on January 9, sailing away into the teeth of a gale. Ellsworth soon found himself in such a desperate fight with the ice that it took eleven days for the *Wyatt Earp* to reach Deception Island. There, the men dismantled the *Polar Star* and stored her in the ship's hold for the trip north. It was a tense voyage, made worse by an explosion of the rat population that had plagued them since leaving New Zealand. As they sailed north, the warmer weather brought the rats out of hiding and they swarmed all over the ship. They even killed and ate one of the cats. When Ellsworth saw a huge rat clinging to his cabin ceiling one morning, he declared war. Using sticks and clubs, the men killed 169 rats on deck and dozens more below.

But Ellsworth had more important issues than rats on his mind as he sailed away from Deception Island. Frustrated twice, he was now unsure about his future plans. He had already spent over $150,000 and traveled 43,000 miles with nothing to show for it. He made up his mind soon after reaching the United States. He wrote, "In New York it was not an encouraging pat on the back but a well-meant attempt to dissuade me that sent me back to the Antarctic once more. . . . That advice, which represented the sensible, comfortable world, crystallized my determination. I was not to be defeated—I *would* cross Antarctica by air."[12]

Ellsworth had been bitterly unhappy with Balchen. Even so, he considered giving the pilot another chance. Balchen, as it turned out, was no longer interested. Wilkins then found two outstanding replacement candidates: Canadian Airways pilots Herbert "Bertie" Hollick-Kenyon and James Harold "Red" Lymburner. Ellsworth hired both so he could play them off against each other, to forestall a repeat of his experiences with Balchen.

The *Wyatt Earp* reached Deception Island for the second time on November 4, 1935. After a week spent there for his men to assemble the *Polar Star,* Ellsworth sailed to Dundee Island, where something he had seen the previous summer suggested there might be a good flying field. He was right. It was so good that Wilkins told Ellsworth, "If I had known five [sic] years ago what this island was like, you would not now be trying to be the first to fly across Antarctica."[13] Not only was there a good snow runway, but the weather was also cooperating. After both pilots test flew the plane, Ellsworth gave the nod to Hollick-Kenyon, the more experienced of the two.

Ellsworth and Hollick-Kenyon took off from Dundee Island on November 20, again heading for Byrd's Little America. This year, however, Byrd would not be there to greet them. He had left in February, completing an expedition that had made significant Antarctic land discoveries from the air and successfully used overland motor transportation for the first time in the south. The effort had also been noteworthy for Byrd's solitary five-month winter stay 100 miles south of Little America. Because Byrd was gone, Ellsworth loaded the *Polar Star*—already laden with over 3,700 pounds of fuel in her topped-off tanks—with 150 pounds of food in addition to his survival supplies. That should be enough, he hoped, to see him and Hollick-Kenyon through until the *Wyatt Earp* arrived to collect them. (Ellsworth also somehow found room for some eccentric Americana—Wyatt Earp's cartridge belt, a Mickey Mouse doll, and an 1849 ox shoe he had found in California's Death Valley years earlier.)

Less than two hours after takeoff, a fuel gauge began to leak badly and then threatened to burst completely. That prospect was dangerous enough for even Ellsworth to agree they should turn back.

The next day, repairs complete, Ellsworth and Hollick-Kenyon tried again. The weather was thickening a few hours later when the *Polar Star* passed Wilkins's 1928 farthest south along the Peninsula. Ellsworth saw a substantial mountain range below just as Hollick-Kenyon was beginning to climb above the clouds. Excited, he named it the Eternity Range. He was sure that it was only the first of many discoveries to come. But an hour and a half later, Hollick-Kenyon turned back. A distraught Ellsworth yelled over the engine noise to ask why. Hollick-Kenyon responded with a note saying there were too many clouds ahead. Ellsworth wrote later, "All the way back I tried to decide what I would do about this fiasco, scarcely observing anything from the plane. . . ."[14]

When the *Polar Star* landed at Dundee Island, Ellsworth stormed off without talking to anyone. Later in the day, he told Wilkins that he was going to try again the next day, but not with Hollick-Kenyon. He would take Lymburner instead. Lymburner, however, had been working for 36 hours straight and was a physical wreck. Wilkins thus argued for Hollick-Kenyon, the fresher of the two pilots. Ellsworth finally agreed, with the proviso, scarcely enforceable, that this time Hollick-Kenyon would go on, no matter what.

On November 22, 1935, Ellsworth and Hollick-Kenyon took off again. Just over eight hours out, the radio quit. Ellsworth did not. He had been in constant contact with Wilkins prior to this moment, and his last message had been that all was well. He was confident that Wilkins would realize that his silence indicated nothing more than a radio problem.

Ellsworth dropped an American flag an hour or so later to claim territory for the United States, because he estimated that they had reached 80° W, the western boundary of the British Falkland Islands Dependencies claim. Four more hours on, about 13 hours from Dundee Island, the elapsed flying time at Ellsworth's estimated speed put the plane near Little America. The position of the sun indicated otherwise, and when Ellsworth took a rough navigation reading, the results really shook him. They were hundreds of miles short of where his dead reckoning had placed them.

An hour later, Ellsworth decided to land to check their position more accurately. This was the first landing, by them or anyone else, in the interior of the Antarctic Continent, and both Ellsworth and Hollick-Kenyon were understandably anxious. Would the surface be smooth enough? Would it be solid enough? To their relief, they had no trouble at all. Ellsworth wrote, "We climbed out of the plane rather stiffly and stood looking around in the heart of the Antarctic. There we were—two lone human beings in the midst of an ice-capped continent. . . ."[15] They had landed at about 79° S, 104° W. The Bay of Whales was still more than 700 miles away.

Ellsworth and Hollick-Kenyon spent 19 hours on the ground before taking off. Once more, they faced a tense and critical test, another first. Would they be able to take off from the polar plateau? Again, there was no problem, but they were aloft for only 30 minutes before bad weather forced them down for three days. Their next flight was nearly as short, because a fierce blizzard assaulted them less than an

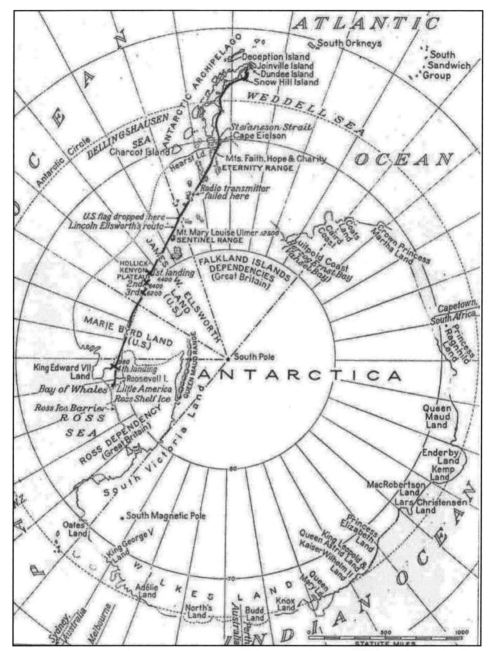

Lincoln Ellsworth's map of his and Hollick-Kenyon's route on their November 22–December 5, 1935 flight across the Antarctic Continent — *The vagueness of the shape of the Antarctic Continent and its coasts as shown here illustrate how much was still unknown about Antarctica at the time Ellsworth made his flight. Note that this map labels the Antarctic Peninsula the "Antarctic Archipelago" and shows it as a group of large islands, reflecting Ellsworth's accepting, and, he thought, confirming, Wilkins's report of straits cutting through the Peninsula. (Map from Ellsworth, "My Flight Across Antarctica," 1936 (Route darkened from the original. "X"s along the track indicate locations where the* Polar Star *landed.))*

THE FIRST AVIATORS ARRIVE: 1928–1936

Ellsworth's *Polar Star* resting on the ice of the Antarctic polar plateau following her first landing — *Ellsworth and Hollick-Kenyon erected their tent here, as they did at all their landings, for shelter while outside the plane. (From Ellsworth, "My Flight Across Antarctica," 1936)*

hour after takeoff. This time they spent a week on the ground, the first three days with screaming winds blasting their tent and piling up snow that took them days to remove. Then, even before they had finished reloading the plane, another blinding snowstorm hit. When they finally did manage to get off in the face of still threatening weather, it turned out that the bad conditions were local. They spent almost four hours in the air before landing again to check their position. They had reached the Ross Ice Shelf. Little America was near. Ellsworth decided to camp for the night and finish the flight the next day.

The *Polar Star* took off for the last time on December 5. Just over an hour later, Ellsworth saw the Ross Sea in the distance. Then the engine started sputtering. Out of fuel after a total of 20 hours and 15 minutes in the air, the *Polar Star* simply glided down to her last landing on a 2,300-mile flight. Ellsworth and Hollick-Kenyon spent the rest of the day securing the plane and preparing to sledge to Little America. Their dead reckoning said that it was only 4 miles away, but in what direction? On December 13, after days of false starts and stumbling retreats, they at last reached the seaward edge of the Ross Ice Shelf. Two days later, they found Little America. It had taken them ten days and 100 miles of sledging to get there—to a place that was in fact 16 miles from the plane.

At first, all they could see of Byrd's abandoned base was a thicket of poles. Then Ellsworth spotted stovepipes poking through the ice. The two men dug about and found a skylight, pried it open, and climbed down into Byrd's radio shack. Once inside, Ellsworth had a surprise for his pilot. For three years, he had been carrying two tiny bottles of Napoleon brandy his wife had given him to celebrate the crossing. The moment to open them had come at last. He wrote, "It contained the best brandy I ever tasted. . . . Hollick-Kenyon took a sip and really smiled."[16]

Confident that Wilkins would not fail them, the two men settled down to wait for the *Wyatt Earp*. They soon scavenged ample coal for the stove and more than enough food to supplement their own provisions. The days dragged, however, especially for Ellsworth, who was bedridden much of the time with a frostbitten foot, an injury suffered while sledging from the *Polar Star* to Little America. Equally frustrating, he had left his glasses behind at the plane and could not read. Hollick-Kenyon, in contrast, thoroughly enjoyed the tattered detective novels Byrd's men had left behind.

Making matters worse, the two men were ill suited as companions, with almost nothing in common other than the flight they had shared. Hollick-Kenyon was a notably silent man whose fastidious habits grated on Ellsworth. The American wrote, "He always shaved every morning, no matter how hard the circumstances and every night before supper, if there was even a chance to melt snow, he always sponged down.... So careful was he of his clothes that at the end of the expedition his trousers still retained their crease."[17] For his part, Hollick-Kenyon found Ellsworth far too casual about nearly everything. As the days passed, both men became increasingly testy as they waited, and wondered, when Wilkins would arrive.

Wilkins was indeed on his way and, for better or worse, the whole world knew it. As Ellsworth had assumed, the radio silence had not worried the Australian because of that last reassuring message that all was well. Unfortunately, the fact that Wilkins had been sending hourly press reports on Ellsworth's progress meant that he had to let the world know that he had lost contact. That included Mrs. Ellsworth, who immediately sent a plane south for Wilkins to use should a search become necessary. When that craft crashed en route, fortunately with no injuries to the crew, she obtained a second one. Wilkins sailed north from Dundee Island, collected the new plane in Punta Arenas on December 22, and returned south across the Drake Passage. He then sailed west around the continent for the Bay of Whales.

Another ship was also heading there. Despite Wilkins's own confidence, the *Polar Star*'s fate had become a worldwide concern, especially in Australia. The prime minister contacted the Discovery Committee, and its directors agreed to send the *Discovery II* to the Bay of Whales to look for Ellsworth—precisely what Wilkins was doing with the *Wyatt Earp*. Wilkins protested that his countrymen's efforts were unnecessary, but the Australians were determined.[18] The *Discovery II* entered the Ross Sea pack four days before the *Wyatt Earp*. Since the two ships were in radio contact, each knew where the other was, and a race mentality developed.

The *Discovery II* won. She reached the Bay of Whales on January 15, picked up Hollick-Kenyon that day and Ellsworth, who had stayed behind in the radio shack tending to his frostbitten foot, the next. The *Discovery II* people were rather disappointed to find that the men they had come to rescue seemed to be just fine, except for Ellsworth's foot. Hollick-Kenyon was an especially disappointing rescuee. One of the *Discovery II* scientists wrote, "We had conjured up in our imagination thin features, covered by a matted growth of beard and wasted by weeks of starvation diet in a twilight cell buried beneath the snow. But his cheeks shone from a recent shave and his stalwart frame in a check shirt showed no signs of anything, but abundant health and vitality."[19]

Wilkins arrived three days later, a clear demonstration of how unnecessary the *Discovery II*'s mission had been. Still, Ellsworth recognized that the Australians had made a generous effort on his behalf. After discussing the matter with Wilkins, he returned to Australia on the *Discovery II* to thank them personally. Hollick-Kenyon remained with Wilkins and the *Wyatt Earp* to retrieve the *Polar Star* before sailing home to New York.

Ellsworth and Hollick-Kenyon had done it. The first transcontinental Antarctic flight. That the flight was successful was the result of a combination of a well-chosen plane, an excellent polar pilot, and a great deal of luck. Hollick-Kenyon never returned to Antarctica, but Ellsworth, Wilkins, and Red Lymburner did. In the summer of 1938–39, Ellsworth would make his fourth, and last, expedition south, this time to East Antarctica. Lymburner was his pilot, and Wilkins was once again his manager and technical adviser. Other than one survey flight that probed several hundred miles south from the coast, the expedition accomplished little, both because of problems finding good takeoff surfaces and because the time in the south had to be cut short due to a serious injury to one of the *Wyatt Earp*'s crew.

As for the *Polar Star*, Ellsworth wrote, she had "traveled something like 65,000

Ellsworth (at left) and Hollick-Kenyon (at right) relaxing aboard the *Discovery II* shortly after her arrival at the Bay of Whales — *The two disappointing rescuees, particularly Hollick-Kenyon, who was in splendid health, clean, and freshly shaven when picked up. Ellsworth, however, had cleaned up before this photo was taken. When he came aboard ship, he had precisely the "matted growth of beard" that the men of* Discovery II *were expecting to see. (Photo from Ommanney,* South Latitude, *1938)*

miles in order to fly a critical distance of 2,300 miles. She was as good as new but far too dear a relic in my eyes to be permitted to grow old and go to aviation's boneyard."[20] In late 1936, he donated her to the Smithsonian Institution in Washington, D.C. She remains there today, a proud exhibit in the Golden Age of Flight gallery at the Institution's Air and Space Museum.

Chapter 16

The Wintering Explorers Return: 1934–1941

Lincoln Ellsworth had company in the Peninsula Region skies in the summers of 1934–35 and 1935–36, but the expedition led by John Riddoch Rymill, a 29-year-old Australian, was quite a different one. Not only did Rymill explore on the ground as well as from the air, but his men also spent two uninterrupted years in the south.

JOHN RYMILL AND THE BRITISH GRAHAM LAND EXPEDITION

Rymill's Antarctic plans changed several times before he settled on a program for what became known as the British Graham Land Expedition (BGLE). In 1934–35, he would sail south along the west coast of the Antarctic Peninsula to Marguerite Bay, taking a plane for aerial survey and dogs for sledging. There he would remain for two entire years, through the summer of 1936–37, undertaking the first exploration of the area since Charcot discovered the bay in 1909.

Five members of the BGLE team had already wintered together on a British expedition in Greenland: Rymill himself; W. E. Hampton, second-in-command and the plane's pilot; Quintin Riley, meteorologist; Alfred Stephenson, chief surveyor; and Edward Bingham, the doctor and chief dog man. Three other members of the nine-man shore party had also been to the Arctic. Six more men, none with polar experience, constituted the ship's contingent. All were amateur sailors except for the captain, R. E. D. Ryder, and H. Millett, the chief engineer, both on loan from the British navy. The ship's party grew to seven in the Falklands when Duncan Carse, a member of the *Discovery II* crew, joined up.

The total BGLE budget was £20,000. Such limited funds meant that Rymill had to economize. He began with his ship, a 32-year-old French fishing schooner he renamed *Penola*, after his South Australia hometown. Although small, the *Penola* was strong and sound. Her twin auxiliary engines were another story. Rymill briefly considered replacing them but decided against the expense—a decision that would cost him dearly. The tight budget also affected the flying plans. Rymill had only one small plane with a limited flying range—a single-engine de Havilland Fox Moth, equipped with both skis and pontoons. He did spend money, however, for a tractor to supplement his dog teams' motive power.

The *Penola* left London for the Falklands on September 10, 1934, with twelve members of the party. Bingham, who had stayed behind to wait for additional dogs, would follow on a commercial vessel. Hampton and Stephenson had departed a few months earlier on another ship with the plane and much of the expedition's cargo, because the *Penola* was too small to carry everything. When Rymill reached Stanley in the Falklands on November 28, Hampton, Stephenson, Bingham, and about 50 dogs were there to greet him. So was the *Discovery II*, ready to take on most of the expedition equipment, the plane, and the dogs, and deliver them to Port Lockroy as prearranged. Rymill had chosen this location, near the southern end of the Gerlache Strait, because it offered a secure harbor where the *Discovery II* could safely leave the cargo until the *Penola* arrived to pick it up.

The *Discovery II* sailed south a few days later. Hampton and Bingham were aboard to care for the dogs at Port Lockroy until the *Penola* arrived. It was fortunate they were, because the ship's violent motion in heavy seas played havoc with the pen holding the dogs on deck. Rymill's men soon found themselves in a constant scramble to catch escaped huskies who went anywhere and everywhere, including happily sacking out on untended bunks.

Rymill and the *Penola* did not leave Stanley until December 31. The delay had resulted from several weeks of sweating over the engines, which had malfunctioned constantly on the voyage from England. When the engines broke down again only a few hours out of Stanley, an exasperated Rymill anchored in a small cove on the east coast of the Falklands to see what he could do. The news was bad, presenting Rymill with two choices. He could stay in the Falklands for more repairs. That might take so long that it would be impossible to reach the Antarctic until the following summer. Alternatively, he could rely on the *Penola*'s sails for the time being and fix the engines during winter. He decided to sail.

Rymill finally reached Port Lockroy on January 22, 1935. Hampton and Bingham were waiting, though they had been there alone for only a few days. They too had had problems. When the *Discovery II* had reached the Gerlache Strait, her captain had found it so choked with ice that he had retreated and taken Rymill's entourage to Deception Island. Then he had gone off to survey the South Shetlands. The two men and their dogs spent a month at Deception Island, living at the abandoned whaling station until the *Discovery II* returned.

Now Rymill had to find a place to winter. The *Penola*'s engine problems ruled out Marguerite Bay that year, because, Rymill felt, it would be too dangerous to go that far south with only sails. He did, however, have the plane. With that, he could look for a nearby winter site. Rymill surveyed the area from the air on January 27 and concluded that the Argentine Islands, about 30 miles south of Port Lockroy, looked the most promising. He used the motor launch the next day to investigate and was delighted with what he found there—a good place for the hut and an excel-

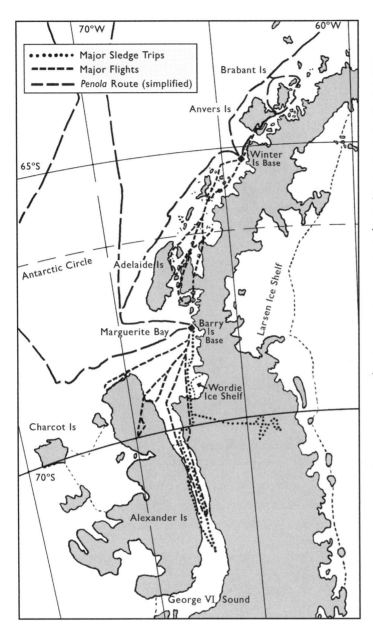

British Graham Land Expedition voyage, sledge, and flight routes — *The BGLE's* Penola *voyages, many flights with the small plane, and extensive sledging program covered a great deal of territory from the Palmer Archipelago in the north to George VI Sound in the south. This map shows a simplified version of the* Penola's *tracks, only the most significant of the flights, and just the tracks of the two most important sledging trips. Not only did the plane make many more flights than shown here, but there was also far more sledging than this map displays, particularly in the vicinity of the Winter and Barry Island bases.*

lent harbor. The *Penola* made two trips from Port Lockroy to transfer everything, and then Hampton flew the plane down. On February 14, the ship eased into her anchorage at what Rymill would name Winter Island.

Rymill and Hampton made their first significant survey flight two weeks later. When they reached 7,000 feet, they could see a few Peninsula mountaintops poking through the clouds. That was all. Lower down, however, Rymill had seen enough to conclude that there was no reasonable route to the Peninsula heights in their vicinity. Wintering just a few miles south of where Charcot had spent the 1909 winter, he was finding himself in much the same situation as the Frenchman.

Domesticity in the hut at Winter Island. *(Photo from Rymill,* Southern Lights, *1938).*

His aerial reconnaissance, however, gave him advance information unavailable to Charcot. And so, unlike his predecessor here, Rymill decided to concentrate his sledging efforts on heading south along the coast rather than spending time in a futile effort to penetrate the Peninsula interior.

Months passed before the sea ice around the Argentine Islands was safe for sledging. In the meantime, nine men lived comfortably in the hut on shore and the other seven resided equally snugly aboard ship. All worked together and enjoyed a busy and happy winter. One man, however, was unable to participate fully: Brian Roberts, the shore party ornithologist, was experiencing recurrent bouts of appendicitis. His illness would have significant consequences for the entire expedition.

Rymill finally began his sledging program on August 18. That day, he and eight others set out with three dog teams. All, humans and animals alike, were excited as they headed toward the rugged mainland coast to the south, country that had previously been seen only from ships. Four men peeled off the second day, when they were about 18 miles south of Winter Island, and sledged eastward to work in a large bay. By August 22, the remaining southbound sledgers were almost 60 miles south of Winter Island. There they marked out an emergency landing strip for the plane and established a depot to support the sledging program. Then Rymill and two others turned back, leaving a final pair of Bingham and R. E. D. Ryder to continue south, toward Adelaide Island, where they planned to put in another landing strip.

The next morning, Bingham and Ryder met real trouble—thin ice that creaked ominously under their sledge. Two days of frighteningly dangerous sledging advanced them a few miles farther south, but

Returning over broken sea-ice from a sledging trip. *(From Rymill,* Southern Lights, *1938)*

that was as far as they could go. Adelaide Island loomed before them, as impossible to reach as it was frustratingly close. Ryder radioed the bad news to Winter Island, and Rymill flew down to consult. After a discouraging conversation, Rymill abandoned much of his spring sledging plan.

Lummo, the BGLE expedition cat, on the snow at Winter Island — *The spelling discrepancy from the cat's namesake reef resulted from a difference of opinion between the ship's crew and the scientists over the proper spelling. (Photo from Rymill,* Southern Lights, *1938)*

The defeat of his sledging program from Winter Island made Rymill all the more eager to move on to Marguerite Bay. But he had a problem. The hut boards, swollen and cracked from exposure to damp weather, were unusable for a new base, and he had no extra building materials because he had originally planned to spend both winters in the same place, in the same hut. When Bingham and Hampton remembered having seen scrap lumber at the abandoned whaling station on Deception Island, Rymill decided to begin the *Penola's* summer season with a scavenging trip. Deception Island turned out to have just what Rymill needed. Captain Ryder spent two weeks there loading lumber and a small stock of coal the men had spotted. On his return south, Ryder added a few names to the map. These included Penola Strait, for the channel between the Argentine Islands and the mainland, and Lumus Reef (now Lumus Rock), for a small group of rocks named in honor of the ship's cat.

On February 17, 1936, several weeks after returning from Deception Island, the *Penola* left Winter Island for Marguerite Bay. Hampton and Stephenson remained behind with the plane. They would fly south when the call came. Rymill sailed slowly down the coast as the men surveyed and made a few landings at places they had originally hoped to reach by sledge. The *Penola* sailed into the far north of Marguerite Bay on the 24th, and Hampton and Stephenson flew down a day later. Before leaving Winter Island, they had boarded up the hut and posted a sign on the door reading, "To Let, for the season 1936–37."[1]

Hampton and Rymill took off on February 27 in search of a base site. They soon spotted the mountains of Alexander I Land, about 80 miles away. And they could see other peaks crowning the Peninsula, about 100 miles distant. These were probably the same ones Ellsworth had named the Eternity Range three months earlier, but Rymill, unaware of the details of Ellsworth's results, excitedly counted them as his own expedition's first significant land discovery. Most important, the flight succeeded in its practical objective when they spied a good base site in Marguerite Bay. It was about 50 miles southeast of the *Penola* and perhaps 100 miles north of what appeared to be the southern limit of the huge bay.

Rymill's plane on the ice at a sledging depot — *The de Haviland Moth used pontoons when taking off from water, on flights where she would not be landing until she returned. Once a runway had been prepared near the Barry Island hut, the plane was able to use skis, making it possible to land in the field. (From Rymill,* **Southern Lights***, 1938)*

The next day, the *Penola* reached the chosen base area, a group of small islands a few miles from the mainland coast. Rymill christened them the Debenham Islands in honor of Frank Debenham, an Australian geologist on Robert Falcon Scott's 1910–13 expedition. He selected Barry Island, named for one of Debenham's children, as the place for his hut.[2] Unloading began at once, and the wintering party moved onto shore on March 12. They would make their home here in Marguerite Bay, the place that Charcot had named for his wife and had so longed to winter at. Now Rymill's men *would* do it. Their work from here would not only be the first significant exploration in this area, but also the first major land surveys in the entire Peninsula Region.

The *Penola* departed the following day. She was taking Brian Roberts north to civilization, because Rymill had decided that the appendicitis the ornithologist had experienced at Winter Island made it too dangerous for him to spend another winter in the south. Colin Bertram, a biologist with the ship's party, remained behind to assume Roberts' duties. The rest of the *Penola* group, however, would lose their second Antarctic winter, since it would be impossible to return south before winter set in.

Rymill's early aerial reconnaissance in the Marguerite Bay area had made it clear that the surrounding country bore little resemblance to Wilkins's descriptions. But the reality of Stefansson Strait, reportedly south of where Rymill's plane had flown, was still an open question. If it did exist, it might offer an easy route to the east coast of the Peninsula. If not, the aerial survey indicated that Rymill would need to travel at least 100 miles south of Barry Island, to the far south of Marguerite Bay, to reach a place where he could climb into the mountains. It would be months, however, before he could carry out the ground reconnaissance he needed to finalize his sledging plans. Once again, he was camped on an island, unable to leave until the sea ice froze.

The new hut, built with the scrap lumber from Deception Island, proved just as comfortable as the one they had left behind at Winter Island. With months to wait before they could begin sledging, the nine men at Barry Island settled into a busy base routine. Local surveys, geological and biological study, and preparing for sledging easily filled their days. In the evenings, despite the fact that these men had already lived together for more than a year, there was no lack of conversation. Unlike their wintering predecessors, they had radio contact with the outside world, and that, Rymill observed, provided ample conversation topics. Indeed, he said, he and most of his companions took more interest in world affairs than they had at home.[3]

Reconnaissance from a flight in early June seemed to indicate that the ice covering Marguerite Bay was finally strong enough for sledging. Seven men set out several days later. Rymill and four others were driving dog teams, while Hampton and Riley had charge of the tractor, towing two sledges. Their goal was to establish a provision depot on a small ice shelf about 100 miles to the south. Rymill had seen it on a survey flight and named it the Wordie Ice Shelf, after James Wordie, the geologist on Shackleton's *Endurance* expedition. But only 40 miles south of base, a violent storm churned the sea ice into a maze of pressure ridges and broken floes. With no way to maneuver the tractor through the chaos of ice and open water, the men abandoned the machine, and then barely escaped to a nearby island with the dogs. It was such a relief to reach solid ground that Rymill named the bit of land they camped on Terra Firma. They spent five days trapped there before they could retreat to the hut on Barry Island. The tractor and its two sledges remained behind, never to be seen again.

Rymill had learned his lesson. The ice needed more time to firm up. In the meantime, Hampton made more reconnaissance flights. On the most significant of these, Rymill realized that what he had originally thought was a small fjord at the southeast corner of Marguerite Bay was probably the entrance to a large sound. It might be a channel to the Weddell Sea in the east or, alternatively, one that separated Alexander I Land from the mainland.

The main sledging work finally got under way on September 5. That day, two teams set out southbound over the sea ice with the dogs. Stephenson was in charge of a three-man party that was to search for Stefansson Strait and head east through it if it existed. Rymill and Bingham planned to explore Alexander I Land to the west. The five men began by traveling together. After three weeks of difficult sledging, they were just over 100 miles south of Barry Island and still short of the point at which the teams had planned to part. They had made such glacial progress that Rymill changed the plans. He handed his own supplies over to Stephenson, and then he and Bingham turned back north to re-supply themselves while Stephenson's team continued south alone.

Stephenson crossed a pass on the slope above the eastern side of the possible sound on September 30. The next day, men and dogs raced down to the ice on the sound. Then it was on south, the early days brutal going because of pressure ridges and crevasses. Once past the difficult area, however, the sledging surface was marvelous. Stephenson was ticking off the distance at a rate of more than 20 miles a day, but there was no sign of a strait to the east. Instead, "As each day passed, fresh promontories would loom up on the east side of the sound, and to the west the 'last point' remained as far ahead of us as ever."[4] On October 19, at 72° S, Stephenson turned back, because he had gone as far as his supplies would permit. It was a difficult decision. He was sure that he was near the southern end of the sound, only a few miles from proving that it separated Alexander I Land from the mainland. He was right on the most important count. Alexander I Land, or Alexander Island as we know it today, is indeed an island. But he was wrong about how close he was. The end of the sound was still more than 150 miles away.

One evening during the return trip, Stephenson had a remarkable experience. He had gone outside and was tuning the radio to pick up a time signal in order to determine their position. He wrote, "I knelt down by the [radio] set and was soon completely absorbed in listening to a broadcast of a public meeting in Europe. The speech was highly oratorical and the audience was completely carried away, cheering wildly, with the result that for five minutes I was back in Europe."[5] And then he glanced up and remembered where he and his companions were, "utterly alone, hundreds of miles farther south than anyone else in the world."[6] It was only months later that he realized he had been listening to Adolf Hitler address a rally at Nuremberg.

Stephenson's team set foot on Alexander Island several times on their return. Recognizing the rock about them as sedimentary, they set out in search of fossils, and "with all the eagerness of the prospector we scrabbled up scree slopes and frantically picked up piece after piece of rock. . . ."[7] And then they found what they were seeking, dozens of fossils, with clear impressions of shells and plant remnants. The shores of Alexander Island were a geologist's dream, an exhilarating place to explore. But the men were also concerned with the rising temperatures of spring, which had transformed the sledging surface into a slushy mess. Their progress slowed to a crawl, forcing Stephenson to cut the men's rations and kill the seven weakest dogs to eke out the animals' food. Then, on November 11, when they were nearing the southern edge of Marguerite Bay, they saw two black specks approaching. Rymill and Bingham, beginning their own trip, had taken a route that would bring them to meet Stephenson's party.

After regaining Barry Island on September 29, Rymill had immediately begun organizing a new plan. He would sledge south along the possible sound to some place where he could cross west over Alexander I Land to Charcot Island. But was the new plan practical? When the two groups met, Stephenson had the answer: it was not. It would be impossible for Rymill and Bingham to reach Charcot Island by the proposed route unless they stayed out far longer than planned and not only duplicated but exceeded Stephenson's trip. Rymill had news of his own to convey, sad news. Stephenson wrote, "From where we stood . . . we commanded a magnificent view to the north of Adelaide Island and Marguerite Bay, and it was a fateful coincidence that in this spot we should . . . [learn] of the death of Dr. J. B. Charcot, the pioneer explorer in the country upon which we were then looking."[8]

Once the meeting ended, Stephenson's party continued homeward. They reached Barry Island on November 19, concluding a hugely successful, near 700-mile trip, by far the most extensive ground exploration in the Peninsula Region to date. They had mapped more than 500 miles of coastline, most of it never seen before, placed the first footsteps on Alexander I Land, found marvelous fossils there, and determined that this landmass was almost certainly an island. (Rymill would later name the icy highway separating Bellingshausen's 1821 discovery from the mainland George VI Sound, in honor of Britain's new king.) These men, who had so thoroughly enjoyed their time on the trail, now delighted in the comforts of the hut, washing away grime, eating off plates, and sleeping in real beds. As for the dogs, Stephenson wrote, "For eleven weeks they had been away from dirt; they had slept each night in the snow, and were beautifully clean . . . now they were reveling and rolling in the dirtiest and most 'blubbery' places they could find. . . . [They] were just as happy as we were. . . ."[9]

In the meantime, Rymill and Bingham sledged off with their own dog teams. They had abandoned the Charcot Island goal after hearing Stephenson's report, and revived the plan to head east and cross the Peninsula. Another of Stephenson's findings—that there did not appear to be a convenient strait cutting through the Peninsula to the east—affected that idea as well: Rymill would have to do it the hard way, climbing into the mountains and then somehow descending to the Weddell Sea coast. The ascent out of Marguerite Bay was an unrelenting struggle, as men and dogs spent days relaying loads and battling storms. But at last, on November 24, Rymill and Bingham reached a plateau at 7,500 feet and became the first people to stand atop the Antarctic Peninsula. It was a triumphant moment. They reached the east side of the plateau the next day and then spent several days looking for a way down. By the 30th, they had dropped 1,700 feet and reached a glacier they thought might provide a descent route. Instead, the glacier proved to be a bedeviling labyrinth of treacherous crevasses. Two days later, after the loss of Bingham's lead dog and many near misses for the men, Rymill gave up.

Once back on the plateau, Rymill set out in search of Stefansson Strait—whose existence he now seriously doubted—from above. After two weeks of sledging first south, then north, with no sign of a sea-level passage through the Peninsula highlands, Rymill concluded that Wilkins had been mistaken. Stefansson Strait did not exist. The Peninsula was not an archipelago. That finding

Rymill (left) and Bingham (right) after returning from their historic sledging trip to the top of the Antarctic Peninsula. *(From Rymill,* Southern Lights, *1938)*

ended Rymill's explorations. On January 5, 1937, he and Bingham were back at Barry Island, their historic 615-mile trip over.

The *Penola* arrived on February 23, and on March 12 the shore party left Barry Island, justifiably proud of their accomplishments. Rymill's small expedition had achieved a great deal, especially during the time at Marguerite Bay. In addition to important natural history studies and significant geological work, hundreds of miles of sledging on the ground and 110 hours of survey from the air resulted in substantial revisions to the Peninsula map, including reinstating the Peninsula as part of the mainland and tentatively separating Alexander I Land from it. It was historic work, the most extensive land exploration so far in the Peninsula Region, carefully carried out using a coordinated combination of aerial survey and ground exploration.

But as work decades later would demonstrate, sometimes even coordinated ground and aerial work are insufficient to determine the true geography of the Antarctic. In fact, both Wilkins and Rymill were wrong—Wilkins that his Stefansson Strait cut through the Peninsula and Rymill that the Peninsula was truly part of the Antarctic Continent. Although no strait bisects the Peninsula anywhere near where Wilkins placed Stefansson Strait, seismic reflection soundings of the ice covering the Antarctic Peninsula have now determined that that ice is thousands of feet deep in places, sometimes extending significantly below sea level.[10] In short, were the ice to disappear, the Peninsula would indeed be an archipelago rather than an extension of the main Antarctic Continent. So long as that deep ice is there, however, for all practical purposes the Peninsula is a part of the Antarctic Continent.

Despite its truly important achievements, the BGLE never received the recognition it deserved. By the time Rymill's team returned home, the imminent possibility of war was absorbing the public's attention. There was another reason as well, perhaps equally important. The BGLE, like Bruce's *Scotia*, produced no heroic drama for the press or exciting achievements such as Ellsworth's flight across the continent. It was simply a well-executed and notably productive piece of work.

For three years following Rymill's departure, whalers and the scientists of the Discovery Investigations were the only humans in the Peninsula Region. Then the explorers returned, men who followed Rymill's lead and wintered at Marguerite Bay. Unlike the general public, these new arrivals were acutely aware of the value of the BGLE's work.

THE UNITED STATES ANTARCTIC SERVICE EXPEDITION

By the early 1930s, Argentina, Australia, France, Great Britain, New Zealand, and Norway had all made claims to some part of the Antarctic regions. In 1938–39, Nazi Germany dispatched a summer expedition to Queen Maud Land, in East Antarctica, and used seaplanes to drop thousands of swastika-shaped darts over the ice as sovereignty emblems. Norway quickly responded by asserting her own ownership of the area. This was also the summer that Lincoln Ellsworth once again took the *Wyatt Earp* south and, at the secret request of the U.S. government, airdropped claim documents over a part of East Antarctica claimed by Australia. As for the Peninsula Region, Argentina hinted at an expansion of her earlier claims when she created a National Antarctic Commission in 1939.

Other than making its confidential request to Ellsworth, the United States was conspicuously absent from the list of claiming nations. Indeed, it had not only made no public claims, but it took the position, first stated in 1924 by Secretary of State Charles Evans Hughes, that no one's Antarctic claims were valid because no one had established actual settlements.[11] In 1939–40, the United States would set out to become the first country to create such settlements, permanent bases that would allow the country to assert its own sovereignty on the terms she had demanded.

In 1938, the U.S. government began considering sending a modest expedition to Marguerite Bay or a bit south. Richard Byrd was simultaneously developing plans for a third Antarctic expedition. Late that year, President Franklin Roosevelt suggested combining the two proposals into one government-sponsored effort, with Byrd as the overall leader. The resultant United States Antarctic Services Expedition (USAS) quickly grew into an ambitious program. There were to be two bases—East Base in the vicinity of Alexander Island and West Base in the Ross Sea area. Both stations were to be permanent, with work to be continued over the years by a series of teams.

The simmering tensions in Europe boiled over on September 1, 1939, when Germany invaded Poland, launching World War II. The United States, however, remained neutral, and in November, when the war in Europe was already two months old, the USAS expedition ships, the *Bear* and the *North Star,* left for the south. Byrd was aboard to serve as expedition commander during the summer season. The ships unloaded the West Base contingent at the Bay of Whales in the Ross Sea during the latter part of January. The *North Star* then returned north to collect

Building the USAS East Base — *A group of huts assembled from prefabricated sections designed by the U.S. Army Corps of Engineers. The main building, 60 feet by 24 feet, had five curtained cubicles on each side of a center aisle with two bunks in each cubicle. The kitchen was at one end, the sick bay and the base leader's cubicle at the other. The floor was built with two layers, 16 inches apart, so that warm air from the kitchen could circulate between them. This design feature successfully prevented snow and ice from accumulating on the floor. (U.S. Government National Archives / National Geographic Stock)*

additional stores and equipment for East Base while the *Bear* sailed east toward the Antarctic Peninsula.

The two ships reunited in Marguerite Bay on March 5. Three days later, they found a site for East Base. It was on a low rocky islet only a few miles from Rymill's Barry Island. Byrd named it Stonington Island in honor of the Connecticut home of the American sealer Nathaniel Palmer. One reason Byrd selected it was that a mainland glacier's end rested on the island's edge. This would provide a bridge to the Peninsula, relieving the men of East Base from dependence on the sea ice for access to the mainland.

The two ships departed on March 21, leaving 26 men and about 75 dogs to winter. Two veterans of Byrd's 1933–35 expedition to the Bay of Whales were in charge—Richard Black in command and Finn Ronne as his second. For transport, they had a twin-engine Curtiss Condor biplane, considerably larger and with a much longer range than Rymill's small aircraft; one light army tank; a light artillery tractor; and the dogs. Tents still housed the men when the ships left, but by using every possible hour of daylight, they finished assembling the main living quarters sufficiently to move in

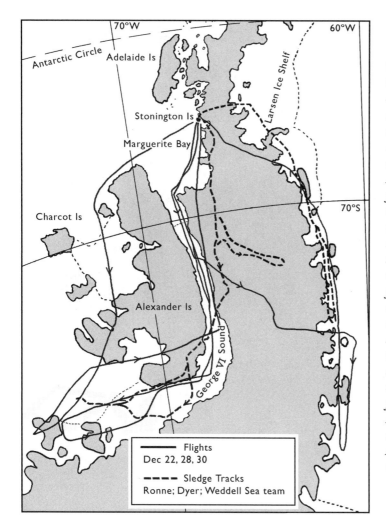

USAS East Base field of operations — *The USAS East Base worked in much the same area the BGLE did during its second year in the Antarctic. Multiple flights and sledge trips set out from the base on Stonington Island. This map shows only the three most important flights and the three most significant sledge trips. A map showing all the flights would be covered with a complex spider web of tracks, especially for plane flights over Marguerite Bay and to the south, over Alexander Island and George VI Sound.*

a week later. It took until the end of April, however, to complete the elaborate five-building camp designed by the Army Corps of Engineers.

The first reconnaissance flight took off on May 20, and the pilot returned with an unexpected bonus. Despite Rymill's conclusion that there was no practical route to the Peninsula highlands anywhere near, the USAS plane had spotted a possibility just north of Stonington Island. Another flight the next day went south. This one landed on the Wordie Ice Shelf to establish a supply depot. Significantly, the men covered the depot with a tarp topped by a U.S. claim sheet. This was the first claim an American government expedition had made anywhere in the Peninsula Region, and it fell well within the British Falkland Islands Dependencies. Unfortunately, the plane was damaged on landing when it returned. This mishap, plus bad weather, ended flying until August.

With flying suspended, the men settled in for the winter doing meteorological, biological, and geological studies about the base. Their main objective, explo-

The men of the USAS East Base celebrate mid-winter — *Going back to the first expeditions of the Heroic Age, mid-winter has been the great celebration of the year for those wintering in the Antarctic. (U.S. Government National Archives / National Geographic Stock)*

ration, would have to wait for spring. But in the meantime, Black was impatient to follow up on his pilot's enticing report of a possible route inland. He sent out three winter sledge trips to investigate.

The first two scouting trips were encouraging. The third one was much more ambitious, an all-out effort to scale the Peninsula mountains. Black himself led the 10 men and 55 dogs who set out on August 6. They proved that it was indeed possible to reach the Peninsula ridgetop from near Stonington Island, but it was not easy. The ascent took three days of backbreaking effort, including navigating a stretch where they had to drag the sledges one at a time up a sheer slope of glare ice. When everyone at last reached the top, at 5,500 feet, their success was rewarded with what Black described as the worst wind and blizzard he had ever experienced. Hurricane-force gales blasted their tents for three days before the weather finally relented and allowed them to return to East Base.

Black had planned to have the sledging and flight operations support each other. A good plan in theory, but weather made executing it impossible much of the time. On September 3, Black wrote in his journal, "I think I will have a rubber stamp made before coming to this area on another expedition, reading, 'Overcast—snowing—wind. No flying today.'"[12]

Although he could do nothing to change the weather, Black thought he might at least be able to improve his ability to forecast it. He thus decided to establish a meteorology outpost on the plateau. Men and dogs left Stonington Island on October 23 with 1,300 pounds of equipment and supplies, including two heavy cylinders of hydrogen that would fill 25 weather balloons. By then, Black's men had found a much less difficult route inland, and it took only two days to reach the selected site, the same place where the blizzard had hit the August group. The two men who occupied the camp then remained there for more than two months, living in tents shielded from the frequent violent storms by a wall of snow blocks. This first-ever high-altitude Antarctic weather station contributed significantly to the success and safety of the major exploratory flights to come. And equally important, the observations there, nearly 300 in all, produced a treasure trove of data on high-altitude air pressure, air temperatures, and wind patterns.

On November 6, five men under Finn Ronne plus a two-man support team hit the trail with 55 dogs to begin the southern sledging program. The plan was for all to travel together for a few days. Then the supporting team would turn back, and a few days later, the remaining five men would split into two groups. Ronne and Carl Eklund, the expedition ornithologist, would explore George VI Sound while the other three men, under Glenn Dyer, were to head toward Ellsworth's Eternity Range.

The men began sledging in cloudy weather with poor visibility, traveling south first overland and then across the Marguerite Bay sea ice for several days. They then struggled for four days, Ronne wrote, "through the most dangerous crevassed area that, I believe, can ever be encountered in Antarctica. Huge open crevasses, over which we crossed on narrow snow-bridges, were most common; but the hidden crevasses were also numerous. . . ."[13] Once past this horror, the supporting duo turned back. The other five men reached 7,000 feet above the eastern flank of George VI Sound on November 21 and then divided into the two planned groups.

Ronne and Eklund set out with fifteen dogs and supplies for two-plus months. They abandoned the original plan to sledge high along the plateau when Black radioed them to descend into George VI Sound to check an aviation cache that had been established to support the aerial exploration program. On December 3, they reached the cache and the sound, a few miles north of where Stephenson had turned back. Before them lay the same vast highway of ice, beckoning them on south to complete the BGLE's exploration of 1936.

It was December 21 when wildly disturbed ice blocked Ronne and Eklund's southward progress. They climbed a nearby slope to look for a way through and instead saw large icebergs drifting in an open blue sea. They had arrived at the end of George VI Sound. Now there was no doubt. Alexander I Land was indeed an island. That night, Ronne radioed Black that he was turning back; he was running out of both time and food.

Things went well for the first few days of the return trip. Then the dogs began struggling with slushy snow, the effect of the bright mid-summer sun. Ronne switched to night travel, but that created a new problem, because the slush froze at night and the icy bits shredded the dogs' paws. Eight became so crippled that Ronne shot them. On January 6, Ronne and Eklund staggered into the depot where they had first reached George VI Sound just over a month earlier. There, Ronne radioed East Base to report that they were camping for a few days to give the surviving dogs a rest. Unfortunately, their radio transmitter quit shortly thereafter. Their receiver, however, was still working. Black, hoping that their receiver *was* still functioning, continued to send messages about plans to relieve them.

Ten days later, Ronne and Eklund resumed sledging. The rest had revived the dogs. That, combined with the canvas booties the animals now wore, made the still dreadful surfaces much less of a problem. At the last camp, just before leaving the sound, Ronne heard on the radio that the plane had been damaged on takeoff and that there was much concern about his party. He wrote, "Here we were sitting in the tent, listening to one emergency after the other, but with no way of telling the worried base personnel that we were absolutely safe."[14] On January 27, only 22 miles from Stonington Island, Ronne and Eklund met Black, on his way to rescue them.

Black had tried to send help earlier, first by air, but weather delayed takeoff for days. Then there was the accident to the plane. Finally, Black himself had set out with a sledging party. They were pushing south on the 27th when they met Ronne and Eklund. Both men, Black wrote, "looked badly burned and very tired, but they certainly did not look to be in need of rescue."[15] All were safely back at East Base less than a day later. Ronne's very productive 1,264-mile trip was over.

Glenn Dyer's three-man team had also had a successful trip, though theirs was much shorter. After leaving Ronne, they had gone southeast. A week later, they reached a small peak at the north end of the Eternity Range. Here they built a rock cairn, raised the U.S. flag, and deposited yet another claim sheet. Dyer's team was back at East Base on December 11.

Ronne's trip had fulfilled one of Black's goals. Another was to cross the Peninsula and sledge south along the Weddell Sea coast to explore that entirely unknown area. Black gave the mission to the geologist, Paul Knowles, who had already cached supplies on the Peninsula plateau and scouted a way down its east side in September. Knowles left Stonington Island on November 19 with two companions and two dog teams. After collecting supplies at the cache, men and dogs worked their way down a heavily crevassed glacier to the Larsen Ice Shelf at the edge of the Weddell Sea. They then turned south and sledged to almost 72° S. The outbound trip ended there on December 22. When they reached East Base on January 17, the three men had traveled more than 800 miles, carried out the first ground survey of the Peninsula's east coast south of the Antarctic Circle, and made the first two land crossings of the Antarctic Peninsula—one in each direction.

While the sledging parties were on the trail, the planes were in the air. East

Base's last exploratory flight was on December 30, across the Peninsula and then down the Weddell Sea coast to well beyond the southern limit of Knowles's sledge trip. Since Knowles had turned back eight days earlier, the plane was entirely on its own, with no hope of help should anything go wrong. The men aboard had this stark fact brought vividly home when both engines stopped for 20 to 30 seconds during the descent to the Weddell Sea. Black, who was aboard, commented years later, "this was a long time—there is no more complete silence this side of the grave."[16] All breathed a huge sigh of relief when they realized that the engines were simply shifting to fresh fuel tanks. Black turned back at 74° 37′ S because of deteriorating weather. With the gathering clouds blocking his view to the south, he had no way to know that he was only a few miles short of the southeast corner of the Weddell Sea.

The East Base field work was finished when Ronne's team returned, and the entire USAS expedition ended shortly thereafter. From an American perspective, the international situation had grown much worse since 1939, and the government decided to evacuate the two bases rather than relieve their personnel as originally planned. Closing up West Base proceeded easily. Not so, East Base.

The *Bear* and the *North Star* reached the vicinity of Marguerite Bay on February 17, 1941, only to be stopped by impenetrable ice 60 miles from Stonington Island. The ships then sailed north, anchoring in the Palmer Archipelago to wait for the ice to open. When it was still solid in mid-March, a desperate Byrd decided to evacuate East Base by air. The men would fly themselves out in their own plane.

The *North Star* sailed to Punta Arenas on March 20 to drop off the West Base men and collect supplies for a second year for East Base should the air evacuation fail. Were that to happen, Byrd would airdrop the supplies to the stranded men. A day after the *North Star* left, the *Bear* landed men on Mikkelsen Island, an islet in the Biscoe Islands group about 120 miles north of Stonington Island. There they created a landing strip and then radioed Black that he should begin the air operation as soon as weather permitted. The East Base plane would fly men, scientific records, and the most crucial scientific collections out in two flights. Everything else would have to be abandoned—equipment, all other scientific specimens, most of the men's personal possessions, and the dogs.

Ashley Snow, the chief pilot, lifted off from East Base early on the morning of March 22 with his co-pilot and the first twelve passengers aboard. Back at Stonington Island, the remaining twelve men crowded anxiously around the radio in the main building while Black scribbled the flight log on the wall with a red crayon so that everyone would know what was happening. They cheered when the *Bear*'s radio operator reported that the plane had landed safely on Mikkelsen Island. Then, while awaiting the plane's return, the East Base men secured their buildings and Black completed a letter that he would leave in the main hut. The letter asked the next arrival, whoever that might be, to collect all the scientific and personal items possible and send them to the United States. Finally, the men tackled the most painful task—dealing with the dogs. They began by shooting most of them. Then they staked out 28 at the airstrip, surrounded with charges of dynamite wired to an alarm clock. This ghoulish arrangement was essential, because the men would need these dogs if the second evacuation flight were to fail. They set the clock to allow time to cut the wire and save the dogs if that proved necessary.[17]

The plane came back to East Base shortly before noon the same day and picked up the rest of the men. Looking back as they flew away, they could see the dogs lying patiently below in their harnesses, a heart-wrenching sight for the men these animals had served so well. But one dog did survive. He was a tiny puppy, smuggled aboard the last flight by Harry Darlington, the youngest member of the expedition. Darlington took him home as a pet.

When the second flight landed safely at Mikkelsen Island, Black's men left the plane unsecured there, in hopes that a gale would lift it into the air, whence it would plunge it into the sea in a quasi-Viking funeral. The *Bear* sailed north as soon as all the East Base men were safely aboard.

The USAS, a highly ambitious undertaking, had failed in its primary goal of establishing permanent Antarctic bases because of the intensification and spread of World War II. Even so, a great deal had been accomplished. At East Base, that included the first crossing of the Antarctic Peninsula and the sledging trip down the Weddell Sea coast; conclusively determining that Alexander I Land was an island; extensive aerial survey; and the two-month occupation of the first high-elevation Antarctic meteorological station.

The U.S. government extended the life of the expedition for six months and employed a few people to transcribe journals and prepare maps and nonscientific reports. That was the end of the official USAS, leaving the scientists to work up their data on their own, though few had sufficient time for the job because of the demands of war work. As a result, many of them never formally published their scientific work. In the end, what was published came out in a single volume with 40 articles, *Reports on the Scientific Results of the United States Antarctic Service Expedition 1939–41,* a large soft-cover work published in April 1945 as a volume in the Proceedings of the American Philosophical Society. Given the tremendous amount of work that had been done, this was very little, leaving the USAS, as one historian called it, "the most poorly reported major expedition in Antarctic History."[18]

Chapter 17

World War II, New Bases, and Political Conflict: 1940–1955

World War II affected Antarctica differently than World War I had. This time, whaling nearly ceased after the early years of the conflict, and the war itself brought non-whalers, including naval vessels, south. After early 1941, when a German raider captured a Norwegian whaling fleet off the coast of East Antarctica, virtually all of this took place in the Peninsula Region.[1]

The War Years and Operation Tabarin

When the war began, Great Britain established limited defensive positions on South Georgia in the form of two elderly four-inch guns, one each at the Grytviken and Leith Harbor whaling stations. The idea was that a volunteer force of whalers would operate them should it become necessary. Years later, one of those supposed volunteers remembered his experience with the gun at Leith. He recalled,

> I did not even know that they had put in a big gun up on the side of the hill above the station but I was put down for going on this gun. . . . The gun had been all in pieces when they hauled her up the hill and . . . when we got her all put together, we had to fire a shot. We got the motor boat men to take a raft made with barrels out to the far end of the harbour at the entrance into Leith Harbour about a mile away. We all got together and fired this shell. It went over the station down below and smashed all the windows in the buildings but we were only a few yards from the raft so we did a very good aim and if it had been a German ship I think we would have hit her.[2]

No Germans showed up, but the gun remained there for years, slowly rusting away.

By 1941–42, Antarctic whaling had collapsed to limited activity out of Grytviken and Leith Harbor. A few whalers remained there throughout the war, but pelagic whaling ceased entirely because the combatants converted nearly all the world's factory ships to war service as oil tankers. As for the catcher vessels, they became minesweepers and subchasers.

Nearly all the whalers may have left, but people with other interests did venture south, in particular the men of the British and Argentine navies. They were a sporadic presence in the Peninsula Region during the war years for two reasons: the British were looking for German raiders rumored to be in the area, and both Britain and Argentina were pursuing political goals.

The first naval vessel on the scene was British, a former passenger liner named *Queen of Bermuda* that the Admiralty had fitted out as an armed merchant cruiser. She sailed to the Antarctic Peninsula early in 1941 to protect the few Norwegian and Scottish whalers still there. Despite being completely unsuited for the ice, she nearly reached the Antarctic Circle in the Weddell Sea. She also visited Deception Island, and while there, her men destroyed the fuel installations at the abandoned whaling station to prevent potential German use. They left the other buildings intact.

The *Queen of Bermuda* in her passenger-carrying role — *Capt. Geoffrey Hawkins, her naval captain, was ordered to visit Deception Island when he was at the South Shetlands and there destroy anything that German raiding vessels might find useful. First, however, he had to navigate this former luxury passenger liner through Neptunes Bellows, the narrow entrance into Port Foster. No one had ever taken such a huge ship through, and Hawkins later commented, "The entrance is very narrow with a pinnacle of rock in the middle of the channel. We had no chart of the entrance except for a two-inch square let-in on a large chart, but the captain of one of the factory ships gave me a drawing and advice on how to get in. It was not funny. When we arrived off the island the entrance was blocked by a small iceberg and we had to wait until it drifted clear before we went in. It was extremely difficult."* [3] *(Photo used with permission of Furness Withy Chartering Ltd.)*

Wartime political moves had begun the year before. Great Britain and Argentina had already made formal claims to parts of the Peninsula Region decades earlier. Chile, however, had only flirted with an interest in Antarctica.[4] In 1940, Chile, like Argentina, was technically a neutral country in the war, and she saw this as an opportune time to step in, given that Britain seemed to be going under to Germany. Thus her government made its first formal claim to Antarctic territory, to "All lands, islands, islets, reefs of rocks, glaciers [pack-ice], already known or to be discovered, and their respective territorial waters, in the sector between longitudes 53° and 90° West. . . ."[5] This was a region that overlapped much of Britain's 1908 Falkland Islands Dependencies (FID) claim.

Chile took no further immediate action after issuing her claim. Argentina, however, employed another tack. In November 1941, she declared the Laurie Island base in the South Orkneys an official Argentine post office. Two months later, she dispatched the naval vessel *1° de Mayo (Primero de Mayo)* south to establish an expanded claim. Chilean government representatives were aboard, a reflection of the one thing

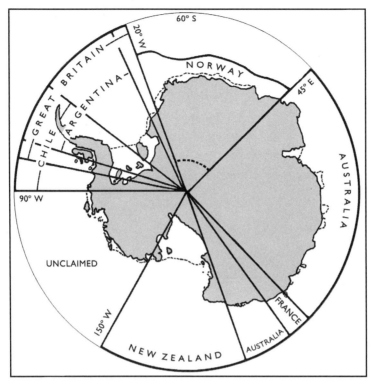

Territorial claims in the Antarctic Regions south of 60° S (the Antarctic Treaty region) — *Seven countries have claimed territory in the Antarctic regions. The four claims outside the Antarctic Peninsula Region—by Australia, France, New Zealand, and Norway—adjoin one another but do not conflict. In the Peninsula Region, however, the claims made by the three countries asserting sovereignty—Argentina, Chile, and Great Britain—overlap to a large degree. Britain also claims South Georgia and the South Sandwich Islands, both north of 60° S. Argentina counters this with a claim of her own to these islands, although she lost an armed fight for them during the 1982 Falklands/Malvinas War (See Chapter 19). The Antarctic Continent from 90° W to 150° W remains unclaimed.*

regarding Antarctica that Chile and Argentina agreed upon, that a "South American Antarctic exists and the only countries with exclusive rights of sovereignty over it are Chile and Argentina."[6]

The *Primero de Mayo* called at Deception Island in early February 1942 and left a bronze cylinder containing a document that formally claimed for Argentina everything below 60° S between 25° W and 68° 34' W (later expanded to 74° W). Her men also painted the Argentine colors on a wall of one of the vacant whaling station buildings. The ship then sailed south, down the west coast of the Antarctic Peninsula. On February 20, a group of men landed on one of the Melchior Islands (a small group in the Palmer Archipelago), raised the Argentine flag, and installed another bronze claiming tablet. A final landing, flag raising, and claim followed at Rymill's hut on Winter Island. Argentina had now fully joined Chile in claiming much of Britain's FID. For the time being, however, she kept silent about her actions.

But word soon leaked out about the *Primero de Mayo*'s voyage, and what ensued became rather a game of tit for tat. In January 1943, the British sent their own warship, the *Carnarvon Castle,* to Deception Island to check on the rumors. The sight of the painted Argentine flag immediately confirmed the reports. The *Carnarvon Castle*'s men painted over the flag, removed the bronze cylinder and all other traces of the *Primero de Mayo*'s visit, and then raised the Union Jack. The captain also nailed a notice to a building stating that the whaling station had become British government property when the company's lease had lapsed. The British then formally returned the *Primero de Mayo*'s Deception Island claim document to Argentina.

Argentina responded in turn. In February 1943, the *Primero de Mayo* sailed again, first to the Melchior Islands and Port Lockroy to leave new claims. The ship then continued south to Marguerite Bay, where her men spent two days at the abandoned United States Antarctic Services (USAS) base. Although the Argentines found the huts well stocked and in excellent shape, things were quite different by the time they left, because they had removed so much. Indeed, they took far more than Black's letter had asked be sent back to the United States. (Eventually, the Argentines did follow through with Black's request for portions of what they had gathered. The first items reached the United States in December 1944.) On her return voyage, the *Primero de Mayo* anchored for a few days at Deception Island. There her men removed the British emblems left by the *Carnarvon Castle* and repainted the Argentine flag, this time on a whale oil tank. They also installed a bronze claiming tablet to replace the cylinder the British had taken.

Britain's counterpunch was more substantial. In May 1943, her War Cabinet decided to send an expedition to establish permanent bases in the FID to reinforce British claims. Initially highly secret, the venture was code-named Operation Tabarin after a famous Paris nightclub, the Bal Tabarin. (The name was chosen by two of the organizers because "when this operation started . . . we had to do a lot of night work and the organization was always so chaotic just as the club. . . ."[7] Expedition members, however, had a caustic explanation of their own for the choice, that it was because "like a night-club audience, the expedition members were kept continually in the dark by the organizers."[8]) The Admiralty appointed James Marr, the Boy Scout who had sailed with Shackleton's *Quest* expedition and had later served as a biologist with the Discovery Investigations, as expedition commander. Marr was to establish two bases—one at Deception Island and the other, the more important one, at Hope Bay on the Peninsula mainland.

The Norwegian government-in-exile in London supplied the expedition ship. The British rechristened her H.M.S. *Bransfield*, a highly symbolic act, given that Bransfield's 1820 claims were a key underpinning of Britain's claim to the FID. The expedition began badly when damage from a wild shipboard party delayed the departure by two months. Then, when the *Bransfield* at last sailed in November, she sprang a fatal leak almost immediately and barely reached safe harbor before sinking. Finding a replacement in wartime posed a serious problem until the Falklands governor came to the rescue. He offered the *William Scoresby*, the Discovery Investigations ship, then based at Stanley as a minesweeper. Since she was too small to carry everything, Operation Tabarin also chartered the S.S. *Fitzroy*, a vessel owned by the Falkland Islands Company (the dominant organization in the islands). Military troop ships transported the expedition personnel and gear to Stanley.

The *Scoresby* and the *Fitzroy* left the Falklands on January 29, 1944, with the fourteen men of Operation Tabarin. All had signed on for two plus years in the south. They reached Deception Island five days later, after sailing the entire way with lights out and radios silent in case German raiders were nearby. No Germans greeted them, but they did find the flag the *Primero de Mayo*'s men had painted ten months earlier. Tabarin's men raised the Union Jack in response, leaving the Argentine colors to be painted over the following summer. Establishing the base at Deception Island was simple: the five-man wintering group just moved into an old whaler dormitory.

The two ships then headed for Hope Bay. Although the *Scoresby* pushed through the belt of pack ice guarding the bay almost casually, the captain of the *Fitzroy* flatly refused to take his ship in. That left Marr no choice. He had to find an alternative site. He eventually settled his nine men at Port Lockroy, the same locale Charcot had rejected in 1904. Marr, too, thought the place less than ideal, but his reasons were political rather than scientific. The Port Lockroy base was on an island, not the mainland, and as Andrew Taylor, Marr's second-in-command, observed years later, "If you're going to claim a continent, you want to be on it, not 10 miles away from it."[9]

The first winter, both bases did a little science, primarily meteorology, and several men at Port Lockroy made a sledging trip on Wiencke Island. Otherwise, they were simply there, in place for political reasons.

The problems with ice at Hope Bay in February 1944 led to a search for a stronger ship to join the *Scoresby* and *Fitzroy* for the second season. The British found her in Newfoundland, a sturdy sealer named S.S. *Eagle*. Labrador provided another important addition to the expedition, 25 sledge dogs. These animals would inaugurate nearly 50 uninterrupted years of British dog sledging in the Peninsula Region.

The *Scoresby* delivered fresh provisions to both bases in early November 1944. When she returned to the Falklands, Marr went along to consult with the expedition organizers about the second year. His plan was to move most of the Port Lockroy men to new bases at Hope and Marguerite Bays. He would personally head up the latter station in addition to continuing as overall Operation Tabarin commander. Taylor would be in charge at Hope Bay.

The *Scoresby, Fitzroy,* and *Eagle* left the Falklands in late January 1945. Aboard were Marr, returning to rejoin the men in the south, several new men, the dogs, and the Falklands governor. After the ships reached Deception Island on January 27, the *Scoresby* sailed on to Port Lockroy and collected Taylor and most of the base's 1944 wintering party. Everything was moving along smoothly until February 8. Then things unraveled. That day, the Falklands governor stunned Taylor by asking him to assume Marr's role as overall expedition leader. Marr, he said, was resigning because of ill health and would be returning north. The plan for the Marguerite Bay base was canceled, and everyone who was to have gone there would join the Hope Bay team.

Three days after the governor's bombshell, the *Eagle* left Deception Island to establish the Hope Bay base. The small sealer was crammed full, so much so that Harold Squires, the 20-year-old wireless operator, wrote, "We looked more like a replica of Noah's Ark than a ship on a secret Admiralty mission, especially with the 25 dogs barking and howling."[10]

Ironically, Hope Bay was largely free of ice that year. After Taylor located a good site about half a mile from the ruins of J. Gunnar Andersson's 1903 hut, his team spent more than a month building three new huts. When they were finished, they took a wooden pole from the muck about the Swedes' structure and mounted it on one of their own building as a flagpole.

The Operation Tabarin base at Hope Bay, still a work in progress — *The flag on one of the smaller buildings flies from a pole salvaged from the ruins of Andersson's hut. (Photo by I. Mackenzie Lamb, 1944. Reproduced courtesy of British Antarctic Survey Archives Service. Ref. no. AD6/19/1/D165/26. Copyright: Natural Environment Research Council)*

The *Eagle* remained offshore during most of the construction. Storms and floating ice made her position precarious at times, but her captain, Robert Sheppard, coped successfully until mid-March. Then a violent storm hit just after the ship returned from Deception Island with a final load of supplies. The wild weather snapped first one anchor line and then the other as ice battered the ship and smashed her bow. Squires radioed the shore party, "We have . . . almost given up hope of saving the ship. . . . The skipper is thinking of beaching her. . . . Will you please go to 'Handy Cove' with ropes and stand by. . . ."[11] David James, the young Hope Bay assistant surveyor, wrote,

> . . . what a desperate expedient! It was blowing gale force, 'Handy Cove' is but a rock-strewn beach and the ship would be certain to rip her bottom out. It would be hard to get anyone ashore in that wind and sea and if she broke up before the gale moderated, those on board would stand a very poor chance indeed. . . . [sic] The first sledge had just started when [the radio operator] came up again . . . to say that the Captain had decided to run for [the Falklands]. . . . Slowly the gallant little ship steamed by, her sides deep coated with ice and her bow stove in.[12]

The *Eagle* did make it back to Stanley, but it was a very near thing.

Operation Tabarin now consisted of three bases—Deception Island and Port Lockroy, each with four men, and Hope Bay, with thirteen men and all the dogs. The second year was quite different from the first. Not only did the men do much more scientific work and exploration, but after Germany surrendered in May 1945, communications with home became significantly more open and more frequent.

On August 8, Taylor, James, and two others left Hope Bay with the dogs for a journey down the east coast of the Peninsula. Only Taylor had any sledging experience, and none of the men was a dog expert. Learning as they went, they found the trip far more difficult than they had anticipated. They reached a point a few miles south of Snow Hill Island before turning back on the 23rd. Several days later, they stumbled on a small depot from Nordenskjöld's expedition. The 43-year-old provisions were a most welcome supplement to their own dwindling supplies. And

The battered bow of the S.S. *Eagle* with makeshift repairs made at sea en route to Stanley — *After limping away from Hope Bay, the ship just survived the night before the storm at last abated. Then the men set to work on temporary repairs. Squires, the radio operator, wrote, ". . . we collected every spare blanket and sleeping bag and stuffed them into the gaping hole the iceberg had punched in our bow. Chippie, the carpenter, was lowered over the bow in a bo'sun's chair to try and stretch a canvas shield over the damaged area. Twice, as the* Eagle *dipped into the waves, Chippie was thrown back onto the deck, but undaunted by the waves, went over the bow again to continue his work. Before the day was over, the canvas shield was in place . . ."*[13] *(Photo from Squires,* S. S. Eagle . . . , *1992)*

then they reached the Snow Hill Island hut itself. There was, James wrote, "a touch of unreality about actually visiting the spot—rather as if one were taken to have tea with Sherlock Holmes in Baker Street—and apart from this aspect, we were looking forward to the good eats and comforts which we were sure awaited us." What they found instead was a great disappointment, "as though one had been invited to a wedding and gone to a funeral by mistake. Over everything hung an air of indescribable gloom. . . ."[14] Although the building was structurally sound, the windows were gone; ice filled the inside; and there was no food anywhere.

A few days later, Taylor's men located a depot the *Uruguay* had left on Seymour Island. Unlike the hut, the depot was full of supplies. The irony was splendid. This inadequately provisioned British team, attempting to establish sovereignty over an area also claimed by Argentina, was taking food from a depot laid down decades earlier by an Argentine party. Taylor's men, however, saw it simply as another link to Nordenskjöld's saga. And besides which, they needed the food. When they reached Hope Bay on September 7, they had traveled about 270 miles and learned a great deal about sledging, the hard way.

After World War II

Explorers returned to Antarctica almost immediately when World War II ended. This time, the governments led the way, and the war-inspired technology they brought with them—long-range aircraft, helicopters, icebreakers, effective overland vehicles, and improved communications equipment—revolutionized Antarctic exploration.

In the Peninsula Region, Operation Tabarin was officially terminated in July 1945. The bases remained occupied, however, because the Falkland Islands government assumed responsibility for them under a new structure, the Falkland Islands Dependencies Survey (FIDS). Although they had lost their military purpose, these stations were still to be symbols of British occupation. And now, much more than under Operation Tabarin, they were to be centers for exploration and scientific work. Continuous, uninterrupted human habitation of the Peninsula Region beyond the South Orkneys and South Georgia had begun.

FIDS relieved the three Operation Tabarin bases in 1945–46 and established two new stations. One was at Stonington Island, a few hundred yards from the old USAS base. The other was on Laurie Island in the South Orkneys, not far from the existing Argentine operation. Edward Bingham, who had been Rymill's doctor and chief dog man, was in overall command as well as base leader at Stonington Island. On the way to Marguerite Bay, he stopped at Winter Island and briefly considered reoccupying the old hut. Luckily, he decided against it, because the structure would disappear by the next summer, probably swept away by the same tsunami that was recorded at nearby Port Lockroy on April 2, 1946.

Argentina and Chile each took careful note of British activity in their mutual claim area. In the summer of 1946–47, both countries sent their first major expeditions south. Argentina opened her second Antarctic base, a year-round station in the Melchior Islands in the Palmer Archipelago. Chile established her first Antarctic base, on Greenwich Island in the South Shetlands.

Argentina had already dispatched many summer voyages south, but for Chile, the 1946–47 venture was the first since the *Yelcho* had rescued Shackleton's Elephant Island party in 1916. After constructing the Greenwich Island base, the Chileans continued on to Stonington Island, landing a party there on February 21. When the Chileans visited the British, the two groups exchanged formal letters protesting each other's presence, and then partied together. Regrettably, many members of the Chilean crew also looted the vacant USAS huts, much to the anger of the British, who had worked hard cleaning them up after their long abandonment and the 1943 Argentine depredations.[15]

OPERATION HIGHJUMP

Another country was also in the area in February 1947. The U.S. government's Operation Highjump, the most massive expedition ever sent to Antarctica, had sailed south in November 1946 with thirteen ships (including the first two modern icebreakers to reach the Antarctic), numerous aircraft, and 4,700 men. Although the effort concentrated on areas other than the Peninsula Region, three expedition ships under George Dufek did sail east through the Bellingshausen Sea and reach Alexander Island in early February. Expedition seaplanes made several survey flights over Alexander and Charcot islands, during which the Americans dropped claim papers over land that the Argentines, British, and Chileans had already claimed. Dufek also attempted to reach the shore of Charcot Island in a motorboat. Drifting sea ice forced him to turn back 500 yards out. Several days later, high winds defeated an attempt to enter Marguerite Bay. When the fleet meteorologists forecast a massive storm for the entire Antarctic Peninsula area, Dufek ordered his ships to head north, round the Peninsula, and sail into the Weddell Sea. There, he hoped, he could push far enough south for his seaplanes to reach the southern coast of the Weddell Sea. Only the men of Filchner's expedition had seen any part of it, and that merely a few miles to the west of Vahsel Bay. From there to the Antarctic Peninsula was still a mystery, one of the longest unseen coasts remaining in Antarctica.

The storm developed into a real monster. As the ships rounded the Peninsula, screaming winds and furious seas slammed the small fleet, but after several hellish days, the ships at last reached the South Orkneys on February 19. From there, Dufek headed south, but when the ships reached 66° S, still at least 100 miles north of where he wanted to be for his planes to operate, the ice was hopelessly dense. And so he turned east. By the time he finally found a way past the Antarctic

Circle, the fleet was at 0° W, far from the area Dufek had hoped to fly over. Still, on February 28th, after sailing back a bit westward, he put his two planes in the water. Only 50 miles after takeoff, clouds blanketed the sky, but the pilots flew on, trying to rise above the murk. One climbed to 14,000 feet but could see nothing below. Eventually both pilots gave up and returned to the ships. Despite Dufek's best efforts, the gap on the map at the base of the Weddell Sea remained. Someone else would have to fill it in.

As it happened, the man who would succeed where Dufek had failed was already on his way to Antarctica as the Operation Highjump ships headed north in early March. And he was someone whom George Dufek knew well, with whom he had wintered on Richard Byrd's 1933–35 expedition and had shared a ship with on the 1939–41 USAS expedition.

Finn Ronne Returns to Stonington Island, Bringing Two Women for the Winter

The man who would at last see the entire southern border of the Weddell Sea was Finn Ronne, the USAS East Base veteran who had explored George VI Sound in 1940–41. In March 1947, Ronne arrived at Stonington Island with his own Ronne Antarctic Research Expedition (RARE), intending to reoccupy the old USAS buildings.

RARE was a private expedition with substantial government backing. The U.S. air force lent two pilots and three airplanes, and the Office of Naval Research contracted for scientific work. The navy also provided an oceangoing wooden tug, rechristened by Ronne the *Port of Beaumont,* and the army contributed two weasels, all-terrain tracked vehicles developed during World War II.

The U.S. government supported Ronne in another fashion as well. The British, unhappy to learn that Americans planned to reoccupy the USAS base, officially protested to the U.S. State Department. Not only was the USAS base in territory claimed by Britain, they insisted, but the resources of Stonington Island were inadequate for two expeditions. Ronne, however, had no money for a new base elsewhere, and the American government, backing him up, refused to forbid him to use the USAS huts. (Chile was far less troubled by the thought of Americans temporarily occupying an area she had claimed. Indeed, she saw it as an opportunity to reassert her own rights, at least symbolically, and granted Ronne's party official visas for the Chilean Antarctic.)

Ronne, now on his third Antarctic expedition, had several other Antarctic returnees with him. These included East Base veteran Harry Darlington, as RARE's chief pilot, and Darlington's large pet husky, Chinook, the former puppy young Harry had rescued in 1941. In all, there were 23 people in the party, two of them women who would be the first of their gender to winter deep in the Antarctic.

Ronne's breaking the Antarctic gender barrier came about almost accidentally. His wife, Edith (Jackie), and Darlington's wife, Jennie, a new bride, had accompanied the expedition to Chile, both of them planning to return home from there when their husbands continued on to Stonington Island. At the last minute, Ronne decided to take Jackie south to help with press dispatches. Some of Ronne's men were so upset at the idea of having a woman accompany the expedition that they threatened to quit. But when they realized that they were in the minority, they compromised, saying that two women would be better than one. Ronne accepted the proposal, and Darlington, who was himself opposed to taking any women, told Jennie that she had been drafted. She and Jackie then quickly organized themselves for an unanticipated winter in the Antarctic.

When RARE reached Stonington Island on March 12, 1947, the British welcome was far warmer than Ronne had anticipated. In fact, Kenelm Pierce-Butler, leader of the FIDS base there, was on the shore to greet the first boat in from the *Port of Beaumont*. In contrast, the state of the USAS buildings, so recently visited by the Chileans, was a nasty shock. It took weeks for Ronne's people to make them habitable. In the meantime, all lived aboard ship.

Unfortunately, one of Ronne's first acts flew directly in the fact of Pierce-Butler's cordial welcome: he raised a large American flag on the old USAS flagpole. That roused British hackles once again, leading to a formal note asking whether the flag implied a territorial claim. If so, the British leader said, he would have to lodge a protest. Ronne replied formally in turn. His note ended disingenuously, "As an American expedition reoccupying this base on Stonington Island, we have reflown the American flag on the American-built flagpole at the American camp."[16] For some time after this exchange, relations between the two parties were distinctly cool. Common sense gradually took over, though, and the RARE and FIDS groups developed a cooperative working relationship—one much to the benefit of both, since each had weaknesses that matched the other's strengths.

Ronne had planned to rely on dogs for sledging. Sadly, almost half his huskies had died en route to Chile, and his efforts to find replacements in Punta Arenas had garnered only a ragbag of unsuitable mutts. Jennie Darlington described the British reaction to these dogs in colorful terms: "When the hardy Britishers, dog lovers all, caught sight of our motley collection of undernourished Chilean canines, including a Spanish spaniel Corgi, an unknown variety of sheepdog, and the joyless 'nude' whippet . . . , they were convinced that the American expedition was a new variety of traveling polar sideshow."[17] The British, for their part, had numerous superb dogs. What Ronne had, in addition to his two surviving good dog teams, were the two weasels and three planes—equipment that was an invaluable complement to Pierce-Butler's one small Auster aircraft.

Regrettably, the growing accord between Ronne and the British was in stark contrast to relations within RARE. Ronne was a man who reacted strongly to anything that he perceived as a challenge to his authority, and severe personality conflicts soon erupted at the old USAS base, particularly between Finn Ronne and Harry Darlington, a sometimes difficult man himself. A few months after the RARE group reached Stonington Island, Ronne excluded Darlington from all expedition work. That forced the rest of the party to choose sides. Each wife naturally supported her own husband, and before long, the only two women in Antarctica were barely speaking to one another. The situation became so extreme that Jennie, who realized in late spring that she was pregnant, pointedly did not tell Jackie. Months passed before either Ronne learned of Jennie's condition.

Despite these interpersonal tensions, Ronne went ahead with an ambitious program. He began with several short sledge trips during winter. One was nearly

Two women and their husbands in the Antarctic — Top: *Edith "Jackie" and Finn Ronne, working in the RARE office at Stonington Island. Finn's first language was Norwegian, and Jackie was an immense help to him, not only in typing but also in writing and editing his press dispatches. (Photo from Ronne, F.,* Antarctic Conquest . . . , *1949)* Right: *Harry and Jennie Darlington. After Finn Ronne fired Harry from his position as chief pilot, the Darlingtons were left with a minimal expedition role while at Stonington Island. (Photo from Parfit,* Reclaiming a Lost Antarctic Base, *1993)*

fatal. In mid-July, following the USAS example, Ronne led a six-man team up to the Peninsula plateau to establish a meteorological station. It took two days of hair-raising travel in the winter darkness to reach the upland ice. Then a wild storm hit. Finally, by the eighth day out, the weather improved enough for Ronne and three other men to return to Stonington Island. Harris-Clichy Peterson, the expedition meteorologist, and Robert Dodson, the assistant geologist and surveyor, stayed to man the new camp.

Ronne lost radio contact with the two men on the plateau several days after he got back to base. Worried, he sent a plane to check. After the pilots failed to see the two men, Ronne decided to mount an air and ground search. Before he was ready to start, however, Dodson stumbled into the base alone. The storm had continued after Ronne's departure, he said, and the violent winds had shredded the tent. In desperation, he and Peterson had abandoned the camp to return to base. Then Peterson had plunged into a crevasse. When Dodson realized it would be impossible for him rescue Peterson—who was wedged in 100 feet down—without help, he left flags and Peterson's skis, to mark the spot. Dodson then collected himself, made his own tense traverse of what they would later name "Pete's Crevasse," and hurried down to the base for help. When he arrived with his story, he found the RARE team entertaining the British for Saturday movie night. Pierce-Butler, the FIDS leader, immediately offered help from his experienced crevasse rescue teams. While the British were getting organized, Dodson turned right around and headed back to Peterson with two RARE men. The weather had improved, but shadows cast by the pale moon and star light made it difficult to pinpoint Peterson's location. Finally, however, the rescuers spotted Dodson's markers, and, looking down into the depths, could see Peterson, still jammed in the same position he had been in when Dodson had left him hours earlier. Together the RARE and FIDS men tied a rope to Richard Butson, the British doctor, who was the smallest man in the group and carefully lowered him down to the trapped American. Butson then tied a rope to Peterson, and those above hauled both men out. Peterson, who had been in his terrifying icy prison for ten hours, amazingly suffered no lasting injury from the experience.

Another sledging party eventually established the plateau weather station at the end of August. From then on, the station operated continuously until the completion of RARE's aerial work. Ronne also created an advance base at Cape Keeler on the east side of the Peninsula, about 100 miles southeast of Stonington Island, to support field sledging and flying operations.

Establishing the Cape Keeler base was difficult, and as with the weather station its beginnings included a near disaster. On September 15, the British Auster took off with three FIDS men—Reginald Freeman, Bernard Stonehouse,

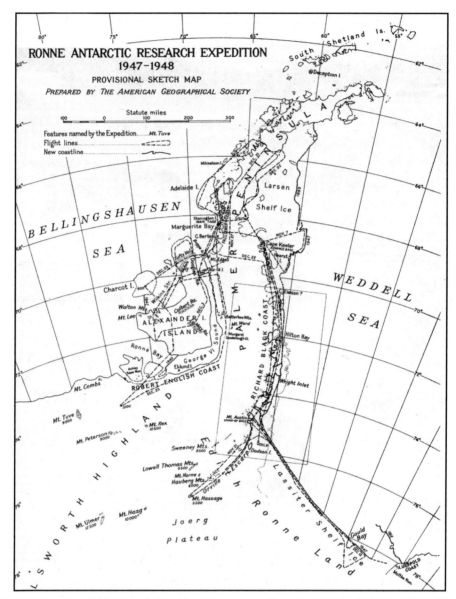

The Ronne Antarctic Research Expedition major exploration/survey flights — *This official expedition map vividly displays the complexity of the expedition flight operations, but even the intricate web of flight tracks shown here is a highly simplified record. In all, there were 86 aircraft landings in the field, half in virgin territory. The three expedition aircraft flew 37,000 miles, surveyed more than 250,000 square miles of previously unseen territory and returned with 14,000 aerial photos covering both sides of the Antarctic Peninsula, Alexander Island, and the Weddell Sea ice shelf area. Particularly notable among the flights were two lengthy exploratory ones that started on the east coast of the Antarctic Peninsula. The first of these headed far south, the second south and then southeast along the never-before-seen far southern reach of the Weddell Sea. The most important sledging trip—not shown on this map—was the joint RARE/FIDS operation that crossed the Antarctic Peninsula and traveled well down the west coast of the Weddell Sea. (Map from* Ronne Antarctic Research Expedition 1946–1948, *U.S. Air Force Report, Washington, D. C., 1948)*

Tommy Thomson—to lay out a landing strip for Ronne's much larger Norseman plane, which was carrying the base supplies. Unfortunately, the planes lost radio contact in flight. When Ronne's men arrived at the appointed rendezvous and saw no Auster, they returned to Stonington Island. No one there knew where the British plane was.

Over the next week, RARE pilots James Lassiter and Charles Adams made numerous unsuccessful flights searching for the Auster. Then, on September 22, Lassiter shifted his search to the west side of the Peninsula and at last spotted the missing men, trudging along on the Marguerite Bay sea ice, about 30 miles south of Stonington Island. They were starving and on the verge of collapse but otherwise unhurt when Lassiter set down and heard their story. They had landed farther inland from Cape Keeler than originally agreed, and the Norseman had flown right over them. When the FIDS men realized that the Americans had missed them, they took off to return to Stonington Island. Then the pilot, Thomson, lost his way in a blizzard and crash-landed after reaching Marguerite Bay. With no way to radio for help because the accident had wrecked their transmitter, the men decided to walk back. The next week was dreadful. They had almost no food, minimal survival equipment, the weather was brutal, and the surfaces underfoot were worse. When Lassiter found them, they had covered barely 20 miles.

Ronne needed the Cape Keeler base for his ambitious aerial exploration plans, but his sledging program got underway first because the same uncooperative weather that had plagued the USAS had forced him to postpone his long exploratory flights. The first sledging team, using RARE's dogs, spent nearly three months traveling some 450 miles about Marguerite Bay and on Alexander Island for geological work, accompanied part of the time by two FIDS men.

The other sledge trip was a joint RARE/FIDS operation. The British supplied all of the sledge dogs, the Americans the air support. A four-man advance team crossed the Peninsula using the route the USAS had pioneered. Pierce-Butler then took over at Cape Keeler and led a team of two FIDS men and two Americans south, picking up supplies from depots laid by RARE's planes. The weather gods smiled on them, and the sledging surfaces were nearly as friendly. On December 13, the men turned back just short of 75° S, a bit over 200 miles beyond the USAS Weddell Sea sledging party's ultimate point. Men and dogs reached Stonington Island on January 22, 1948, after a round-trip of about 1,200 miles.

In the meantime, Ronne had begun his aerial exploration. On November 21, he made the first of several long flights. It began with the cargo-carrying Norseman heading south from the Cape Keeler advance base with five drums of fuel. The Beechcraft followed. The two planes landed several hundred miles to the south, where the crew transferred half the Norseman's fuel cargo into the Beechcraft. Then Ronne, with Bill Latady operating a camera and Lassiter at the controls of the Beechcraft, took off into the unknown, following the east coast of the Peninsula south and beyond to about 77° 30' S, 72° W, over the main body of the Antarctic Continent.

Ronne made his second important flight nearly three weeks later. On December 12, again employing the Norseman to advance the starting point well south of Cape Keeler, he flew the Beechcraft southeast, tracing the ice shelf at the southern edge of the Weddell Sea for hundreds of miles to its eastern edge at Vahsel Bay. In addition to an ice shelf, he thought he saw a substantial stretch of land below him. He named the area surveyed—from the farthest west seen on November 21 to the farthest east on December 21—the Lassiter Ice Shelf, for his chief pilot, and Edith Ronne Land, for his wife, Jackie. (There is, in fact, no land in much of the area Ronne flew over. The name Edith Ronne Land has now disappeared from the map, replaced by the single name Ronne for the ice shelf that fills much of the area he saw. Lassiter's name was transferred to a major stretch of the mainland coast.)[18]

RARE's final significant flight took place on December 23, to the southwest down and well beyond the southern end of George VI Sound. Previously unknown mountains loomed in the distance and Ronne dished out names as he plotted them on his charts.[19] Then, after landing on the ice at about 74° S, 79° 35' W, he turned back and flew northwest to Charcot Island for another landing. This was the first time anyone had set foot there.

Ronne originally planned to leave the south in mid-March. The Marguerite Bay ice, however, remained stubbornly solid and even thickened in February. But this time, there was an alternative to an air evacuation. Icebreakers had arrived in the Antarctic. Two from America's Operation Windmill, an expedition working principally outside the Peninsula Region, came to the rescue and freed the *Beaumont* in late February. The expedition's escape from Marguerite Bay was a major relief for one expedition member, Jennie Darlington, who by then was visibly pregnant. (Her daughter, Cynthia, would be born in the United States in July 1948.[20])

As he sailed away, Ronne knew that his privately funded expedition, which had worked so cooperatively with the British, had accomplished a great deal. His three planes had made 86 landings in the field, half in virgin territory, and had surveyed some 250,000 square miles of never-before-seen territory, including the southern border of the Weddell Sea, one of the largest remaining gaps on the Antarctic coastal map. The sledging had also been highly productive, as had the more localized scientific work. But despite all these important results, the RARE expedition is probably best remembered for its two pioneering women, neither of whom had sought the honor.

Politics and Conflict

Ronne's departure in February 1948 left Peninsula Region exploration to the three claimant countries, none of which had accepted the others' positions. In

July 1947, Argentina and Chile signed a joint declaration on the rights of both countries in the south. A few months later, the British ambassadors in Buenos Aires and Santiago delivered formal protest notes against Argentine and Chilean "Acts of Trespass" in the FID in the 1946–47 season. The notes reasserted British sovereignty and invited Argentina and Chile to refer the dispute to the International Court of Justice at The Hague. Britain agreed to accept whatever decision the Court made. Chile and Argentina flatly rejected the proposal, both then and later, when Britain repeated the offer. In the meantime, all three claimants went on adding bases.

In the summer of 1947–48, Argentina threw naval maneuvers in the far south into the mix. It was a huge operation, with fifteen ships, five admirals, newspaper reporters, and multiple special guests. A smaller fleet reinforced the Argentine presence during the 1948 winter.

Chile also upped the ante that summer. When her president, Gabriel González Videla, paid an official visit to Chile's claim, he was the first head of state to see the Antarctic. The official party included Videla's wife and daughters, several government ministers, the commanders-in-chief of the Chilean navy and air force, senators, and a flock of journalists and press photographers. On February 17, 1948, this group inspected Chile's first Antarctic base, on Greenwich Island. The next day, President Videla formally opened Chile's second station, on the tip of the Peninsula mainland; personally reiterated Chile's claim to Antarctic territory; and proclaimed the name of the Antarctic Peninsula to be Tierra O'Higgins, after Chile's first president, Bernardo O'Higgins. (Argentina had its own name for the Peninsula, Tierra San Martín, after her own hero.)

Argentina's naval maneuvers, Chile's presidential visit, and Britain's repeated protests ratcheted up the tension in the Peninsula Region. None of the claimants was willing to back down, but neither did any of them want a shooting war to develop. In late 1948, the three governments formally agreed that they would not send naval vessels south of 60° "apart . . . from routine movements such as have been customary for a number of years."[21] This tripartite agreement would be renewed every year after 1948 until 1959, when the Antarctic Treaty rendered it superfluous. But to no one's surprise, the interpretation of this declaration would be loose. Both Argentina and Britain sent individual warships into the region nearly every year after 1948, and some of these voyages were far from routine.

The closest the three countries came to armed conflict in the area was an event that took place in early 1952. Argentina had begun construction of its Esperanza base at Hope Bay that summer. When a FIDS team arrived at the beginning of February and landed near the new Argentine base to reestablish its own Hope Bay base, closed in early 1949 after a fatal fire had destroyed the main building, the Argentine commander informed them that he had been ordered "to prevent you from building a base here, using force if necessary."[22] He then commanded his men to shoot over the heads of the "invading" British and ordered them to return to their ship.

The unarmed British reluctantly retreated and then, once aboard their ship, radioed the news to the Falklands. The Falkland Islands governor Miles Clifford ordered the captain of the British ship to stay right where he was. Clifford also advised the Colonial Office in London of the situation, and the British government immediately responded with a strong protest to the Argentine government. The latter quickly countered that it was all a mistake, blaming the Esperanza commander for overreacting and exceeding his orders. In the meantime, Clifford, without waiting for directions from his superiors, had already summoned a naval vessel and sailed for Hope Bay. When he arrived, he sent a boatload of marines ashore to confront the Argentines. Fortunately, the Esperanza men had already received orders to back down and had retreated inland. That took the steam out of the situation. The British went about reestablishing their base, and Argentina soon replaced their scapegoated base leader.

Another potentially dangerous incident occurred the following summer. This time, it was the British who took aggressive action in defense of an established base. In January 1953, both Chile and Argentina installed new refuge huts on Deception Island, only a few hundred yards from the British base. Since Argentina already had a year-round station on the island, its practical reason for the new building was unclear. Chile, however, built its structure with a view to later expansion. When it learned about the huts, the British government ordered Clifford, still the Falklands governor, to remove them and deport their occupants.

Without warning Argentina or Chile, Clifford dispatched the warship *Snipe* to Deception Island. A magistrate, two Falklands constables, and fifteen marines landed there on February 15 and dismantled the huts. They also arrested the two Argentines in residence. The *Snipe* then sailed to South Georgia, where the magistrate turned the prisoners over to an Argentine ship bound for Buenos Aires. At that point, the British ambassadors at Buenos Aires and Santiago presented formal notes to the Argentine and Chilean governments detailing the *Snipe*'s actions and protesting what Britain called an infringement of her sovereignty. Both Argentina and Chile replied on February 20 asserting *their* claims to the territory and protesting the British actions. After all three governments agreed to hush up the entire matter, the conflict went no further.

The events of the 1951–52 and 1952–53 summers had had explosive potential, but cooler heads had eventually prevailed. After that, the three competing nations reached an uneasy accommodation, even as they added more bases. By 1954–55, Argentina and Britain had eight bases each in the Peninsula Region and Chile had four. Argentina was the most forthright about her intentions. In April 1954, President Juan Domingo Perón announced that his policy was to "saturate

Antártida Argentina with Argentine occupants."[23]

The most populated place in the region, just as in the early days of the whalers, was Deception Island. Although neither Argentina nor Chile rebuilt their refuge huts, Argentina and Great Britain remained firmly planted at their year-round bases on the island, and Chile established the island's third full station in 1954–55. It was only a few miles from the FIDS base, but this time the British did nothing more than register a formal complaint. Despite the continued official political wrangling, the men at the three bases got along well, exchanging social visits, trading food and drink, and helping each other out when the need arose. And for years, they personally settled Deception Island's sovereignty with a series of soccer matches and darts championships that entitled the winning team to claim the island in its country's name. Until the next match.

CHAPTER 18

The International Geophysical Year and the Antarctic Treaty: 1955–1959

While the Peninsula Region was becoming home to more and more bases during the early 1950s, an event occurred elsewhere that would have a major impact throughout the Antarctic. In 1950, several American and British scientists proposed that the world's nations undertake a third International Polar Year (IPY). The international scientific community enthusiastically embraced the idea, and countries around the world agreed to participate. The second IPY had taken place in 1932 and 1933, 50 years after the first IPY had brought a German expedition to South Georgia. Because of the Great Depression, the second IPY had been a very lean, almost entirely Arctic, program. In contrast, this new undertaking would focus its polar region work on Antarctica. Formally called the International Geophysical Year (IGY), it would extend for a year and a half, from July 1, 1957, to December 31, 1958. Twelve countries would participate with Antarctic region bases: the three already well established in the Peninsula Region—Argentina, Chile, and Great Britain—plus Australia, Belgium, France, Japan, New Zealand, Norway, South Africa, the Soviet Union, and the United States.

IGY field preparations began in the summer of 1955–56, when several countries set up new Antarctic bases. This included the United States in the Ross Sea Region and the Soviet Union on the coast and interior of East Antarctica. More countries arrived in 1956–57, and additional bases sprang up, among them the U.S. Amundsen-Scott Station, at the South Pole itself. In all, the twelve participating countries operated more than 50 bases on or near the Antarctic Continent, approximately half of them in the Peninsula Region. By historical standards, Antarctica was experiencing a human invasion. From late 1955 to the end of 1958, literally thousands of men (and a minuscule number of Russian women) added their footprints to the Antarctic ice—more noncommercial visitors than had called in all the years since James Cook circumnavigated the continent in the early 1770s.

Although work connected with the IGY dominated most scientific/exploratory activity in Antarctica during the latter half of the 1950s, this was far more true of other parts of the Antarctic than in the Peninsula Region. There, Argentine, British, and Chilean bases predating the international effort simply added IGY tasks to existing programs. Only two significant Peninsula Region bases were established explicitly for the IGY, one by a British scientific organization independent of the FIDS program and the other by the United States. Both were deep in the Weddell Sea, as was Argentina's General Belgrano station, established in the summer of 1954–55 for both political and IGY purposes.

THE IGY IN THE PENINSULA REGION

Britain's Royal Society established the British IGY-specific base in the summer of 1955–56, a year before the IGY began, to allow time for everything to be readied for the official July 1, 1957, start of scientific work. The Royal Society chartered the *Tottan*, a World War II vintage former trawler and sealer, to transport the advance team. She left South Georgia the day after Christmas 1955 and headed for the Weddell Sea, with the goal of reaching as far south along the east coast of the Weddell Sea as possible. She reached northern Coats Land with little difficulty and then pushed south along the coast. On January 6, dense ice stopped her well short of where the expedition organizers had hoped to be. They settled for establishing the base at 75° 36' S, 26° 41' W, at a location they named Halley Bay, in honor of the British scientist who had touched the fringes of the Antarctic in 1700. No rock was visible in the area, but they thought the chosen location was on an ice piedmont, or at least a grounded ice shelf. They were wrong, as they soon learned. The Halley Bay base had been built on a floating ice shelf, with a surface that was advancing slowly but steadily toward the Weddell Sea.

Ten men spent the 1956 winter preparing the base for the work to come. The main 21-man scientific team arrived in early 1957, relieved the advance party, and began the IGY program of meteorology, geomagnetism, glaciology, and ionosphere observations.

In 1959, FIDS took over the Halley Bay base, originally planned to close at the end of the IGY, because the location was excellent for studying atmospheric phenomena. This work justified continued operation of what turned out to be a difficult base to maintain. Indeed, in later years, the British had to rebuild it several times as the steadily shifting ice sheet crushed each successive structure. By 2011, the base at Halley Bay—by now simply called Halley—was the sixth version there.

The summer after the British established the Halley Bay base, the United States set up its own Peninsula Region IGY station in the Weddell Sea. One of seven U.S. IGY bases, it was named Ellsworth Station in honor of the first man to fly across the Antarctic Continent. Establishing this base was not easy. In mid-December 1956, the cargo vessel *Wyandot* and the icebreaker *Staten Island* sailed into the Weddell Sea, bound for the chosen base site far south along the eastern coast of the Antarctic Peninsula. The *Staten Island* led the way as the ships first pushed southward into

The icebreaker *Staten Island* leads the cargo ship *Wyandot* through the dense pack ice of the Weddell Sea en route to establish Ellsworth Station. *(U.S. navy photo, from reproduction in Behrendt,* Innocents on the Ice, *1998)*

the Weddell Sea and then slowly, sometimes beset for days at a time, battled the ice west toward the Peninsula. It was a dangerously difficult voyage, especially for the vulnerable *Wyandot*. The ice punched a hole in her hull, bent her frame, and sheared the tips off all four of her propellers. Even the *Staten Island*, a vessel designed for the ice, lost an entire propeller blade. Despite the damage, the two battered ships penetrated the ice to within only a few miles of the junction of the Ronne Ice Shelf and the Antarctic Peninsula. But back at home, U.S. IGY organizers grew increasingly concerned, and thus ordered their captains to retreat before they reached the planned base site. Eventually, the Ellsworth Station team found an alternative site on the Filchner Ice Shelf, less than 80 miles west of Vahsel Bay.

Finn Ronne led the 39-man wintering party. Just as his RARE team had, this group experienced severe interpersonal problems. The disputes between Ronne and the scientific staff were particularly rancorous. Some of the men found themselves very much in sympathy with Jennie Darlington when they read *My Antarctic Honeymoon,* her recently published account of her experiences at Stonington Island. Ronne, for his part, was furious when he discovered them devouring it.

Unlike the British at Halley Bay, who focused their time and work on the immediate vicinity of their base, the Americans spent significant time in the field, both on the ground and in the air. Ellsworth Station boasted several aircraft. Ronne used them for aerial surveys, including a flight on October 21, 1957, during which he flew over mountains to the south of the Filchner Ice Shelf and dropped a claim paper, the last claim made by an American in the Antarctic.[1]

The most significant overland trip was what the participants called the Filchner Ice Shelf Traverse. On October 28, 1957, five scientists set out with two Sno-Cats. The purpose was to study the ice sheet itself, make magnetic and gravity measurements, and carry out seismic investigations. In addition to this scientific work, which was part of the official IGY program, they planned to explore and study the geology of the newly discovered mountains that Ellsworth Station flights had seen south of the ice shelf. When seriously crevassed areas made it difficult to find a way forward, the traverse team called for help, and planes from Ellsworth guided them from above. The trip, which ended on January 17, 1958, was highly productive. In the course of about 1,200 miles, the participants reached the land and mountains at the far south of the Filchner Ice Shelf, carried out the first important study of the ice shelf structure and the seabed beneath it, and discovered Berkner Island, an 85-mile-wide, 200-mile-long, ice-covered island embedded in what had been thought to be a continuous stretch of shelf ice at the head of the Weddell Sea. As a result of this last discovery, the portion of the ice shelf to the west of Berkner Island would be renamed the Ronne Ice Shelf, while the section east of the island would continue to be called the Filchner Ice Shelf, after its original discoverer.

Forty new men replaced Ronne's first-year team in January 1958 and continued with the base's IGY work in meteorology, geomagnetism, and ionosphere studies, among others, as well as carrying out another major overland traverse. At the end of the IGY, the Americans turned Ellsworth Station over to Argentina, leaving the three claimant countries once again the only ones with bases in the Peninsula Region.

Argentina's Belgrano station had been established two summers earlier than Ellsworth. Also on the Filchner Ice Shelf, it was situated about 35 miles east of the U.S. base. It had officially opened in mid-January 1955 and focused on meteorology during its first year. In 1955–56, the relief ship delivered a second aircraft. With the greater safety that two planes offered, the Argentines expanded their efforts to include aerial exploration. On the most significant of their flights, which lasted five hours and probed a significant distance inland, the pilot spotted two entirely unknown mountain ranges. Unfortunately, the Argentines never documented their sightings for the outside world. As a result, the discovery of what would be named the Theron and Shackleton mountain ranges by the British have typically been credited to the men from Vivian Fuchs's Commonwealth Trans-Antarctic Expedition, who saw these mountains on a flight of their own in early February 1956.[2]

Fuchs's expedition was responsible for the fourth new Peninsula Region base that participated in the IGY, one that cooperated actively with both the British at Halley Bay and the Americans at Ellsworth. Rather than a dedicated IGY operation, however, it was the Weddell Sea staging point for a non-government expedition that set out to make the first crossing of the Antarctic Continent. An abbreviated account of this complex expedition is included here because of its Weddell Sea location and its links to the *Deutschland* and *Endurance* expeditions.

THE COMMONWEALTH TRANS-ANTARCTIC EXPEDITION

Vivian Fuchs, the expedition leader, had been the FIDS leader at Stonington Island in the late 1940s. While stormbound on a sledging trip during the winter of

1949, he had come up with the idea of a much more ambitious journey. He would lead the first expedition to cross the Antarctic Continent overland. Fuchs eventually developed a plan to travel from the Weddell Sea to the South Pole and then on to McMurdo Sound in the Ross Sea. He would use motor vehicles to pull the loads, dogs to scout the way, and aircraft in support. A second team operating from McMurdo Sound would reconnoiter a route to the polar plateau and lay supply caches for the crossing party. As Fuchs readily acknowledged, this plan was essentially a modern version of Shackleton's original *Endurance* expedition program.

Most of the financial backing came from the four British Commonwealth governments that had an interest in Antarctica—Australia, Great Britain, New Zealand, and South Africa. Each made a monetary grant as well as providing logistical support. Fuchs's role was analogous to Shackleton's. He was overall commander, in charge of operations on the Weddell Sea side, and leader of the crossing party. New Zealand assumed responsibility for the Ross Sea team and appointed Sir Edmund Hillary, the 1953 conqueror of Mount Everest, to head the support effort.

Fuchs scheduled the crossing trip for the summer of 1957–58, in the midst of the IGY, but the expedition began operations two summers earlier to allow time to establish staging bases. In early November 1955, the expedition's chartered ship *Theron* left England carrying the 1956 Weddell Sea wintering party. Ken Blaiklock, who had been with Fuchs at Stonington Island, would be in charge. Fuchs and Hillary were along for the summer voyage. Vahsel Bay was the goal, a place that Fuchs, who had studied the *Deutschland*'s reports, was convinced he could reach. He knew, of course, that the *Endurance* had been lost in a similar effort. The *Theron*, however, had a far more powerful engine than her two Heroic Age predecessors, and she carried a seaplane for aerial reconnaissance. Fuchs put his trust in this new technology to get him through the ice.

The *Theron* met pack ice only one day after leaving South Georgia on December 21, 1955. Five days later, John Lewis, Fuchs's chief pilot, took off from the water to scout the ice to the south. The leads he spotted served the expedition well until strong winds closed both the leads and the open pools needed for aerial operations. After that, the ice controlled the *Theron*'s movements for nearly a month. She finally escaped using information from reconnaissance flights by planes from a British naval vessel that had sailed to the northern edge of the Weddell Sea pack to help Fuchs. On January 20, the *Theron* at last clawed her way to enough open water for her own plane to operate. Three days later, she was sailing through looser and looser pack. Having an eye in the sky had indeed made a difference.

Fuchs arrived at the just established Halley Bay base on January 27. Although it was farther north than he wanted to be, he was willing to consider stopping here for the sake of sharing the existing base. Lewis's reconnaissance flights, however, found heavily crevassed ice a few miles inland. That tipped the balance. Fuchs sailed on for Vahsel Bay.

Fuchs reached Vahsel Bay on January 30, 1956, 45 years less a day after Filchner had discovered the place. Both leaders faced the same key questions. Was there a base site here? And was there a reasonable route to the south? Filchner had never had a chance to investigate the second issue. Fuchs, however, had a plane that could look for answers from above. This time, Lewis returned with good news. The way south appeared passable for motor vehicles, and there was an excellent base site on the Filchner Ice Shelf just a few miles west of Vahsel Bay.

Unloading for what became Shackleton Base began on January 31. Unfortunately, weather and gathering sea ice defeated Fuchs's plan to spend

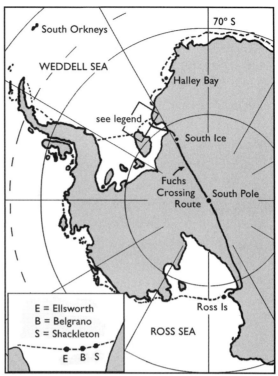

Vivian Fuchs's Commonwealth Trans-Antarctic Expedition route across the Antarctic Continent and the location of the four bases operating in the Weddell Sea area during the International Geophysical Year.

several weeks helping Blaiklock's advance team establish the station. Although everything had been unloaded when the *Theron* sailed away on February 7, the eight men who remained to winter still had a great deal to do. Not only were most of the stores still sitting on the bay ice, but the living hut remained unbuilt. Rather than spend time erecting the hut right away, Blaiklock's men fitted out a Sno-Cat crate as a temporary shelter. They used it for work during the daytime hours and slept in tents at night, all through the frigid winter until mid-September, when they at last finished the hut sufficiently to move in. As for the stores, the men worked hard to move them to safety, but before they could finish, disaster struck when a storm destroyed the bay ice and carried away 25 tons of coal, the materials for a wooden workshop, a tractor, 240 barrels of liquid fuel, and two tons of chemicals for the scientific program. Despite barely surviving a horrific winter, the advance party had accomplished a great deal by the time Fuchs returned with the full team the following summer. They had completed the hut; installed the radio and electrical equipment; trained the dogs; pioneered a route toward the south; and even found time for some scientific work.[3]

Fuchs's full Weddell Sea team spent much of the 1956–57 summer organizing at Shackleton Base and setting up South Ice, an advance base 300 miles to the south at an elevation of 4,430 feet. Fuchs established South Ice entirely by air with a small plane that carried one ton of supplies and gear on each of twenty trips. Blaiklock led the three men who wintered at South Ice in 1957. Fuchs headed up the sixteen at Shackleton Base.

Preparing for the trip to come and for scientific work kept the men busy during the winter. When spring arrived, Fuchs turned to his main objective, crossing the Antarctic Continent. On October 8, 1957, he and three other men set out to reconnoiter a surface route to South Ice. Reaching the advance base overland proved to be very difficult. Crevasses seemed to be everywhere, and two of the four vehicles had to be abandoned en route before the team completed its 37-day, 400-trail-mile trip. The men then flew back to Shackleton Base in less than three hours.

The crossing journey proper set out from Shackleton Base on November 24. That day, Fuchs and nine companions departed with two weasels, three Sno-Cats, a Muskeg

Fuchs's Sno-Cats and dog teams reach the South Pole with flags flying. *(U.S. Government National Archives / National Geographic Stock)*

tractor, and dog teams that went ahead to scout the way. Only 30 miles into the trip, the men found themselves once again in "a maze of crevasses," Fuchs wrote, "that could not have been more diabolical if they had been traps deliberately set."[4] Fuchs, who had devoted more than a month to pioneering an overland route, now faced an entirely new battle, because rising summer temperatures had weakened the crevasse bridges. By the time he reached South Ice for a second time, on December 21, it had taken him almost a month, and he had nearly lost the Sno-Cats in crevasses.

Fuchs left South Ice on Christmas Day 1957 with eight vehicles and twelve men. The South Pole was 555 miles away, along a route entirely over virgin territory. Fuchs's vehicles soon met another challenge—sastrugi, hard wind-created ridges of compacted snow that studded the plateau surface. As Fuchs's convoy struggled to advance even a few miles a day, Hillary, working from the Ross Sea side, completed his depot laying and went on to the South Pole. Fuchs reached the Pole just over two weeks later.[5] On January 19, 1958, he wrote, "we topped a snow ridge, and suddenly there it was—a small cluster of huts and radio masts: the United States Amundsen-Scott IGY Station at the South Pole."[6] One of several flags that Fuchs flew briefly while at the South Pole was that of the *Scotia*, raised in honor of William Speirs Bruce, the man who had made the first proposal for an overland crossing of the Antarctic Continent.[7]

The crossing party spent four days at the South Pole enjoying hot showers and iced beer—a surreal experience, given where they were. The remainder of the trip to McMurdo Sound was challenging but relatively uneventful, other than one medical emergency, dealt with by help from American aircraft. Fuchs reached Scott Base on March 2. There,

> as the Sno-Cats thundered and weaved between the ridges, escorted by a wide variety of vehicles, scores of figures stood, cameras clicking. . . . An improvised band from [the U.S. McMurdo station] . . . did their worst with our national airs, ending up with what we were told was 'God Save the Queen.' The band had been formed the night before by calling for all who thought they could play an instrument. 'It doesn't matter if you can play,' they were told, 'but you gotta be able to play loud.' And they certainly did.[8]

Fuchs had completed the first overland crossing of the Antarctic Continent, fulfilling Shackleton's dream. His expedition, however, was also a last, the final (quasi) independent, individual-leader-oriented, great Antarctic expedition. A new era was beginning, one in which government programs would dominate Antarctic activity. The change resulted from the costs of the new technology introduced to the Antarctic after World War II. And the situation intensified during the IGY, because that massive program introduced more and more complex technology and logistics to Antarctic work, technology that few but governments could afford. After the IGY, individuals interested in the Antarctic would have to find their niche with small private or semi-private expeditions.

THE ANTARCTIC TREATY

The International Geophysical Year changed humanity's relationship with Antarctica. The international effort, a huge success scientifically and cooperatively, produced an immense body of scientific work that laid the foundation for Antarctic research for years to come. The logistical techniques developed during the period made the continent far more accessible. Several governments that had established bases specifically for the IGY, notably the Soviet Union and the United States, decided to stay. And most important, the countries that had participated realized that they needed a long-term framework if they were to continue working together.

In May 1958, the United States invited the twelve countries with Antarctic bases to a conference. Despite the tensions of the Cold War, all attended and created what became the Antarctic Treaty. It was signed in Washington, D.C., on December 1, 1959, and came into effect in June 1961 when Chile, the final signatory country to do so, ratified it.

The advent of the Antarctic Treaty was a watershed event for the human story in Antarctica. For the first time, visitors to the south, no matter their purpose, would have to deal with Antarctica's challenges within a structure of laws that reflected a recognition that people had come to stay. The original Antarctic Treaty was a short and deceptively simple agreement with fourteen brief articles that applied to all land and attached ice south of 60° S. Its key points were that human activity in Antarctica should be restricted to peaceful purposes, and that scientific cooperation among national programs should continue. The Treaty negotiators sidestepped two potentially deal-breaking issues. The first was the question of mineral exploitation. The original Treaty dealt with it by ignoring it. The other was the question of sovereignty, of particular significance for the Peninsula Region with its competing claims. In effect, the Treaty froze things as they were. Article IV stated that the Treaty required neither the renunciation or recognition of any existing claims. It concluded,

> No acts or activities taking place while the present Treaty is in force shall constitute a basis for asserting, supporting or denying a claim to territorial sovereignty in Antarctica or create any rights of sovereignty in Antarctica. No new claim, or enlargement of an existing claim, to territorial sovereignty in Antarctica shall be asserted while the present Treaty is in force.[9]

Although the Treaty had no expiration date, Article XII stated that any member could request a review 30 years after ratification. When 1991 arrived, no one did, and the Treaty, which by then had evolved into a highly complex system incorporating supplementary conventions, annexes, and protocols, remained in force. As of 2011, it still does, now with a greatly expanded membership of participating countries. The nearly 50 countries that are now members of the Antarctic Treaty system are home to about 80% of the world's population.

CHAPTER 19

The Antarctic Treaty Era: Antarctica After 1959

The years following the signing of the Antarctic Treaty saw ever-increasing human activity throughout the Antarctic. In the Peninsula Region, Argentina, Great Britain, and Chile continued to operate multiple bases. Science was the official reason, but politics remained important. Despite the Treaty's freezing of claims, each of these three claimant countries wanted to be ready to support its position should the Treaty end. Only the British, however, responded to the Treaty with significant political action. In 1962, they split the Falkland Islands Dependencies (FID). The portion below 60° S, i.e., within the Treaty area, became the British Antarctic Territory. That left only South Georgia and the South Sandwich Islands within the FID. Simultaneously, the Falkland Islands Dependencies Survey became the British Antarctic Survey (BAS). (Further changes would take place in later years).

Another official nomenclature change, at least on some maps, was made two years later. In 1964, the United States and the British Commonwealth countries mutually agreed on a common name for the long, thin finger of land stretching north from the main body of the Antarctic Continent. The Americans had been calling it Palmer Land or the Palmer Peninsula since the 1820s. British maps had designated it Graham Land for nearly as long. Now, the English-speaking countries formally agreed to adopt a name already in common use, the Antarctic Peninsula.[1] (Argentina and Chile officially continue to use the names Tierra San Martín and Tierra O'Higgins respectively for the northern portion of the landmass, but their maps now typically call the whole the Antarctic Peninsula.)

Although only Argentina, Chile, and Great Britain were operating permanent bases in the Peninsula Region as the 1960s opened, other countries were also working there. The Americans, in particular, were active, especially in the far southern reaches of the Antarctic Peninsula. Shortly after the Peninsula name agreement, the United States and the Soviet Union became the first new countries to establish permanent bases in the Peninsula Region since Chile in the summer of 1946–47. The United States had had year-round bases in the region earlier—the USAS East Base on Stonington Island from 1939–41 and Ellsworth Station on the Filchner Ice Shelf during the IGY. The Americans had also established Eights Station on the extreme southern icecap of the Peninsula as a staging place for seismic land traverses and glaciology in late 1961. Eights Station, however, was never intended as a permanent base and was closed in late 1965. What became America's first truly permanent Peninsula Region base, Palmer Station, was established on Anvers Island, initially in temporary quarters, in 1964–65. Three years later, in 1967–68, the Soviet Union, which had never had any base in the Peninsula Region, opened its Bellingshausen station on King George Island. Other countries would follow, on King George Island and elsewhere in the Peninsula Region, a few years later.

The essence of the story of the government bases following ratification of the Antarctic Treaty is that their numbers grew, they did important work, and people, including more and more women, lived in some of them year-round. Humans had come to stay. But that did not mean that people had conquered the Antarctic. It remained a dangerous place where things could go wrong with little warning. During the years after 1959, government expeditioners fell fatally into crevasses, were lost in blizzards, were stranded when sea ice broke out, ... And then there was the volcano.

DECEPTION ISLAND BLOWS UP

In the mid-1960s, people had known for well over a century that Deception Island was an active volcano. Still, it had never erupted seriously enough while people were in residence to raise lasting concerns, and by the mid-1950s, Argentina, Britain, and Chile all had year-round bases on the island. It was simply too attractive a location to worry about the seemingly remote danger of an eruption. That was a mistake.

Deception Island started rumbling in early 1967. Then, on the evening of December 4 that year, steam burst through the ice in Telefon Bay and rose thousands of feet in the air. Hot ash and hailstones began raining all over Port Foster, while lightning and fiery volcanic bombs filled the air. The explosions blew out the power supply at the Chilean base, and the 27 men there took refuge in a basement. When the walls and ceilings began collapsing, the desperate men inside managed to restore enough power to make contact with their supply ship, the *Yelcho*. Then they headed out into the volcanic storm toward the British base, 4 miles away.

Eight men occupied the British base, which was at Whalers Bay. When the eruption began, their leader wrote,

> As a man we raced to the base hut, grabbed our cameras and rushed out again to take pictures. Having done the most important thing, we set about the secondary things like saving our skins. . . . We had just completed preparations when the ash started to fall. This was black, and like coarse sand, and fell for some minutes while we took shelter in a small hut. Walking became difficult because the snow was melting from below from the warmth of the ground, and

Soot and ash explode into the air during the 1967 Deception Island eruption. *(Photo by M. J. Cole, 1967. Reproduced courtesy of British Antarctic Survey Archives Service. Ref. no. AD6/19/3/Sg19. Copyright: Natural Environment Research Council)*

because the ash and hail had fallen on top to a depth of some inches. All the time it was getting darker and darker, even though in summer the night was normally never more than twilight. When the storm broke we made a dash for the base unit amid continuous thunder and lightning crashing round the hills. There were little flashes of fork lightning in the air at eye level . . . , looking for all the world like tinsel from a Christmas tree in mid-air.[2]

A British attempt to radio the Chilean base had failed because the Chileans had no power. The British then sent out a general distress call. About two and a half hours after the initial eruption, the BAS men learned that the nearest ship was the Chilean *Piloto Pardo*. She was coming as fast as possible. So were the British R.R.S. *Shackleton* and Argentina's *Bahía Aguirre*. A few minutes later, the BAS contingent received word that the Chileans were on their way to join them. The refugees staggered in an hour or so later after a nightmare walk. The British and Chileans finally learned the fate of their Argentine neighbors just after midnight. Wild winds had kept the ash from engulfing their buildings, and everyone there was safe, but they had left the base and were all making their way to the outer coast where the *Bahía Aguirre* would pick them up.

A foot of ash covered the BAS base by the time the *Piloto Pardo*'s helicopters flew in early the next morning to evacuate the waiting men to the Chilean ship. When the *Shackleton* arrived, the British transferred to her. Two days later, a Chilean reconnaissance flight over the island observed that volcanic debris had completely buried the Chilean base. The *Shackleton* carefully sailed into Port Foster that same afternoon to inspect the British base. Finding the buildings undamaged under a coating of ash, the men secured the station to await reoccupation. The British moved back in at the end of 1968, because they believed the eruption had been an isolated event. They were wrong.

The eruptions resumed early the following year. Dawn was breaking on February 21, 1969, when a violent earthquake hurled the five men at the British base, the only humans then on the island, out of bed. After two more strong quakes, they radioed the *Shackleton* and arranged to walk to the outer coast of the island to meet the ship. They were well on their way when the volcano began erupting. Blinding steam and flying debris assaulted them and ruined their trail radio. The men took shelter behind rocks before they managed to retreat to a hut at the old whaling station. When the ashfall eased off, they ripped iron sheets from the walls of the building. Holding these over their heads, they stumbled back to the base to use the main radio. What they saw on arrival there astounded them. Only a few hours before, hills had separated the whaling station from the BAS building. Now those hills and much of the whaling station itself were gone, swept away by an ice-choked flood of meltwater. As for their own main structure, the flood had split it in two, and the insides, mud-caked and ice-clogged, were a shambles. The cemetery that had stood at Whalers Bay since 1908 had suffered even more; mudflows had completely buried what they had not washed away.[3]

The men's decision to leave the base had saved their lives, but they were still stranded, and a volcano was erupting about them. Since their residence hut was uninhabitable, they took refuge in the still-intact aircraft hangar. Then, with both the trail and base radios gone, they lit a signal beacon in a drum of aircraft fuel. Helicopters from the *Piloto Pardo* once again came to the rescue, this time flying through thick snow and ash on a very dangerous mercy mission. That evening, the BAS men overheard the Chilean captain muttering, "One time—yes. Two time—yes. Three time—by God—not at all."[4]

There were no eyewitnesses when Deception Island exploded again in August 1970. Scientists inferred the fact of this eruption from earthquake recording instruments in the Argentine Islands and the film of ash that drifted over bases in the South Shetlands. An international team of geologists visited the island the following summer to investigate and confirmed the inferences. Since then, Deception Island has rumbled from time to time, but as of mid-2011, there have been no more major eruptions. Although the year-round bases remain abandoned, Spain now operates a Deception Island station each summer and Argentina has had one open there in many recent summers. And the island, given special attention when the 2005 Antarctic Treaty Meeting designated it an Antarctic Specially Managed Area, is one of the most visited tourist stops in the entire Antarctic. But Deception Island's explosive potential is not forgotten. The management plan concludes with an "Alert Scheme and Escape Strategy for Volcanic Eruptions on Deception Island."[5]

The Hunters

While more and more Antarctic scientific bases were being established in the 1960s, the whalers were ending their tenure. Grytviken, the birthplace of Antarctic whaling, ceased operations after the 1964–65 summer. The last shore station, Leith, closed the next year. A few years later, nearly all commercial whaling ceased throughout the Antarctic, because there were so few whales left to hunt.

A hunter of a new sort arrived in the Antarctic at nearly the same time the South Georgia shore stations closed. In the summer of 1961–62, a Russian ship netted 4 tons of krill, the tiny crustaceans that many whales, seals, and penguins feed on around the Antarctic. The results of this experimental voyage were sufficiently encouraging for the ship to return the following summer for commercial operations, and other countries soon joined the hunt. Early krill fishing concentrated on the waters of the Peninsula Region, but soon expanded around the continent. In addition, fishing for finfish, particularly around South Georgia, began about the same time as krill fishing, soon expanding into a large and profitable industry.

Whales, seals, krill, and fish were valuable resources for those looking to reap profits from the Antarctic, but they could be devastated when not exploited with an eye to conservation. The results of the uncontrolled slaughter of Antarctic seals in the nineteenth century and of whales in the twentieth were ample evidence of that. Recognizing this, the Antarctic Treaty parties had adopted conservation annexes to the Treaty in 1964 and 1972 (respectively, the Agreed Measures for Conservation of Antarctic Flora and Fauna and the Convention for Conservation of Antarctic Seals). In early 1977, with the prospect of major commercial fishing in the far south looming, the Antarctic Treaty parties determined to take a proactive approach to controlling the industry before it was too late. Three years of discussions resulted in the 1980 adoption of the Convention on the Conservation of Antarctic Marine Living Resources (CCAMLR), an agreement that by then had grown far beyond an attempt simply to control fishing. The final text applied to *all* Antarctic marine living resources—fish, mollusks, crustaceans, and all other species of living organisms, including birds—from south of 60° S, or from south of the Antarctic Convergence where that ecological boundary lay north of 60° S. It was a groundbreaking agreement, the first to consider the entire Antarctic ecosystem as an integrated whole.

But the South Polar Region also contains other, less wasting, assets. The ice, the landscape, and Antarctic fauna, including many of the animals that hunters regarded as prey, had fascinated explorers and scientists for years. Others, who were willing to pay for the privilege, were eager to see these sights for themselves.

The IGY helped make this possible, because one of its legacies was a logistical infrastructure that facilitated travel to the Antarctic. And that, in turn, made commercial tourism viable.

Tourists, Yachtsmen, and Adventure Expeditions

Tourism in the south had been a long time coming. The earliest proposals dated back to at least 1892, when an Australian company suggested sending sightseers south as a way to promote Antarctic exploration. Nothing developed. From the mid-1920s on, a few tourists did reach South Georgia, the South Shetlands, and the Antarctic Peninsula as paying passengers on supply boats servicing the whaling industry. This was not organized tourism, however, and a prospective visitor needed good connections to gain a berth. The Argentine and Chilean governments took a few tourists to the Antarctic Peninsula in the late 1950s, beginning with a single Chilean sightseeing flight in December 1956. The next summer, Argentina sponsored two ten-day cruises to the South Shetlands and the Antarctic Peninsula. Chile then joined Argentina, each country operating one cruise in 1958–59. After that, these offerings ceased.

A new tour operator appeared on the scene in 1965–66. He was Lars-Eric Lindblad, a Swedish-American who had been running what he called adventure travel trips for years. In January 1966, Lindblad chartered a ship from the Argentine navy and took 57 passengers, nearly all Americans, for a two-week cruise to the South Shetlands and the Antarctic Peninsula. This voyage was the real beginning of the Antarctic tourism industry, because Lindblad did not abandon the effort after just a few trips. And when other tour operators joined Lindblad, Antarctic tourism was off and running. As evidence that this industry had come to stay, seven companies offering Antarctic tours banded together in mid-1991 to form the International Association of Antarctica Tour Operators (IAATO). The following summer, tourist vessels took approximately 7,000 people south. By the 2007–08 summer, the number of tourists reaching the Antarctic annually was approaching 50,000. The vast majority of these people, well in excess of 95%, went to the Peninsula Region. As for IAATO, its membership had grown to more than 100 companies.[6] (Although the worldwide economic downturn resulted in reduced tourism to Antarctica following 2007–08, the numbers remained high, with approximately 34,000 tourists heading south in the summer of 2010–11.)

The first season he took tourists south, Lindblad arranged to have an Argentine naval vessel escort his passenger ship, for safety reasons. Although he dropped the naval escort in 1966–67, his concerns about tourist safety were valid. The Antarctic was as dangerous for tourist ships as for any others, and perhaps more so for their passengers, many of whom were significantly older than the traditional Antarctic visitors. In fact, Lindblad's clients and other tourists at times experienced

Lars-Eric Lindblad in the Antarctic, behind him at top right, the *Lindblad Explorer* — The 250-foot-long Lindberg Explorer *had 50 air-conditioned cabins, each with a private bathroom. She made her first Antarctic cruises in the summer of 1969–70, then operated in the Antarctic for decades, initially for Lindblad Expeditions, then for other operators as she was sold several times. Over the years she grew weary, and, sadly, she met her end in November 2007 when she hit submerged ice in the Bransfield Strait and sank. (Photo from Lindblad, with Fuller,* Passport to Anywhere, *1983)*

unplanned adventures on their voyages, particularly in the early years. Their ships ran aground on several occasions, and weather or other circumstances occasionally stranded passengers on shore for hours or even days. And in November 2007, the M.V. *Explorer*, the very ship that Lindblad had commissioned in 1969 as the first vessel to be built for Antarctic tourism, sank after hitting submerged ice in the Bransfield Strait. Fortunately, all 154 people aboard were safely evacuated to lifeboats and picked up, several frightening hours later, by other tourist ships that responded to the *Explorer*'s SOS calls. Over the years, however, it had often been ships or personnel from government programs that had had to come to the rescue of tourists in distress—a call on scientific resources that hardly endeared such visitors to base personnel. The tourists who descended on the government bases also interrupted the residents' work. But at the same time, these were people who would return home and support funding for Antarctic activity. Thus, over the years, an accord, though sometimes an uneasy one, developed between most of the government programs and the tourist industry.

Tour operators carried paying passengers who came, saw, learned, and appreciated, but staff members always escorted them whenever they left their ships. If they found themselves in challenging situations, it was because of a mishap rather than as part of their itinerary. As the 1960s and 1970s progressed, however, more and more people visited Antarctica just because it did offer challenges—to sail an icy sea, climb a mountain, experience skiing to the South Pole. . . .

The first modern adventure yachtsman to reach the South Shetlands arrived in early 1966. Others soon followed, and by the 1990s, the trickle of yachting trips had grown to a virtual flood, at least in Antarctic terms. Dozens of yachts turned up each year, some simply for the experience of sailing to Antarctica, others to pursue further adventures or for scientific research. A few even spent the winter. The challenge of Antarctica's glacier-covered mountains was a special draw, one that Britain's Robert Falcon Scott had recognized at the opening of the twentieth century. He wrote then, "One wonders when the mountaineer . . . will descend on this lonely region, for here indeed lies a field where the boldness of man might have play for many a year. . . ."[7] The first Antarctic mountaineers were men from exploring expeditions who climbed mountains for recreation or in connection with their scientific and other work. Some of Scott's own men were among them. And Charcot had taken an Italian alpine guide on the 1903–05 *Français* expedition. But it would be five decades before Scott's question was truly answered, before men at last traveled to the Antarctic regions expressly for the sake of mountain climbing.

In the 1950s, several expeditions arrived at South Georgia with adventure mountaineering as an explicit element of their programs. Duncan Carse, the Discovery Investigations man who had joined Rymill's expedition in 1934, led the first of these, the four-summer South Georgia Survey Expedition, which began its work in 1951–52. Although Carse's primary goal was to carry out the first topographical survey of South Georgia, his men bagged a number of first ascents of challenging peaks on the island. And during their 1955–56 season, they also roughly retraced Shackleton's 1916 crossing route, the first time this had been done. Others soon followed Carse, because South Georgia's demanding mountains were very tempting, as was the mystique of the Shackleton story.

Malcolm Burley arrived in 1964–65 leading a ten-man South Georgia Joint Services Expedition group. This venture was one in a long tradition of adventuring parties sponsored by the British armed forces. Burley, who had already been to the island in December 1960 for five days of mountain climbing, was now back for a much longer visit. A British naval vessel delivered Burley's team to South Georgia in mid-November 1964, and a ship's helicopter flew them across the island to King Haakon Bay, where the men easily identified Shackleton's campsite. They then began their own South Georgia trek, attempting to follow Shackleton's crossing route as precisely as possible.

It was an exciting and emotional moment when they reached the northeast side of the island, where Shackleton had stood when he heard the whistle from the whaling station. Like Shackleton, they went on from there and, Burley wrote, "on what was nearly the last leg, a sense of relaxation and complacency set in. South Georgia has a habit of sorting out those who presume easy familiarity, and we were brought to heel with an all-enveloping avalanche. This underlined again how fortunate Shackleton, Crean, and Worsley were. . . ." After descending what Burley believed was the same waterfall that Shackleton had scrambled down, they reached "the silent, derelict remains of Stromness whaling station, now haunted only by ghosts of the past."⁸ The men followed the crossing with the mountain-climbing phase of their expedition. That effort, equally successful, included the first complete ascent of 9,625-foot-high Mount Paget, South Georgia's highest peak.

Burley returned to the Peninsula Region in 1970–71 with another Joint Services Expedition. This time he went to Elephant Island. Prior to his arrival, virtually nothing was known about the locale beyond its coasts. But Burley already knew one thing related to the place he was about to visit, something that almost no one else did. After Shackleton had dramatically discarded Queen Alexandra's Bible before the *Endurance* party's first failed sledging attempt, Thomas McLeod, one of the seamen, had secretly retrieved it and carried it with him to Elephant Island. When the *Yelcho* rescued the Elephant Island party, he took the Bible to Punta Arenas. There, he presented it to the family he stayed with, as thanks for their hospitality. In November 1970, just before leaving for Elephant Island, Burley gave a public lecture in Buenos Aires about his 1964–65 South Georgia expedition. The daughter of McLeod's host family was in the audience, and after Burley's talk, she told him the story and gave him the Bible. Burley took it with him to Elephant Island. (He later presented it to the Royal Geographical Society in London.)

The ship that delivered Burley's fourteen-man team to Elephant Island in December 1970 was another link to Shackleton. She was H.M.S. *Endurance*, named in honor of Shackleton's ship. Her walls sported a number of Hurley's photographs, and three of Shackleton's men had even set foot on her decks. Earlier in the year, Charles Green, the original *Endurance*'s cook, Lionel Greenstreet, her first officer, and Walter How, a member of the crew, had toured the new *Endurance* and given her their blessings.

Burley's men spent most of their time on surveying and other scientific work, but when they had time, the climbers among them tackled many of Elephant Island's virgin mountains. They also visited Cape Valentine, the place where Shackleton had first landed. Burley was even more eager to see Point Wild, the bleak gravel beach where 22 men had lived under upturned boats for months while waiting for Shackleton to return. Burley's entire party eventually visited the site, in three separate groups. The first, including Burley himself, reached the historic location in mid-February. Their overland trip to Point Wild, a struggle with crevasses, glaciers, and rock falls, made it clear why Shackleton's ill-equipped men had found it impossible to move beyond their beach. Burley used copies of Hurley's photographs to locate the site of Wild's hut, but there was no sign of the actual structure. Instead, a boisterous crowd of penguins covered the spot.

This first group to camp at Point Wild since 1916 had problems finding a site large enough for even two small tents. The only possible place was a narrow gravel spit occupied by about 50 fur seals. Burley reported,

A naïve attempt to shoo the fur seals off it met with seal-like derision at the ludicrous idea of surrendering their territory. . . . An uneasy truce was then tacitly accepted and we pitched our tents. . . . Two yards to landwards, the penguins kept up an incessant and indescribably raucous clatter, pecking periodically at the guys. From one side of the spit, the waves sluiced up to within two feet of the tents . . . and on the other, the brash ice crunched relentlessly a mere two yards away. Throughout all of this, the fur seals cavorted around the tent, shuffling and snuffling with inexhaustible supplies of energy. . . .[9]

Fur seals were one problem that neither Shackleton's men nor anyone else had had to cope with in the South Shetlands for many decades. Indeed, it was only thirteen years earlier, in January 1958, that a FIDS party visiting Livingston Island had seen the first group of fur seals anyone had reported in the area since the nineteenth century. Burley was facing the advance guard of an amazing population recovery, more successful than the men of the Discovery Investigations could have dreamed of in 1933–34 when they found that tiny colony of the near-extinct animals on Bird Island off South Georgia. Today, literally millions of these animals crowd the South Georgia beaches during the breeding season, and tens of thousands more occupy shores farther south. The animal the sealers had nearly wiped out in the nineteenth century has come back, and more.

Fur seals crowding a beach at South Georgia — *In the left center, a bull fur seal watches alertly to defend his territory and females from other males. This photo was taken in January 1995, at a time when it was still possible, if only barely, to land on this beach during breeding season. Since then, the South Georgia fur seal population has increased so much that it is probably now impossible for humans to set foot here safely in January. (Author photo)*

Burley was in the vanguard of a special category of adventurers, people who wanted to visit celebrated locales and recreate exploits of the past, particularly Shackleton's from the *Endurance* expedition. From 1994 to 2000, three separate groups set out to duplicate the voyage of the *James Caird* in differing degrees of exactitude. Arved Fuchs, a German who had skied across the continent as one of a two-man team in 1989–90, led the most successful effort. From January to February 2000, he and three companions sailed from Elephant Island to South Georgia in a custom-built replica of Shackleton's small boat. Their thoroughly miserable trip brought them to South Georgia in roughly the same time it had taken Shackleton, only to have to endure a terrifying night in a storm, as Shackleton had, before they could slip into King Haakon Bay. Then one man from their support vessel joined them, and the party of five crossed South Georgia following Shackleton's route as closely as possible.[10]

Retracing Shackleton's mountain crossing was on the way to becoming an adventure icon when Fuchs's team did it. Carse had tried to trace the route in the 1950s; Burley had done it in the 1960s; and a group from the British garrison on South Georgia had followed Burley's path in 1985. Around the end of the twentieth century, tourist companies started offering guided trips along all or parts of Shackleton's South Georgia route.

Fishermen, tourists, adventurers. Like government scientists, all were becoming a regular part of the summer Antarctic scene. As the 1970s ended, Argentina and Chile added yet another element to the population mix.

Antarctic "Settlers"

Both Argentine and Chilean heads of state had visited their countries' Antarctic bases over the years. Argentina took this a step further in 1974. That year, President Isabel Perón formally designated the country's Marambio base on Seymour Island a provisional seat of government. She then flew down with her entire cabinet to conduct business for the day. Chile's president, Augusto Pinochet, made his own trip south in 1977, sailing all the way to Marguerite Bay, where he deposited an urn containing soil from every region of Chile. It was a symbolic gesture, emphasizing that Territorio Antártico Chileno was part of the rest of Chile.

After President Perón's visit, Argentina took another step to demonstrate that some of her bases were true outposts of the homeland. Late in 1977, the Argentine program began bringing families to its Esperanza base at Hope Bay. The first two arrived in November 1977. One was the de Palma family—Captain Jorge de Palma, the base officer-in-charge, his seven-months-pregnant wife Silvia, and their three young children. On January 7, 1978, Silvia gave birth to a baby boy, whom she named Emilio. Argentine President Jorge Videla

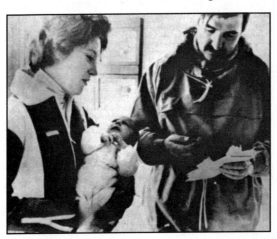

Silvia de Palma holds her two-week-old son, Emilio, while her husband examines gifts sent to the baby by the Argentine president. *(Associated Press, from reproduction in Anon., "Antarctica's First Baby...," 1978)*

hailed the arrival of this first child born on the Antarctic Continent as "a reaffirmation of the inalienable role of Argentines in these far lands."[11] In 1983–84, following Argentina's precedent, Chile established its own settlement at its King George Island station.

Presidential visits, families, and babies all conformed to the rules of the Antarctic Treaty, even when their presence was at least partially politically motivated. But military action was not allowed. The Treaty, however, applied only to the Antarctic regions south of 60° S.

War on South Georgia

North of 60° S, Argentina and Britain had both long claimed South Georgia and the South Sandwich Islands. More seriously, Argentina had never accepted British sovereignty over the Falkland Islands, or Islas Malvinas, as she called them. In February 1976, Lord Shackleton, Sir Ernest's son, visited the area on a fact-finding mission for a report on the British future in the Falklands. While he was at South Georgia, his father's namesake ship, the R.R.S. *Shackleton,* encountered an Argentine destroyer near the Falklands/Malvinas. The Argentine captain, incorrectly assuming that Lord Shackleton was aboard, fired a warning shot across the *Shackleton*'s bows to discourage his lordship's visit.

Argentina made a more substantial move nine months later. In November, she secretly established a base on Thule, one of the southernmost islands in the South Sandwich group. Air force personnel staffed the base, and the facilities included a fuel tank farm, an aircraft hangar, and accommodations for more than 100 people. After Britain learned about it when a helicopter from the *Endurance* flew over on a routine survey flight in mid-December, her ambassador to Argentina immediately delivered an official protest. Although the British said little else about the matter, they did replace the *Endurance*'s helicopters with ones capable of launching missiles.

The simmering tensions over the three disputed island groups—the Falklands/Malvinas, South Georgia, and the South Sandwich Islands—exploded into outright war in 1982. Although most of the action took place in and around the Falklands/Malvinas, beyond this book's scope, the war extended to South Georgia and the South Sandwich Islands as well. The latter two island elements are the part of the tale covered here.[12]

In December 1981, an Argentine naval vessel anchored at South Georgia's abandoned Leith Harbor whaling station for several days. She maintained radio silence the entire time and ignored the British requirement to check in with the magistrate at King Edward Point. Officially, the ship was there for a benign reason. She was transporting Constantino Davidoff, a Buenos Aires salvage contractor who had secured an option to collect scrap material from the derelict whaling stations. Davidoff had come to examine his proposed operation. On his return to Buenos Aires, he apologized to the British Embassy for his actions, or rather lack of them. He then received permission for another trip to South Georgia for the actual salvage work.

Davidoff returned to Leith Harbor in mid-March 1982 aboard the *Bahía Buen Suceso,* another Argentine naval vessel. Once again, the Argentines ignored the formalities with the magistrate. It was not long before nearly 40 scrap-metal merchants and ten naval personnel landed, raised the Argentine flag, occupied a British refuge hut, and started shooting reindeer in defiance of the animals' protected status. A BAS field party that had been alerted to the Argentine arrival reached Leith Harbor on March 19. They saw what was happening and radioed the magistrate at King Edward Point, who passed the news on to the Falklands governor. The governor's reply directed the BAS men to inform the Argentine ship's captain that the salvage party had landed illegally; Argentine military personnel were forbidden on South Georgia; they must report to the magistrate, lower the Argentine flag, and cease killing reindeer. After the BAS men delivered the message, the captain removed the flag, but ignored all the other directives. Moreover, when the *Bahía Buen Suceso* sailed away on March 22, he left the salvage workers and a few military personnel behind.

Other than the message delivered to the *Bahía Buen Suceso,* the initial British reaction was limited. The Admiralty ordered the *Endurance* to South Georgia to land a 22-man marine detachment at King Edward Point. The ship's captain, Nick Barker, was then simply to observe the Argentines. In the meantime, Britain advised Argentina that she was willing to grant retroactive authorization to Davidoff's salvage party.

Argentina declined the offer, instead broadcasting an adament assertion of her

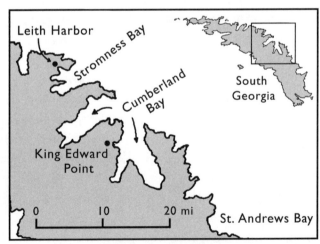

The 1982 Falklands/Malvinas war zone on South Georgia — *Nearly all the action in the hostilities that took place on South Georgia occurred along a small section of the northeast coast of the island. The war here began at the abandoned Leith Harbor whaling station, then expanded to King Edward Point in Cumberland Bay, where the British had their administrative and scientific headquarters on the island. All of the serious fighting took place in the vicinity of the latter location. The British special forces who had to be rescued by helicopter from a storm in the interior of the island were 10–15 miles to the west of Leith Harbor.*

own South Georgia sovereignty. This was a signal that war was imminent. In fact, about half the country's navy was already at sea, most of it headed for the Falklands/Malvinas. One ship, the *Bahía Paraíso*, was bound for South Georgia. She sailed into Leith Harbor on March 25, landed about fourteen more sailors and marines, and re-raised the Argentine flag. The Argentines then began to establish what was clearly going to be a military position, but they restricted their activity to the vicinity of Leith Harbor for the time being.

Captain Barker left the British marines at King Edward Point when the *Endurance* was ordered back to the Falklands/Malvinas at the end of March. At that point, BAS men were scattered about the island. Thirteen of them were at the King Edward Point base; several other groups were working in the field; and three men had just hiked to St. Andrews Bay because of concern for two British wildlife photographers—the only women then on the island—who had been living there in a tiny hut for months. On April 2, the same day the Argentine navy landed in the Falklands/Malvinas, the *Bahía Paraíso* sailed into Cumberland Bay and approached King Edward Point. The weather made it impossible for her to launch boats or helicopters, but before she left, her captain, Ismael Jorge García, radioed the British that he would deliver a message the next morning. The marines, well aware that war had begun in the Falklands/Malvinas, spent the day preparing defensive positions around King Edward Point. Simultaneously, most of the BAS men not in the field took refuge in the nearby Grytviken church. Two other men left to warn a field party in the South Georgia interior that they should not return to base.

The *Bahía Paraíso* returned the next day along with a second, newly arrived, Argentine naval ship, the *Guerrico*, and delivered the promised message: Argentina had taken the Malvinas and now South Georgia. The British must surrender. The magistrate refused, and after a fruitless radio exchange, the *Guerrico* opened fire. To the Argentine's surprise, the vastly outnumbered British shot back. Rockets blasted the *Guerrico*'s side and slammed her two helicopters, one of which crashed into a hillside across the bay from King Edward Point. The other, belching smoke, just managed to land near King Edward Cove. But the Argentine bombardment and 100-plus troops were more than the British marines could handle. It was a costly victory—three Argentines dead and several times that many wounded, to only one injured British marine. The victors gave the captured BAS men less than twenty minutes to gather their belongings before taking them and the marines on board the *Bahía Paraíso*. The British had to abandon all their ongoing data, including the meteorology records that had been kept continuously since 1905. Ironically, Argentina had cited these observations, begun and maintained for years by the Argentine whaling company Pesca, in support of her claim to South Georgia.

The *Bahía Paraíso* immediately took her captives to Argentina. Fifteen days later, they were released and sent back to Britain. Argentina had intended to remove all the British personnel on South Georgia but had been unable to finish the job because of the damage to the helicopters. The BAS field parties and the five people at St. Andrews Bay thus remained on the island. Moreover, the *Endurance* had returned and established an observation post overlooking King Edward Cove in time to witness the closing actions of the April 3 battle. Then, under orders to stay clear of the fighting, Captain Barker hid in remote inlets well away from Cumberland Bay, doing what he could to make his bright red ship look like an iceberg to Argentine radar.

On April 7, the British decided to mount an effort to retake South Georgia. The decision made, they ordered three ships from the Falklands Task Force to rendezvous with the *Endurance*.

The four ships were together in position off South Georgia by April 21. Because the British hoped to keep their presence secret until they were ready to strike, the leader of the special forces unit that had arrived from the Falklands/Malvinas suggested landing at King Haakon Bay. His men would follow Shackleton's route across the island to a position where they could observe Leith covertly. Barker and a BAS adviser who had come aboard the *Endurance* from the St. Andrews Bay group both pointed out how difficult this would be. The special forces commander, however, was convinced that if a starving Shackleton could do it, so could his fit men. Still, he did grudgingly agree to eliminate the first part of the trek by using helicopters to fly to the interior of the island. Within an hour, the troops found themselves pinned down by violent winds in a hopeless whiteout. The situation was so desperate that they radioed for the helicopters to return and pick them up. The first two helicopters to try crashed when they attempted to land. Although all the men aboard escaped serious injury, the machines were total wrecks. After the weather improved a bit, a third helicopter finally managed to evacuate everyone in a single hair-raising trip. Following this debacle, the British changed tactics and tried stealth to approaches to Leith and Grytviken by sea. These efforts were equally unsuccessful.

Argentina strengthened her own South Georgia position three days after Britain began operations. On April 24, a submarine arrived and dropped off fifteen more soldiers at Grytviken. This development, along with the failures of a secret approach, was enough for the British authorities. They authorized their small South Georgia task force to open fire the next day. The ships began by attacking the sub. As she ran for safety, one of the *Endurance*'s helicopters hit her with a missile. After the entire crew had safely escaped the sinking vessel, the British took them all prisoner. (Sadly, one submariner died several days later, shot mistakenly by the British when he was helping to move the sunken hulk. Named Felix Artuso, he was buried with full military honors in the Grytviken cemetery.) The British then bombarded the shore near King Edward Point to demonstrate their firepower,

Casualties of the 1982 Falklands/Malvinas war on South Georgia — Top: *The wreckage of the helicopter shot down when the Argentines took Grytviken and King Edward Point, later used for target practice by the King Edward Point British garrison. (Author photo, 2001).* **Right:** *The grave of Argentine submariner Felix Artuso in the Grytviken cemetery. (Author photo, 2009)*

and the Argentine forces soon surrendered. It was April 25, and Britain was back in control at Grytviken. The Argentine garrison at Leith Harbor surrendered the next day without a shot, though there were some tense moments.

Retaking South Georgia was valuable strategically for Great Britain. Argentina was fighting the Falklands/Malvinas War close to home. Not so the British. But with South Georgia back in her control, Britain had a staging area for her Falklands Task Force. During the next few weeks, as many as 25 warships—the largest the *Queen Elizabeth II*, commandeered as a troop carrier—filled Cumberland Bay as they transferred men and supplies for onward movement to the Falklands/Malvinas. When the Argentines surrendered at Stanley on June 14, the British were once more firmly in control, not only of South Georgia but also the Falklands.

One more disputed location remained, the Argentine base on Thule in the South Sandwich Islands. The *Endurance* arrived on June 18 with a small task force, and the ten Argentines in residence surrendered without a fight. When the *Endurance* and another ship returned to Thule six months later, however, Captain Barker discovered that someone had replaced the British flag with an Argentine one. The

British Admiralty then dispatched another ship to the island to destroy the station.

As for South Georgia, the British repatriated the remaining BAS personnel and the two wildlife photographers at the end of April. They then established a military garrison in the BAS building at King Edward Point. Military personnel occupied the location for nearly 20 years before the British finally closed the garrison in March 2001 and reinstated the scientists.

In the aftermath of the Falklands/Malvinas War, Britain took a careful look at her interests in the Peninsula Region. One result was that the government doubled the BAS budget. This action was taken primarily for political reasons, but the new funds meant that BAS could expand its scientific efforts, rather than scaling back as had been planned before the Falklands/Malvinas War broke out. As for Argentina, she made no substantial changes to her Antarctic program, which continued to operate multiple year-round bases, seven in the summer following the war.

The Ozone Hole and Global Climate Change

Great Britain's renewed commitment to science in the far south bore fruit several years later, when scientists at BAS's Halley base confirmed an ominous discovery. A number of Antarctic bases had been measuring Earth's atmospheric ozone since the IGY. To the surprise of the scientists at Halley, their 1982 data indicated a drop of more than 20% in ozone levels directly above Antarctica. Their immediate reaction was that something must be wrong with the equipment, because, as far as they knew, no other measurements had shown anything of the sort. They were wrong. The Nimbus 7 weather satellite, launched by the United States in 1978, carried instruments that *had* recorded such changes. The levels were so unexpected, however, that the computer analysis program interpreting the Nimbus data had simply noted the readings and flagged them as anomalies for recheck, a task no one had done.

The Halley scientists kept silent about their 1982 findings. They again said nothing when their instruments recorded even lower ozone levels the next year. But in 1984, they employed new equipment and set up a second measuring station at BAS's Faraday base in the Argentine Islands.[13] When the new results confirmed the 1982 and 1983 readings, the Halley team went public. That led to reanalysis of the earlier Nimbus 7 data, which supported the Halley observations. The large and growing seasonal hole in the ozone, scientists concluded, had resulted from humanity's activities well outside the South Polar Region. The culprit was the release of manufactured chemicals, especially chlorofluorocarbons, into the atmosphere.

Another impact of human activity beyond Antarctica, global climate change, also was affecting the far south, most obviously in the Peninsula Region. Average annual temperatures along the west side of the Antarctic Peninsula had increased more than 5° F during the second half of the twentieth century and approximately 90% of 244 measured glaciers along the west coast of the Peninsula had retreated since 1940. That included the very convenient one that had connected Stonington Island to the Peninsula mainland. That connection was long gone by 2000. Nearby, the Wordie Ice Shelf virtually disappeared in the early 1990s, and in the summer of 1994–95, the far northern part of the Larsen Ice Shelf, on the east coast of the Peninsula, collapsed. The Larsen Ice Shelf breakups continued in later summers,

Satellite photo of the Larsen B Ice Shelf, taken March 7, 2002, showing the ice shelf disintegrating into a mass of icebergs and brash ice, with icebergs floating off into the Weddell Sea. *(NASA Jet Propulsion Lab photo)*

including a massive one in early 2002 when about 1,250 square miles disintegrated in a 35-day period. The Larsen Ice Shelf events were the most dramatic, but lesser happenings of this sort were going on all over the Peninsula Region. One of these, in March 2008, provided ominous evidence of a possible acceleration of the impact of climate change in the Antarctic. Over the course of that month, a significant part of the Wilkins Ice Shelf, which lies between Alexander, Charcot, and several other islands, disintegrated. As a scientist with BAS put it, "In 1993, we predicted that this was going to be a vulnerable ice shelf. But we got the time scales completely wrong. We were saying 30 years at the time, and now it's happened within 15 years."[14] And in recent years, winter sea ice has on average been forming weeks later than it had in previous decades and correspondingly, breaking up significantly earlier. Wildlife, too, is responding to the temperature changes. Fur seals are moving farther south, as are elephant seals, and gentoo penguins, whose population is increasing notably in places they were seldom seen only a few years ago. At the same time, the more northerly Adélie penguin colonies are crashing.[15]

Both the ozone hole and global climate change are examples of ways that humanity has been affecting the planet by exploiting its resources outside the Antarctic regions. But Antarctica itself is certainly vulnerable to an exploitation mentality. For two centuries, men had profited from hunting the continent's elephant and fur seals, whales, krill, and fish. And from the mid-twentieth century on, more and more people had begun speculating about other possibilities. Were there commercial minerals in the south? The Antarctic Treaty had finessed the question of controlling such exploitation by deliberately ignoring it. As the 1980s began, the Treaty parties addressed it. This continent-wide issue would have unique consequences for the Peninsula Region.

MINERALS, BASES, AN OIL SPILL, AND AN ENVIRONMENTAL PROTOCOL

In June 1982, the Antarctic Treaty parties began negotiating the Convention on the Regulation of Antarctic Mineral Resource Activities (CRAMRA). Their intent was to establish a regulatory framework that would control minerals activity in the Antarctic Treaty area. A number of non-Treaty member countries took note of what was happening and decided that they wanted to have a say in the discussions.

The original Antarctic Treaty had allowed for the admission of new countries. By the time the CRAMRA negotiations began, 26 countries had signed on to the Treaty, fourteen of them as Consultative Members with full voting rights. These voting members were the original twelve plus Poland and West Germany. The latter two had earned their status by fulfilling the Treaty requirement of having a significant scientific presence in Antarctica. When other countries began looking into ways they could do the same, many concluded that the easiest place to conduct scientific work in Antarctica was in the Peninsula Region. As a result, nine newly arrived countries had established bases in the area by 1988. The largest concentration was on King George Island. This single island was now occupied by four year-round bases that had been there for years—one each for Argentina, Chile, Poland, and the Soviet Union—plus new ones operated by Brazil, China, South Korea, and Uruguay. Other countries established summer-only bases on the island. Forty-four-mile-long King George Island, the place where William Smith had made the first landing in the South Shetlands, in 1819, had become home to an Antarctic version of urban sprawl. (Most of these new bases in the Peninsula Region would remain after CRAMRA was signed, and a few more would join them, on King George Island and elsewhere in the area. As of winter 2010, by which time the numbers had held fairly steady for many years, there were 20 year-round bases in the Peninsula Region, operated by eleven countries. In addition, five more countries participated with one or more summer-only bases.)

The Antarctic Treaty members, including the new adherents, signed CRAMRA in June 1988, but many non-participants were extremely unhappy with the agreement. Conservation groups in particular argued that CRAMRA did as much to open Antarctica to minerals activity as it did to control it. There was also substantial concern over the weaknesses of the agreement's environmental safeguards.

An event in early 1989 dramatically demonstrated the danger that human activity posed to the fragile Antarctic environment. On January 28, the Argentine supply ship *Bahía Paraíso* visited America's Palmer Station for the sake of the 81 tourists aboard. As the ship was leaving, she grounded on an underwater ledge. Passengers and crew safely reached Palmer Station in the lifeboats, and before long, tourist vessels picked them up. Dealing with the *Bahía Paraíso*'s cargo was much more difficult. The ship carried hundreds of thousands of gallons of petroleum products bound for the Argentine bases, and these began pouring out through a huge gash in the vessel's hull. By January 31, currents and tides freed the wreck from the rocks and she drifted to within a mile of Anvers Island, then rolled over and sank. Fuel continued to escape, eventually creating an oil slick that covered nearly 40 square miles. It was a major environmental disaster, the worst in the Antarctic since a similar-sized spill in McMurdo Sound in 1956.

Measures to control the spill began immediately, starting with a makeshift effort by the Americans from Palmer Station. Several days later, Argentina, Chile, and the United States brought in professional containment teams. The cleanup was finally declared complete in January 1993, by which time the price tag for the effort had run into the millions of dollars. Although the cost to the Antarctic environment, as measured in degraded waters and dead fauna, was less severe than originally feared, it too was significant. And it is not yet at an end. As of early 2011, more than twenty years after the spill, the hulk of the *Bahía Paraíso* continues to seep residual bits of oil into the waters off Anvers Island.

The potential for environmental damage from minerals exploration activity had already been amply demonstrated elsewhere in the world, and now, via the *Bahía Paraíso*'s spill, in Antarctica itself. When several Treaty-member countries began proposing changes, CRAMRA was doomed before it even came into force. In 1989, the Antarctic Treaty nations agreed to consider comprehensive environmental protection for the Antarctic Treaty area. Two years of intense negotiation resulted in what became known as the Madrid Protocol. This new agreement, adopted in October 1991, replaced CRAMRA with a comprehensive environmental protection regime that applied to everything humans did in the Antarctic. What CRAMRA had been designed to regulate—mining and other minerals activity—was now banned entirely, subject to review after 50 years. The Madrid Protocol came into force on December 15, 1997, when Japan, the last of 26 nations with voting rights that needed to agree, ratified it. (Sadly, as of 2011 several Antarctic Treaty participants are looking for loopholes in the Madrid Protocol or ways to bypass the Antarctic Treaty that will allow them to explore for minerals, particularly petroleum, in the Antarctic.)

The core minerals issues of the Madrid Protocol were the most contentious, but several other elements of the negotiations were also controversial. One was a relatively minor clause that both prohibited the introduction of non-indigenous animal and plant species without a permit and required the removal of those already there. That included dogs. Many people, especially those with the Australian Antarctic program, objected passionately to removing these animals, which had been with humans in the Antarctic since the days of James Cook. Although government programs had phased them out of their work roles years earlier, they were a treasured part of Antarctica's history. And not just history. In 1989–90, a six-man private international expedition had used dog teams to cross the Antarctic Continent, beginning with a three-month-long trek along the entire 1,000-mile length of the Antarctic Peninsula.[16] As for the governments, only three bases still had dogs in 1991, but all were places where people were extremely attached to their canine companions. The bases were an Australian one in East Antarctica and one base each for Argentina and Great Britain on the Antarctic Peninsula. The Madrid Protocol required that all these animals go by April 1, 1994. Australia raised a great fuss, then conceded and said good-bye to her dogs with much sad fanfare. Argentina removed hers more quietly, although there were rumors that she had actually hung onto one or two past the deadline.

The British had twenty dogs at Rothera base on Adelaide Island, kept there mainly for recreation and companionship. These last BAS dogs left Antarctica in style, beginning with a final field trip titled Lost Heritage. Two men from Rothera—John Sweeney, the last dog handler there, and John Killingbeck, who had worked with BAS dogs 30 years earlier—took a team of fourteen dogs sledging to Alexander Island in early 1994. A film crew went along, traveling on snowmobiles, the dogs' replacements. After the trip, Sweeney boarded a plane with the animals and flew to the Falklands. He and his charges then continued to England on a special RAF flight laid on just for them. From there, they flew to Boston, where man and dogs transferred to a truck and rode to a village in northern Quebec. The final leg of the dogs' odyssey was a sledge trip to their new home, an Inuit village on the northwest shore of Hudson Bay.

Although the Madrid Protocol focused on preventing future environmental damage to Antarctica, it also required Treaty members to repair human-created scars from the past. In particular, the governments had to do something about their abandoned bases. It was left to each country to decide whether to remove completely or simply to clean up, unless a location had designated historic status. If it did, the Protocol, incorporating an earlier Antarctic Treaty action, forbade removal.

In 1972, almost 20 years before the Madrid Protocol was adopted, the Treaty parties had taken steps to protect human history in the south by conferring protected status on designated Antarctic Treaty historic monuments. The initial 43 historic sites and relics had included eighteen in the Peninsula Region. (Many more sites were added after 1972.) Once the list was approved, the sites became the responsibility of those treaty members working nearby.

The governments took these responsibilities seriously. Argentina, for example, instituted a multi-year program to preserve all three huts from Nordenskjöld's expedition. The wooden structure on Snow Hill Island received the most attention. Extensive restoration efforts there ultimately resulted in a site museum furnished in part with artifacts left behind by the Swedes. It opened to the public in 2004. Argentina also has worked to preserve what remains of the *Scotia* expedition's Omond House on Laurie Island, as well as her own Moneta House structure, which had replaced Omond House in early 1905 as the dwelling place for the teams manning the South Orkney's year-round scientific base. Today, Moneta House is a museum.[17]

The British also have mounted a substantial effort in the Peninsula Region, beginning with a survey of their old bases. To carry it out, they created an Antarctic Heritage Trust modeled on a similar New Zealand organization created to preserve historic huts in the Ross Sea Region. When the survey began in 1993–94, BAS had 30 distinct bases or refuge huts, seven of them still in use as year-round or summer bases. The British recommended several additional structures for historic monument status as a result of this review, and in 1995–95, the Trust began conservation work on one of these: Port Lockroy, an original Operation Tabarin base. The objective was to create a place where visitors could see a base as it had been decades earlier. When it opened to tourists in November 1996, the restored Port Lockroy Base was an instant hit.

Farther north, the British had already established a whaling museum at Grytviken, on South Georgia. Located in the old manager's house, the museum opened in early 1992. Several years later, another project restored the Grytviken Church.

Tourists at Port Lockroy, 2003. (*Author photo*)

Unfortunately, efforts to clean up the working parts of the stations could not defeat the progressive effects of South Georgia's harsh climate. As buildings continued to deteriorate, the British put more and more places off limits. In November 2001, they closed public access to most of Grytviken, the last station they had allowed tourists to enter. Four years later, after removing the worst of the derelict buildings, the British reopened all of South Georgia's historic first whaling station to visitors.

❆ ❆ ❆ ❆ ❆ ❆ ❆ ❆ ❆

THE STORIED ICE

Antarctica, the last continent to be discovered and a mystery for so many centuries, is now a place with a recognized and respected, even revered, past. It is today very different in many respects from what it was around 1500 when this narrative begins. People, men and women alike, now live there year-round, albeit only with help and support from the outside world. Others now visit almost routinely. And human actions have affected not only the Antarctic's populations of seals, whales, fish, and birds, but also the landscape itself, especially in the Peninsula Region. Antarctica, however, remains a vast and mysterious place of snow and ice, of danger and challenge, and always, of great beauty and enduring fascination. The Antarctic is rich in so many, many ways, some of them ones that the early explorers never imagined.

Five hundred years ago, men in tiny ships set out in search of a theoretical and theoretically rich Southern Continent. James Cook circumnavigated the Antarctic regions and declared that even if land were there, it was not worth finding. The supposed riches were a chimera. But then men discovered fur seals in the south, and people realized that there actually was wealth to be had there. Later, different hunters, explorers, and other visitors found additional sources of profit in whales and elephant seals, krill and fish, as well as vast stores of scientific knowledge. But they found something else too, something intangible—the magnificence, wonder, and magic of the Antarctic itself. Jean-Baptiste Charcot wrote of this when he tried to explain the lure of Antarctica:

> Why then do we feel this strange attraction . . . , a feeling so powerful and lasting, that when we return home we forget the mental and physical hardships, and want nothing more than to return . . . ? Why are we so susceptible to the charm of these landscapes when they are so empty and terrifying? Is it a delight in the unknown? Are we intoxicated by the struggle and effort needed to travel and survive here? Is it pride in attempting and achieving what others have not done? Or is it the pleasure of being far removed from meanness and pettiness? It is something of all these, but it is something else as well. . . . [These] regions stamp us in some way with religious awe. . . . Here we are in the sanctuary of sanctuaries, where nature reveals herself in all her tremendous power. . . . The man who has been able to enter such a place feels his spirit lifted.[18]

The stories of those who discovered, explored, and braved this vast part of the world, and prized out its secrets, magnify the power Charcot wrote of. Antarctica's human history, including that of the Peninsula Region, has, without question, become an integral part of its riches. The part of the far south covered in this book, like all of Antarctica, is indeed a Storied Ice.

APPENDIXES

Appendix A: Antarctic Timeline

See Glossary for abbreviations of expedition names. Re location names, note that many of the places cited were given the names used here only in later years. From the International Geophysical Year (1957–58) onward, this table does not typically include detail for government expedition activities. The notation (?) following an entry indicates that there is a question regarding whether the event took place.

YEARS	ANTARCTIC PENINSULA REGION	OTHER ANTARCTIC REGIONS	EVENTS ELSEWHERE
1400–1499	1421–22 Chinese fleet to South Shetlands (?)		1421–22 Chinese fleet to Falklands (?) 1492 Columbus discovers "New World" 1493, 1494 Papal Bull and Treaty of Tordesillas divide "New World" between Spain and Portugal 1498 Vasco da Gama reaches India via Cape of Good Hope
1500–1550			1501 Amerigo Vespucci reaches high latitude along South America's east coast (?) 1503 or 04 Portuguese voyage reports sighting Southern Continent (?) 1513 Vasco Nuñez Balboa crosses Isthmus of Panama and discovers "Great South Sea" (Pacific Ocean) 1520 Ferdinand Magellan discovers Strait of Magellan; crew completes first world circumnavigation in 1522
1551–1599	1599 Dirk Gherritz perhaps reaches 64° S and sees land (?)		1578 Francis Drake discovers sea south of Tierra del Fuego 1582 Pope Gregory XIII introduces Gregorian calendar, shifting dates forward about ten days; Christian countries gradually adopt it over following three centuries 1588 England defeats Spanish Armada 1592 John Davis discovers Falklands

YEARS	ANTARCTIC PENINSULA REGION	OTHER ANTARCTIC REGIONS	EVENTS ELSEWHERE
1600–1649			1616 Willem Schouten and Jacob Le Maire discover and round Cape Horn
1650–1699	1675 Anthony de la Roché discovers South Georgia		1699–1700 Edmond Halley makes scientific voyage to Southern Ocean
1700–1749		1738–39 Jean-Baptiste Bouvet de Lozier discovers Bouvet Island, in the far South Atlantic Ocean	1714 British Board of Longitude established, offers reward for creation of accurate seagoing timekeeper
1750–1799	1756 Gregorio Jerez, *León*, rediscovers South Georgia 1762 Joseph de la Llana in *Aurora* reports discovery of Aurora Islands 1775 James Cook lands on and claims South Georgia for Great Britain; discovers South Sandwich Islands Late 1780s Start of sealing at South Georgia	1768–71 James Cook voyages to Tahiti to observe transit of Venus, searches for Southern Continent 1771–72, 1773–74 Summers, Yves-Joseph de Kerguelen-Trémarec discovers Iles Kerguelen 1771–72 Marion du Fresne discovers Iles Crozet, Marion, and Prince Edward Islands 1772–75 Cook circumnavigates Antarctic continent; Jan 17, 1773, crosses Antarctic Circle; Jan 30, 1774, reaches 71° 10' S 1791–92 Whaling/sealing begins at Iles Kerguelen	1753 James Lind's *A Treatise of the Scurvy* published 1761 James Harrison's seagoing timekeeper (chronometer) successfully tested 1760s British and French occupy Falkland Islands/Islas Malvinas; French cede their position to Spain Early 1770s Whaling ships to Falklands 1774 James Watt builds first modern steam engine 1775 American Revolution begins 1776–78 James Cook's third voyage 1780s Sealing begins in Falklands 1789 French Revolution
1800–1809	1800–01 Summer, South Georgia fur sealing peaks	1806 Abraham Bristow discovers Auckland Islands, to south of New Zealand 1810 Frederick Hasselburg discovers Macquarie, Campbell Islands; sealing begins	1803–15 Napoleonic Wars engulf Europe 1807 Robert Fulton builds first commercial steamboat
1810–1819	1819 Feb 19, William Smith discovers South Shetlands Sept, *San Telmo* sinks in Drake Passage; wreckage reaches South Shetlands 1819–20 Dec–Jan, *Espírito Santo* and *Hersilia* make sealing voyages to South Shetlands; *San Juan Nepomuceno* may have sealed in South Shetlands Dec–Jan, Fabian von Bellingshausen surveys South Georgia and South Sandwich Islands, beginning two-summer Antarctic circumnavigation		1812–15 War between U.S. and Great Britain 1816 Argentina declares independence from Spain 1816 Earliest experiments in photography, by Nicéphore Niepce of France 1818 Chile declares independence from Spain

APPENDIX A: ANTARCTIC TIMELINE

1820–1829	1820 Jan–Mar, Edward Bransfield surveys South Shetlands; Jan 30, sights Antarctic Peninsula Nov, Fur seal rush in South Shetlands begins Nov 16, Nathaniel Palmer sights Antarctic Peninsula 1821 Jan 27, Bellingshausen discovers Alexander I Land; Feb, surveys South Shetlands Feb 7, John Davis lands on Antarctic Peninsula Marooned party from *Lord Melville* winters on King George Island Dec 6, George Powell and Nathaniel Palmer discover South Orkneys 1823 Feb 20, James Weddell reaches 74° 15′ S in Weddell Sea Mar, Benjamin Morrell discovers "New South Greenland," in Weddell Sea 1829 Jan–Mar, *Chanticleer* scientific visit to Deception Island	1820 Jan–Feb, Bellingshausen sails along coast of East Antarctica; Jan 27, sights Antarctic continent 1821 Jan 20, Bellingshausen discovers Peter I Island, first land south of Antarctic Circle 1820–21 Summer, Bellingshausen completes Antarctic circumnavigation 1822–23 Summer, Benjamin Morrell reports sailing off unseen coast of East Antarctica at high latitudes, landing on Bouvet Island	1820 World map showing South Shetlands published 1822–25 *Coquille* completes world circumnavigation to study terrestrial magnetism; first attempt to fix position of south magnetic pole
1830–1839	1830 Jan–Mar, *Seraph*, *Annawan*, and *Penguin*, sealing/scientific voyage to South Shetlands; James Eights makes scientific observations 1832 Feb, John Biscoe discovers Adelaide Island; names and claims Graham Land for Britain 1833 Dec, *Rose* and *Hopeful* to South Shetlands, *Rose* crushed in ice 1838 Jan–Mar, Dumont d'Urville to Weddell Sea, South Shetlands, and Antarctic Peninsula 1839 Feb–Mar, Charles Wilkes to South Shetlands, Antarctic Peninsula, Bellingshausen Sea	1831 Feb, John Biscoe discovers Enderby Land 1833 Dec, Peter Kemp discovers Kemp Land 1839 Feb–Mar, John Balleny discovers Balleny Islands and Sabrina Land	1830 Friedrich Gauss calculates position of south magnetic pole 1831 James Clark Ross reaches north magnetic pole Chile makes vague Antarctic claim 1833 Jan, Great Britain re-asserts sovereignty over the Falkland Islands
1840–1849	1841–42 Summer, William Smyley visits Deception Island, reports eruption 1842–43 Summer, James Clark Ross to Weddell Sea, South Shetlands, Antarctic Peninsula	1840 Jan, d'Urville discovers Adélie Land Jan–Feb, Wilkes sails 1,500 miles along coast of East Antarctica, sights scattered land 1840–42 James Clark Ross discovers Ross Sea,	1843 Jun 23, Great Britain asserts formal claim to "Settlements in the Falkland Islands and their Dependencies" 1844 First news transmitted by telegraph

307

YEARS	ANTARCTIC PENINSULA REGION	OTHER ANTARCTIC REGIONS	EVENTS ELSEWHERE
	1846 *Esther* visits South Georgia; her surgeon and four other men buried at King Edward Cove (earliest graves in Grytviken cemetery)	Victoria Land, Ross Ice Shelf; Feb 23, 1842, new farthest south of 78° 10′ S 1844–45 *Pagoda* voyage, follow-up to Ross	1845–49 Irish Potato Famine 1847 Search for Sir John Franklin, lost in the Arctic, begins; absorbs British interest in polar matters for years 1848 Gold discovered in California Dec, settlement at Punta Arenas founded
1850–1859	Several, generally unsuccessful, sealing voyages to South Shetlands	1850 Feb–Mar, Thomas Tapsell, *Brisk*, whaling to Balleny Islands 1853 Jan 26, Sealer Mercator Cooper lands on Victoria Land 1853 Nov 15, John Heard discovers Heard Island; sealing begins 1855	1850s Matthew Fontaine Maury's sailing directions encourage ships to take Great Circle routes through Southern Ocean 1853–56 Crimean War
1860–1869			1861–65 U.S. Civil War 1867 U.S. purchases Alaska from Russia 1869 Suez Canal opens to traffic U.S. transcontinental railroad completed, significantly reduces Cape Horn traffic
1870–1879	Sealing voyages to South Shetlands, South Georgia, and South Sandwich Islands 1873–74 Eduard Dallmann, *Grönland*, whaling and sealing in South Shetlands, Antarctic Peninsula	1874 Feb, British oceanographic vessel *Challenger* dips south of Antarctic Circle 1874–75 Summer, transit of Venus expeditions: France, U.S., Britain, Germany, to Iles Kerguelen, Iles Crozet, Auckland Islands, and Campbell Island	1870 Svend Foyn patents harpoon gun 1870–71 Franco-Prussian war 1874 Gray brothers publish Antarctic whaling proposal 1878 Baron Adolf Erik Nordenskiöld completes first transit of Northeast Passage
1880–1889	1882–83 German International Polar Year (IPY) Expedition to South Georgia		1882–83 First IPY, mostly in Arctic 1884 World adopts Greenwich as common prime meridian 1880s National geographical societies begin to promote Antarctic exploration
1890–1899	1892–93 Summer, Dundee, Scotland, whaling fleet to South Shetlands and Antarctic Peninsula 1892–93, 1893–94 Summers, Carl Anton Larsen, whaling reconnaissance to South Shetlands and	1894–95 Summer, Henrik Bull, whaling reconnaissance to Ross Sea; Jan 24, seven men land at Cape Adare 1898–99 Summer, *Valdivia*, German	1895 Sixth International Geographical Congress urges Antarctic exploration 1896 First Modern Olympics 1897 Britain grants Marconi first patent for

APPENDIX A: ANTARCTIC TIMELINE

	Antarctic Peninsula; significant exploration	oceanographic expedition to Southern Ocean	a radio device
	1897–99 Adrien de Gerlache's *Belgica* expedition	1898–1900 Carsten Borchgrevink's *Southern Cross* expedition; 1899, ten men winter at Cape Adare	1898 Spanish-American War 1899 *Yermak*, first modern icebreaker launched, used in Arctic Russia
1900–1909	1901–03 Otto Nordenskjöld's *Antarctic* expedition 1902–04 William Speirs Bruce's *Scotia* expedition 1903–05 Jean-Baptiste Charcot's *Français* expedition 1904 Feb, Argentina takes over Bruce's base in South Orkneys, maintains it continuously thereafter 1904 Nov, Shore whaling station established at Grytviken, inaugurating Antarctic whaling industry 1905–06 Summer, *Admiralen*, first whaling factory ship, to South Shetlands 1906 Feb, H.M.S. *Sappho* visits South Georgia 1906–07 Whaling based at Deception Island begins 1908–10 Charcot's *Pourquoi-Pas?* expedition	1901–03 Erich von Drygalski's *Gauss* expedition, explores Wilkes Land; 1902, winters frozen in off coast of East Antarctica 1901–04 Robert Falcon Scott's *Discovery* expedition; 1902, 03, winters on Ross Island; first significant probes into Antarctic interior 1907–09 Ernest Shackleton's *Nimrod* expedition; 1908, winters at Cape Royds on Ross Island; Jan 9, 1909, four men reach 88° 23′ S; Jan 16, 1909, three men reach south magnetic pole	Early 1900s Hydrogenation processes developed 1903 Wright Brothers' first powered and controlled heavier-than-air aircraft flight 1904–05 Russo-Japanese War 1905 Roald Amundsen completes first sea transit of Northwest Passage 1908 Jul 21, Great Britain formally claims much of Peninsula Region as "Falkland Islands Dependencies" 1908 Frederick Cook claims to reach North Pole 1909 Robert Peary reaches North Pole (?)
1910–1919	1911 Reindeer introduced to South Georgia 1911–12 Wilhelm Filchner's *Deutschland* expedition 1912–13 Summer, *Daisy* whaling/sealing voyage to South Georgia 1913 Oct 8, Baby born on South Georgia Dec 25, Church consecrated on South Georgia 1914–16 Ernest Shackleton's *Endurance* expedition 1916 Jan, American research vessel *Carnegie* calls at South Georgia during sub-Antarctic circumnavigation	1910–12 Roald Amundsen's *Fram* expedition; 1911 winters on Ross Ice Shelf; Dec 14, 1911, five men reach South Pole 1910–13 Robert Falcon Scott's *Terra Nova* expedition; 1911, 12, winters at Cape Evans; Jan 17, 1912, five men reach South Pole, all die during return 1910–11, 1911–12 Summers, Nobu Shirase's *Kainan Maru* expedition to Ross Sea 1911–14 Douglas Mawson's *Aurora* expedition; 1911 two groups winter East Antarctica; major sledging; 1912, one group winters 1914–17 Shackleton's Ross Sea party; 1915, 16, winters on Ross Island; major sledging 1915–16 Dec–Apr, American research vessel *Carnegie*'s sub-Antarctic circumnavigation	1910 Thomas Cook and Sons proposes tourist cruise to Antarctic for 1911 1912 Alfred Wegener proposes theory of continental drift Apr, *Titanic* sunk by iceberg 1914 Panama Canal opens to traffic 1914–18 World War I 1919 League of Nations formed Treaty of Versailles formally ends World War I

YEARS	ANTARCTIC PENINSULA REGION	OTHER ANTARCTIC REGIONS	EVENTS ELSEWHERE
1920–1929	1920–22 John Lachlan Cope attempts to explore Antarctic Peninsula; 1921, Maxime Lester and Thomas Bagshawe winter at Waterboat Point 1921–22 Summer, Ernest Shackleton's *Quest* expedition; Jan 5, 1922, Shackleton dies at Grytviken 1923 Discovery Committee formed 1925–26 Summer, *Lancing* uses slipway on whaling factory ship, beginning pelagic whaling 1925–26 Summer, First Discovery Committee voyages 1925–27 Summers, *Meteor* scientific voyage, identification of Antarctic Convergence 1927–28 Summer, Lars Christensen's first *Norvegia* voyage, scientific work in Peninsula Region and elsewhere 1928–29 Summer, Ludwig Kohl-Larsen at South Georgia 1928–29, 1929–30 Summers, Hubert Wilkins, first Antarctic airplane flights	1923–24, 1924–25 Summers, Carl Anton Larsen and *Sir James Clark Ross* initiate Ross Sea whaling 1928–30 Richard Byrd; 1929, winters at Bay of Whales (Little America); major sledging, aerial exploration; Nov 29–30, 1929, flight over South Pole 1929–30 Summer, Hjalmar Riiser-Larsen, *Norvegia*, explores East Antarctic coast using seaplanes Dec 26, Plane from whaling ship *Kosmos* lost with both men aboard 1929–30, 1930–31 Summers, Douglas Mawson leads Australian/New Zealand expedition (BANZARE), explores East Antarctic coast, using seaplanes	1923 Britain claims Ross Sea Region for New Zealand as "Ross Dependency" 1924 France claims Adélie Land 1925, 1927 Argentina claims South Orkneys, South Georgia, and South Sandwich Islands 1926 Byrd flies over North Pole (?) Roald Amundsen/Lincoln Ellsworth/Umberto Nobile fly over North Pole from Spitsbergen to Alaska in *Norge*, a dirigible 1927 Charles Lindbergh flies solo across the Atlantic 1928 Hubert Wilkins and Carl Ben Eielson fly across Arctic from Alaska to Spitsbergen 1929 U.S. stock market crashes
1930–1939	1931 Most shore whaling stations close as pelagic whaling takes over 1934–35, 1935–36 Summers, Lincoln Ellsworth attempts, then succeeds, in making first trans-Antarctic flight 1934–37 John Riddoch Rymill's BGLE to west coast of Antarctic Peninsula; major ground and aerial surveys	1930–31 Summer, *Norvegia* circumnavigates Antarctica 1933–34 Summer, Lincoln Ellsworth's first attempt at trans-Antarctic flight, from Bay of Whales 1933–35 Byrd's 2nd expedition; 1934, winters at Little America II, Byrd winters alone inland; major sledging and aerial exploration 1938–39 Summer, Ellsworth, aerial exploration in East Antarctica Summer, German *Schwabenland* expedition to East Antarctica	1930 and after, worldwide economic collapse 1931 International Whaling Convention 1932–33 Second IPY 1933 Britain claims Australian Antarctic Territory on behalf of Australia Macquarie Island declared a wildlife sanctuary 1939 Norway claims Queen Maud Land Sep 1, Germany invades Poland, setting off World War II
1940–1949	1940–41 USAS base at Marguerite Bay; 1940, 26 men winter, extensive sledging and aerial survey 1941 Jan–Mar, *Queen of Bermuda* to South Shetlands, Antarctic Peninsula, Weddell Sea	1940–41 USAS base at Bay of Whales, 1940, 33 men winter; extensive sledging and aerial survey 1940–42 German raiders use Iles Kerguelen as	1940–45 World War II engulfs world 1940 Nov 6, Chile claims much of Antarctic Peninsula Region 1944, 1945, 1946 International whaling

APPENDIX A: ANTARCTIC TIMELINE

	1942 Jan-Feb, *Primero de Mayo* to South Shetlands, Antarctic Peninsula 1943 Jan, *Carnarvon Castle* to Deception Island Feb-Mar, *Primero de Mayo* to South Shetlands, Antarctic Peninsula including Marguerite Bay 1943-45 Operation Tabarin establishes permanent bases 1945 FIDS replaces Operation Tabarin; establishes new bases 1946-47 Summer, Argentina establishes her first new Antarctic base after South Orkneys Chile establishes her first Antarctic base Operation Highjump, survey of Alexander and Charcot Islands, probes of Weddell Sea 1947-48 Finn Ronne's expedition; 1947, 21 men and 2 women winter Stonington Island Summer, U.S. Operation Windmill brings first icebreakers to Peninsula Region	base; Jan 1941, *Pinguin* captures Norwegian whaling ships off Queen Maud Land 1946 Antarctic whaling resumes 1946-47 Summer, Operation Highjump, massive U.S. summer Antarctic program, primarily outside Peninsula Region 1947-48 Summer, Australian Antarctic program begins with Heard and Macquarie Island bases Operation Windmill, U.S. follow-up to Operation Highjump, primarily outside Peninsula Region 1949-52 International (Norwegian-British-Swedish) expedition to Queen Maud Land; 1950, 51, winter near Cape Norvegia. 1949-53 French government expeditions to Adélie Land	conferences 1945 End of World War II 1945 United Nations formed 1946 International Whaling Commission formed 1947 Argentina and Chile reject British offer to refer Antarctic claim disputes to International Court of Justice at the Hague 1948 U.S. proposal to settle Antarctic claim disputes rejected Berlin Airlift; Cold War intensifies 1949 U.S.S.R. says Bellingshausen was first to see Antarctic continent
1950–1959	Throughout decade, Argentina, Chile, and Great Britain establish more and more bases 1951-52, 52-53, 55-56, 56-57 Summers, Duncan Carse's South Georgia Survey 1954-55 Summer, British South Georgia Expedition, mountain climbing 1955-56, 56-57 Summers, FID Aerial Survey Expedition maps South Shetlands, Antarctic Peninsula 1955-58 Commonwealth Trans-Antarctic Expedition; 1956, 57, winters in Weddell Sea 1955-59 British Royal Societies Expedition establishes, occupies IGY base at Halley Bay 1956 Dec, Chilean tourist flight over Peninsula 1956-57 Summer, Great Britain's Prince Philip visits the FID aboard the *Britannia*	1955-57 Multiple governments establish bases for IGY; participants include Australia, Belgium, France, Japan, New Zealand, Norway, South Africa, U.S., U.S.S.R. 1956 Jan, U.S. fuel tanker *Nespelen* spills 140,000 gallons of aviation gasoline in McMurdo Sound Nov 12, Huge iceberg, 208 x 60 miles, observed in Ross Sea 1956-58 Commonwealth Trans-Antarctic Expedition, staging from Ross Island; crossing completed at Ross Island Mar 1958 1958 Vinson Massif, highest mountain in Antarctica at about 16,050 feet, discovered	1950-53 Korean War 1957 U.S.S.R. launches Sputnik satellite 1957-58 IGY, scientific effort in Antarctic regions and elsewhere in world 1958 Scientific Committee on Antarctic Research (SCAR) created 1958-59 Conference to discuss continuation of IGY 1959 Dec, Antarctic Treaty signed

311

YEARS	ANTARCTIC PENINSULA REGION	OTHER ANTARCTIC REGIONS	EVENTS ELSEWHERE
	1956–59 U.S. IGY Ellsworth Station on Filchner Ice Shelf 1957–58 Summer, Argentina offers two tourist cruises to Antarctic Peninsula 1957–59 Argentine, Chilean, British bases participate in IGY		
1960–1969	1960 Dec, British Joint Services Expedition to South Georgia for mountaineering 1964–65 Nov–Mar, British Joint Services Expedition to South Georgia 1964–68 U.S. establishes Palmer Station on Anvers Island 1965 Dec, Last South Georgia whaling stations close 1966 Jan–Feb, Lindblad tourist cruise inaugurates modern Antarctic tourist industry 1967, 69, 70 Volcanic eruptions at Deception Island 1967–68 U.S.S.R. establishes Bellingshausen base on King George Island	Whaling continues at declining level 1961–62 Summer, Nuclear reactor installed at U.S. McMurdo station on Ross Island (closed down, removed 1973) 1964 Jun, Mid-winter flight made from New Zealand to Ross Island	Through decade, most countries begin including women in government Antarctic programs 1964 U.S.–British Commonwealth agreement on name for Antarctic Peninsula 1964 Agreed Measures for Conservation of Antarctic Flora and Fauna adopted, annex to Antarctic Treaty 1965–75 Vietnam War 1969 U.S. moon landing
1970–1979	Trickle of private yacht cruises to region begins, increasing in numbers by end of decade 1970–71, 76–77 Summers, British Joint Services Expeditions to Elephant Island 1976–77 Poland establishes Arctowski base on King George Island, to be followed in the next ten years by other countries 1977–78 Argentina establishes "settlement" at Esperanza base; Jan 7, baby born there	Whaling industry continues at greatly reduced level 1972–73 Summer, U.S. ship *Glomar Challenger* conducts scientific drilling in Ross Sea 1977 Feb, Tourist sightseeing flights over East Antarctica and Ross Sea Region begin 1979 Nov 28, sightseeing flight crashes into Mt. Erebus; all 257 people aboard die; sightseeing flights stop 1979–81 Transglobe Expedition crosses continent on snowmobiles; 1980 winters inland	1971 First International Earth Day celebrated, marking growing worldwide environmental consciousness 1972 Convention for Conservation of Antarctic Seals adopted, annex to Antarctic Treaty 1972 Antarctic Treaty recommendations for Antarctic Historic Monuments 1972 World Conference on National Parks, proposal to make Antarctica an international park
1980–1989	1982 Mar–Jun, Falkland Islands/Malvinas War 1983–85 Brabant Island Expedition; 1984 winters on the island. 1985 Ozone hole over Antarctica confirmed 1986–87 Summer, Peck Range in southern Antarctic	Nearly all commercial Antarctic whaling ends Tourist and yacht cruises increase in number 1983 Nov, Seven Summits expedition to Vinson Massif, start of Seven Summits craze 1985 Adventure Network International formed,	1980 Convention on Conservation of Antarctic Marine Living Resources ratified (CCAMLR); enters into force 1982 1982 Commercial whaling moratorium

APPENDIX A: ANTARCTIC TIMELINE

	Peninsula, last previously unvisited major Antarctic range, explored and mapped 1989 Jan 28, *Bahia Paraiso* oil spill off Anvers Island 1989–90 July–Mar, six-man International Trans-Antarctic Expedition travels length of Antarctic Peninsula (and continues across continent to end at U.S.S.R.'s Mirnyy base)	providing aerial access to Antarctic interior for private individuals 1985–86 Greenpeace Antarctic expeditions begin	1983 Challenge to Antarctic Treaty at United Nations 1987 New Zealand Antarctic Heritage Trust established 1988 Convention on Regulation of Antarctic Mineral Resource Activities (CRAMRA) adopted 1989 Fall of Berlin Wall; break-up of Soviet Union
1990–1999	1992 Whaling Museum opens on South Georgia 1993 United Kingdom Antarctic Heritage Trust begins survey of abandoned BAS bases 1994 Feb, Last dogs removed from Antarctica 1994–95 Wordie and Larsen Ice Shelves begin major disintegration 1996–97 Ice breaks up in Prince Gustav Channel, opening passage between east coast of Antarctic Peninsula and James Ross Island to navigation	1994–95 Sightseeing flights over continent resume 1999 South Pole station doctor becomes ill, treats self during winter before evacuated Oct 16	1991 International Association of Antarctica Tour Operators (IAATO) founded Madrid Protocol adopted Gulf War among Iraq, Kuwait, U.S., etc. 1994 Southern Ocean whaling sanctuary established 1997 Kyoto Protocol (amendment to U.N. Framework on Climate Change) signed; comes into force for signatories 2005
2000 and after	2000 and following, Warming in Peninsula Region continues 2002 Feb, Much of "B" (northern) section of Larsen Ice Shelf collapses 2007 Nov 23, M.V. *Explorer* sinks in Bransfield Strait 2007–09, Mar–Mar, (4th) International Polar Year; existing government bases participate 2008 Mar, Wilkins Ice Shelf partially collapses	2000 Mar, Giant iceberg, 180 miles long, approximately 4,500 square miles in area, calves off Ross Ice Shelf; disrupts wildlife migration for years after 2001 South Pole station doctor become ill, evacuated Apr 24 2006–07 U.S. inaugurates surface route from Ross Island to South Pole for heavy cargo 2007–09, Mar–Mar, (4th) IPY, existing government bases plus one new one (Belgium) participate	2001 Islamic terrorist attack on U.S. U.S. invades Afghanistan 2003 U.S. invades Iraq 2007–09, Mar–Mar, (4th) IPY in the Arctic

Appendix B: Antarctic Firsts

See Glossary for definitions of expedition names and/or acronyms. Events occurring in the Antarctic Peninsula Region are in **bold** type; when an event took place both there and elsewhere, it is also underlined. Names in brackets following entries are for leader of the expedition achieving the first, except in cases where the entry is for an achievement by a single individual or indicated individuals. The notation (?) at the end of an entry indicates that the event itself is in doubt. The same notation inside brackets indicates a question concerning the person responsible.

Many "firsts" involve the first time that a *particular* landmass or part of the Antarctic was seen. There are, of course, a great many such events, and only the most significant are listed. The following also omits most "firsts" of the sort that exploded in the 1990s: the first person from a particular country to do something, the first to use a new route to reach somewhere such as the South Pole, speed records to achieve something, etc.

YEARS	DISCOVERY, EXPLORATION, AND OCCUPATION	COMMERCE AND TOURISM; TECHNOLOGY AND TRANSPORT	SCIENCE AND NATURE	WOMEN AND CHILDREN; POLITICS; MISCELLANEOUS
Prior to 1500	**1422 South Shetlands, Antarctic Peninsula perhaps sighted [Chinese fleet] (?)**		1497 European description of penguin [Vasco da Gama]	About 1370 The word "Antarktyk" used in English-language publication [*The Travels of Sir John Mandeville*]
1500–1549	1502 Far south voyage along South American coast [Amerigo Vespucci] (?) 1503 or 04 Reported sighting of Southern Continent [Cristobal Jacques] (?) 1520 Nov, Sea passage from Atlantic to Pacific [Ferdinand Magellan] 1526 Ship reaches open sea south of Staten Island [Francisco de Hoces, *San Lesmos*] (?)			1515 Maps and globes explicitly showing Southern Continent published [Leonardo da Vinci; Johannes Schöner] 1531 Term "Terra Australis" used on map [Orontius]
1550–1599	1567–69 Expedition sent to search for Southern Continent, in South Pacific [Alvaro de Mendaña] 1578 Oct, Discovery of open sea south of Tierra del Fuego [Francis Drake]		1587 Penguins called by their English name [Thomas Cavendish]	

YEARS	DISCOVERY, EXPLORATION, AND OCCUPATION	COMMERCE AND TOURISM; TECHNOLOGY AND TRANSPORT	SCIENCE AND NATURE	WOMEN AND CHILDREN; POLITICS; MISCELLANEOUS
1600–1649	1599 Sep, Ship reaches 64° S, land seen [Dirk Gherritz] (?) 1616 Jan 29, Discovery and rounding of Cape Horn [Willem Schouten and Jacob Le Maire]			
1650–1699	1675 Apr, Land discovered south of Antarctic Convergence (South Georgia) [Anthony de la Roché]			
1700–1749	1738–39 Summer, Voyage through many degrees of longitude at high latitude [Jean-Baptiste Bouvet de Lozier, South Atlantic]		1700 Jan–Feb, Scientific expedition to Southern Ocean [Edmond Halley]	
1750–1799	1772–75 High-latitude Antarctic circumnavigation [James Cook] 1773 Jan 17, Antarctic Circle crossed [Cook] 1774 Jan 30, New farthest south, at 71° 10′ S [Cook] 1775 Jan 17, Landing south of Convergence [Cook, on South Georgia] 1790s Wintering south of Convergence [sealers, at South Georgia]	1772–75 Use of chronometer on Southern Ocean exploratory voyage [Cook] 1788 Proposal for seal conservation [John Leard] Late 1780s Sealing south of Convergence [at South Georgia]	1764 Feb–Mar, Sealing in near-Antarctic regions [Louis de Bougainville in Falklands] 1772–75 Trained scientists to spend significant time south of Convergence [Johann Forster et al., with James Cook]	1775 Jan 17, Sovereignty claim in Antarctic regions [James Cook, South Georgia] 1776 Published map of land south of Convergence based on observation [Cook, of South Georgia and South Sandwich Islands] 1789 Dec 24, Ship hit and sunk by iceberg in Southern Hemisphere [*Guardian*, in Indian Ocean]
1800–1819	1819 Feb 19, Definite sighting of land south of 60° S [William Smith, South Shetlands] 1819 Oct 16, Definite landing on land south of 60° S [Smith, South Shetlands]	1819 Dec, Sealing in South Shetlands [*Espírito Santo*] or *San Juan Nepomuceno* (??)]	1820 Identification of krill [Fabian von Bellingshausen] Jan, Volcanic activity observed in Antarctic regions [Bellingshausen, at Zavodovski island]	1812 Proposal for U.S. exploratory expedition to Southern Ocean [Edmund Fanning] c1818, Birth in near-Antarctic [daughter, to wife of John Carnell, British sealer, at Iles Kerguelen]
1820–	1820 Jan 27, Sighting of Antarctic		1820–21 Summer, Geological	1820 Jun, Map published

APPENDIX B: ANTARCTIC FIRSTS

1829	continent [East Antarctica, Fabian von Bellingshausen] Jan 30, Sighting of Antarctic Peninsula [Edward Bransfield] 1821 Jan 20, Land discovered south of Antarctic Circle [Peter I Island, Bellingshausen] Feb 7, Documented landing on Antarctic continent [John Davis, on Antarctic Peninsula] Wintering south of 60° S [10 men of *Lord Melville*, in South Shetlands] 1823 Feb 20, New farthest south, 74° 15' S [James Weddell, in Weddell Sea] Mar 15, Claimed landing south of Antarctic Circle [Benjamin Morrell, on New South Greenland] (?)	samples collected from Antarctic [from South Shetlands, Bellingshausen and sealers B. Astor and Donald Mackay of *Jane Maria* and *Sarah*] 1829 Jan–Mar, Purely scientific expedition south of Convergence [Henry Foster, *Chanticleer*, at Deception Island]	showing significant part of Antarctic regions based on discovery [of South Shetlands, on sheets to be used in Adrien Brue's 1820 World Atlas]
1830–1839	1831 Feb 24, Reported sighting of East Antarctica [Enderby Land, John Biscoe] Voyage into Ross Sea [Samuel Harvey, *Venus*] (?) 1839 Feb 12, Definite landing south of Antarctic Circle [Thomas Freeman, off Borradaile Island, Balleny Islands]	1830 Jan–Feb, Fossils found in Antarctic regions [James Eights in South Shetlands, on *Annawan*] Jan–Feb, Flowering plan found below 60° S [Eights] 1830 Calculation of location of south magnetic pole [Friedrich Gauss]	1833 Dec, Ship crushed by ice in Antarctic [*Rose*, near South Shetlands] 1839 Jan 29, Woman south of Antarctic Circle [name unknown, with John Balleny]
1840–1849	1841 Jan–Mar, Documented voyage into Ross Sea [James Clark Ross] 1842 Feb 23, New farthest south, 78° 10' S [Ross, in Ross Sea]	1840 Jan, Emperor penguin egg found [Dumont d'Urville, off Adélie Land, East Antarctica] 1841 Jan, Major ice shelf seen, [Ross Ice Shelf, Ross] 1842 Feb, Deception Island seen	1840 Southern Continent formally called "Antarctic continent" [Charles Wilkes, off East Antarctica]

YEARS	DISCOVERY, EXPLORATION, AND OCCUPATION	COMMERCE AND TOURISM; TECHNOLOGY AND TRANSPORT	SCIENCE AND NATURE	WOMEN AND CHILDREN; POLITICS; MISCELLANEOUS
1850–1859	1853 Jan 26, Landing on Greater Antarctica [Mercator Cooper, on Victoria Land]			
1860–1869		1860s Harpoon gun invented, patented 1870 [Svend Foyn]		1860 Promotion of international Antarctic exploration [Matthew Fontaine Maury] 1866 Southern Hemisphere ice chart [British Admiralty] 1869 Published proposal for Antarctic expedition with deliberate wintering [John Davis]
1870–1879		1873–74 Nov–Mar, Steamship in Antarctic waters [Dallmann, with *Grönland*] 1874 Feb 16, Steamship crosses Antarctic Circle [*Challenger*] Photographs of Antarctic icebergs [*Challenger*] Published proposal for whaling in Antarctic regions [Gray Brothers, Scotland]	1874 Scientific evidence that Antarctica is a continent rather than an archipelago [ocean bottom dredgings, *Challenger*]	
1880–1889	1882–83 Scientific expedition to winter south of Convergence [German IPY expedition, South Georgia]	1882–83 Photographs of land south of the Convergence [German IPY expedition, South Georgia]	1882–83 First International Polar Year (IPY)	
1890–1899	1895 Jan 24, Widely recognized landing on greater Antarctica [seven men led by Henrik Bull, at Cape Adare, Victoria Land] 1898 Wintering south of Antarctic Circle [eighteen men aboard *Belgica*, in Bellingshausen Sea] 1899 Wintering on mainland [ten	1892–93 Summer, Whaling fleets to Antarctic [Dundee whalers; Carl Anton Larsen, *Jason*] 1893 Jan, Photographs of land south of 60° S [Charles Donald, Dundee whalers] Dec 12, Skis used in Antarctic [Larsen, on Christensen Nunatak]	1895 Jan 18, Plant life found south of Antarctic Circle [lichens on Possession Island in Ross Sea, Carsten Borchgrevink, on Henrik Bull's expedition] 1898 Jan 24, Insects found in Antarctic regions [Émile Racovitza, *Belgica*, west side of Antarctic Peninsula]	1899 Oct 14, Known death on Antarctic mainland [Nicolai Hanson at Cape Adare, *Southern Cross* expedition] 1890, Term "Antarctica" used as name for south polar continent [by J. G. Bartholomew, on map in an atlas]

APPENDIX B: ANTARCTIC FIRSTS

		men at Cape Adare, Carsten Borchgrevink's *Southern Cross* expedition]		
		1894 Dec, Photographs from south of Antarctic Circle [Henrik Bull, of the Balleny Islands]	1899 Nov, Insects found south of Antarctic Circle [Herlof Klövstad, *Southern Cross*, near Cape Adare]	
		1897–99 Major photographic expedition record [*Belgica*]		
		1898 Jan 30–Feb 6, Sledging trip in Antarctic [*Belgica* expedition, on Brabant Island] Ship beset for winter [*Belgica*]		
		1898–1900 Dogs used; kayaks used [*Southern Cross*]		
		Attempt to shoot movies [*Southern Cross*]		
1900–1909	1900 Feb 17, Landing on Ross Ice Shelf; sledge journey on Ross Ice Shelf [*Southern Cross*]	1902 Jan, Prior expedition photos used for geographic feature identification [*Antarctic*, using photos taken by *Belgica*]	1902 Oct 12, Emperor penguin rookery seen [Reginald Skelton and two others at Cape Crozier, Ross Island, *Discovery*]	1903 Mar, Permanent scientific base established [Laurie Island, South Orkneys, *Scotia*]
	1902 Major inland exploration [*Discovery* expedition, from Ross Island]	Feb 4, Balloon flight [Scott, *Discovery*]	Nov, Vertebrate fossil found [Nordenskjöld, extinct penguin species, on Seymour Island]	1904 Feb 20, Mail postmarked from Antarctic regions [from Laurie Island, South Orkneys, Argentine and Falklands' cancellations]
	Dec 30, New farthest south, about 82° 16′ S [Robert Falcon Scott, Ernest Shackleton, Edward Wilson, *Discovery*]	Feb 4, Aerial photos [Shackleton, *Discovery*, from balloon on its second ascent]	1903 Nov, Fossils found south of Antarctic Circle [Hartley Ferrar, Southern Victoria Land, *Discovery*]	1906–07 Summer, Women to South Shetlands [whaling wives]
	Dec 31, Polar plateau reached [Albert Armitage and ten others, *Discovery*]	Motor boat used [*Gauss* expedition]	**1904 Feb–Mar, Major oceanographic exploration in Weddell Sea [*Scotia*]**	1908–09 Book published in Antarctic [*Aurora Australis*, *Nimrod*]
	1903 Dec 18, Antarctic dry valley seen [Taylor Dry Valley, by Scott, *Discovery*]	Mar 4, Electricity for lighting [*Discovery*]	1908 Dec 17, Coal found [Shackleton on Beardmore Glacier, *Nimrod*]	**1908 April, Formal proposal for Antarctic crossing [William Speirs Bruce]**
	1908 Mar 10, Ascent of Mt. Erebus, Ross Island, first major Antarctic mountaineering [six men led by T.W. Edgeworth David, *Nimrod*]	Mar 29, Telephone used [Erich von Drygalski, from balloon, *Gauss*]		1908 July 21, Formal claim to Antarctic regions [Great Britain, to "Falkland Islands Dependencies"]
	1909 Jan 9, New farthest south, 88° 23′ S [Shackleton, Frank	Nov, Penguin voice recorded [Adélie penguins, *Gauss*]		1909 Nov 30, Resident
		1903 Sep, Seal voice recorded [Bruce, *Scotia*]		
		Oct, Successful movie (50 feet of a penguin rookery) [Bruce,		

319

YEARS	DISCOVERY, EXPLORATION, AND OCCUPATION	COMMERCE AND TOURISM; TECHNOLOGY AND TRANSPORT	SCIENCE AND NATURE	WOMEN AND CHILDREN; POLITICS; MISCELLANEOUS
	Wild, Jameson Adams, Eric Marshall, *Nimrod*] Jan 16, South magnetic pole reached [T. W. Edgeworth David, Douglas Mawson, Alistair Mackay, *Nimrod*]	*Scotia*] 1904 Nov 16, Antarctic whaling station [Carl Anton Larsen, Grytviken on South Georgia] 1905–06 Summer, Whaling factory ship in Antarctic [*Admiralen*, to South Shetlands] 1908 Feb 1, Motor car used [Shackleton, *Nimrod*]		government official [James Innes Wilson, Stipendiary magistrate, South Georgia] Dec 20 or 23, official mail from Antarctic [from Grytviken, South Georgia]
1910–1919	1911 Dec 14, South Pole reached [Roald Amundsen, Olav Bjaaland, Helmer Hanssen, Sverre Hassel, Otto Wisting, *Fram*] 1912 Ship beset in Weddell Sea for winter [*Deutschland*]	1910 Nov, Significant proposal for tourist cruise to Antarctic [Thomas Cook travel agency, New Zealand] 1910–12 Professional photographer in Antarctic; commercial movie produced [Herbert Ponting, *Terra Nova* expedition] 1913 Feb 20, Two-way wireless communication from/to Antarctic [from/to Commonwealth Bay, East Antarctica, Mawson's *Aurora* expedition]	1911, Jul, winter visit to emperor penguin rookery [Wilson, Cherry-Garrard, Bowers, *Terra Nova* expedition] 1912 Theory of continental drift proposed [Alfred Wegener] 1912 Oct 13, Record wind speed [measured gust of 202 mph, limit of instruments, Mawson, Commonwealth Bay] Dec 5, Meteorites found [Francis Bickerton, near Commonwealth Bay, Mawson's *Aurora* expedition]	1912 Dec 14, Death down Antarctic crevasse [Belgrave Ninnis, with Mawson, *Aurora*] 1913 Oct 8, Recorded birth south of Convergence [Solveig Gundjörg Jacobsen, Grytviken, South Georgia] 1913 Dec 25, Church consecrated in Antarctic [Grytviken, South Georgia] 1913–14 Summer, Jail in Antarctic [Grytviken, South Georgia] 1914 Women known to apply to join Antarctic expedition [*Endurance*] 1914–16 Attempt to cross Antarctic continent [Shackleton, *Endurance*] Ordained clergyman south of Antarctic Circle [Arnold Spencer-Smith, with Shackleton's Ross Sea Party]
1920–1929	1928–29 Inland exploration of South Georgia [Kohl-Larsen]	1920 Proposal for heavier-than-air flight [John Lachlan Cope] 1923–24 Summer, Whaling in	1923 Ongoing scientific investigation begins [Discovery Committee formed]	1920 Polar research institute formed [Scott Polar Research Institute, Cambridge, England]

APPENDIX B: ANTARCTIC FIRSTS

	Ross Sea [Carl Anton Larsen, *Sir James Clark Ross*]
	1924 Tourists [Peninsula Region]
	1925 Dec 11, Slipway used on whaling factory ship [*Lancing*]
	1925–27 Antarctic Convergence identified [*Meteor* and *Discovery*]
	1927 Mar, Time signals received at sea in Antarctic [*Discovery*]
	1928 Nov 16, Airplane flight [George Hubert Wilkins and Carl Ben Eielson (pilot)]
	1928–29 Sep–May, Woman explorer [Margit Kohl-Larsen, South Georgia]
	1928–30 Professional journalist with expedition [Russell Owen of *New York Times*, with Richard Byrd, Bay of Whales]
	1928–30 Major use of motor transport [Richard Byrd, from Bay of Whales, Ross Sea]
	Extended aerial exploration [Byrd]
	Regular radio contact with outside [Byrd]
	1929 Nov 29–30, Flight over South Pole [Byrd, Bernt Balchen (pilot) and two others]
	1929 Dec 26, Aircraft-related deaths [Leif Lier (pilot) and Ingvold Schreiner, of whaling ship *Kosmos*]
	1929 Whaling laws [Norway]
1930– 1939	1930 *The Antarctic Pilot* published [Hydrographer of the (British) Navy]
	1931 International effort to control whaling [League of Nations, International Convention for the Regulation of Whaling]
	1931–33 Antarctic Convergence location plotted around Antarctic continent [*Discovery II*]
	1931–33 Winter circumnavigation [*Discovery II*]
	1932 Feb 24, Marriage in Antarctic regions [Mr. A.G.N. Jones and Miss Vera Riches, at King Edward Point, South Georgia]
	Jun, Official government Antarctic place names committee formed [Great Britain]
	1933–35 Fully successful use of motor transport; regular live radio broadcasts to outside [Byrd, Bay of Whales]
	1934 Emperor penguin and Weddell seal voices recorded [Byrd]
	1934 Mar–Oct, Inland wintering [Richard Byrd, 123 miles south of Bay of Whales, at 80° 08′ S]
	1935 Feb 20, Woman sets foot on Antarctic continent [Caroline Mikkelsen, whaling
	1935 Nov 23–Dec 5, Aircraft flight across continent [Lincoln Ellsworth, Herbert Hollick-Kenyon, from Dundee Island to Bay of Whales]
	Nov 23–24, Plane landing on and taking off from polar plateau
	1936 Nov 24, Men reach Antarctic Peninsula plateau [John Rymill and Edward Bingham, BGLE]
	1939–41 Effort to establish permanent bases beyond South Orkneys [USAS]

YEARS	DISCOVERY, EXPLORATION, AND OCCUPATION	COMMERCE AND TOURISM; TECHNOLOGY AND TRANSPORT	SCIENCE AND NATURE	WOMEN AND CHILDREN; POLITICS; MISCELLANEOUS
		[Ellsworth, at about 79° S, 103° W]		wife, just offshore Vestfold Hills, East Antarctica]
				1937 Jan 27, Woman flies in Antarctic [Ingrid Christensen, whaling wife, as passenger over Vestfold Hills]
1940–1949	1940–41 Nov–Jan, Antarctic Peninsula crossed overland [USAS] 1944 Feb, Permanent bases established in Antarctic beyond South Orkneys [Deception Island and Port Lockroy, Operation Tabarin]	1941 "Wire photos" transmitted by radio from Antarctica [USAS] 1946–47 Summer, Extensive use of aircraft for whale spotting [*Balaena*] 1946–47 Summer, Modern icebreakers used [Operation Highjump] Modern helicopters used [Operation Highjump] Submarine in Antarctic [*Sennet*, Operation Highjump] 1947 Jan–Apr, Location filming for commercial movie [*Scott of the Antarctic*, at Hope Bay] 1947 Dec 13, Plane flight from another continent to Antarctica [Argentine navy, round trip from Piedrabuena, Patagonia, to Adelaide Island; did not land]	1940 Nov–Dec, High-altitude weather station [at 5,500 feet on Antarctic Peninsula plateau, east of Marguerite Bay, USAS]	1941 Jan 14, Act of war in Antarctic regions [German raider, *Pinguin*, captures Norwegian whaling vessels at about 59° S, 2° W] 1947 Women winter south of 60° S [Edith "Jackie" Ronne and Jennie Darlington, at Marguerite Bay, RARE] 1948 Jan 11, Child born south of 60° S [boy named Antarctic, to Aleksandra Akimovna Leonova, waitress on factory ship of Russian whaling fleet] 1948 Feb, Head of state visit [President González Videla, Chile, to South Shetlands and Antarctic Peninsula] 1948 Agreement restricting military action [Argentina/Chile/Great Britain] 1948 Nov 8, Base destroyed by fire; fire-related deaths (two) [Hope Bay base, FIDS] 1949–52 Fully international expedition [Norwegian-British-Swedish, at Cape Norvegia, Queen Maud Land]
1950–1959	1954 Feb, Permanent base established on mainland outside Peninsula Region [Mawson	1954–55 Oct–Mar, Modern adventure expedition [George Sutton and four others,	1952–53 Full-year observation of Emperor penguins [French group of seven at Pointe Géologie	1952 Feb, Modern shots fired in anger [Argentina, in

APPENDIX B: ANTARCTIC FIRSTS

	station, Australia] 1956–57 Base established, wintering at South Pole [Amundsen-Scott Station, U.S.] **Surface exploration of the Filchner Ice Shelf** [Seismic Traverse party from U.S. Ellsworth Station] **1957–58 Summer, Land crossing of Antarctic continent** [Vivian Fuchs, Commonwealth Trans-Antarctic Expedition]	mountaineering on South Georgia] 1955 Dec 20, Flights from outside to mainland other than Peninsula Region [U.S. flights from New Zealand to Ross Island, McMurdo Sound] 1956 Oct 31, Aircraft landing at South Pole [U.S. plane *Que Sera Sera*, Cmdr. Conrad Shinn (pilot) and six others] 1956 Nov, Parachute jump over South Pole [Technical Sgt. Richard Patton] 1956 Dec 22, Tourist overflight [Lan Chile, over Antarctic Peninsula] 1958 Jan–Feb, Tourist cruises [by Argentine government to South Shetlands, Peninsula]	Adélie Land] 1956 Sighted Nov 12, Iceberg size record, 208 x 60 miles [in Ross Sea] **1957–58 Major international cooperative effort in Antarctic** [IGY]	landing at Hope Bay] 1953 Apr 14, Criminal law case beyond South Georgia [Re wildlife protection, heard by FID magistrate on Deception Island] 1955–56 Summer, Woman scientist to work in field on mainland [Marie Klenova, at U.S.S.R. base Mirnyy] 1956 Jan, Major oil spill [U.S. supply vessel *Nespelen*, 140,000 gallons of aviation gasoline in McMurdo Sound] 1957 Oct 15, Women to see South Pole [Patricia Hepinstall and Ruth Kelly, stewardesses on Pan Am VIP flight overflying the pole] 1959 Dec 1, Governance regime for Antarctic [Antarctic Treaty]	
1960–1969	1964–65 Summer, Non-claimant country permanent base established in Peninsula Region [U.S., Palmer Station on Anvers Island]	1961 Apr 8–9, Winter flight lands on mainland [U.S., from New Zealand to McMurdo and Byrd bases and back] **1961–62 Summer, Krill harvesting** [U.S.S.R. fishing fleet] Nuclear power plant installed [at McMurdo Station, Ross Island] 1964 June 27–28, Mid-winter flight from outside to mainland [U.S., New Zealand to McMurdo Station, to evacuate ill winterer] Nov, Retracing Shackleton's South Georgia crossing route	1960–61 Summer, Meteorites found in Antarctic interior [U.S.S.R. and U.S. geologists at several locations] **1961–62 Summer, Discovery of genuine strait separating the Antarctic Peninsula from the main Antarctic continent** [subglacial strait near base of the Peninsula, by U.S. Antarctic Peninsula Traverse Party] 1967 Dec 4–5, Major destructive volcanic eruptions [Deception Island] 1968 Dec 28, Vertebrate animal	1960–61 Summer, Historic hut restoration [New Zealand, on Ross Island] 1964 Conservation protocol to Antarctic Treaty [Agreed Measures for the Conservation of Antarctic Flora and Fauna]	

YEARS	DISCOVERY, EXPLORATION, AND OCCUPATION	COMMERCE AND TOURISM; TECHNOLOGY AND TRANSPORT	SCIENCE AND NATURE	WOMEN AND CHILDREN; POLITICS; MISCELLANEOUS
		[Malcolm Burley and nine others]	fossils found in Antarctic interior [Peter Barrett, New Zealand, and David Johnston, U.S., at 85° 03' S, 172° 19 E]	
		1966 Jan–Feb, Modern tourist cruise [Lindblad Expeditions, to South Shetlands and Antarctic Peninsula]		
		Dec, Adventure mountaineering expedition to the Antarctic continent; Dec 18, ascent of Vinson Massif [American Alpine Club expedition, led by Nicholas Clinch; ascent of Vinson Massif by Barry Corbett, John Evans, William Long, Peter Schoening]		
1970–1979			1970 Nov 10, Complete vertebrate animal fossil found in Antarctic interior [U.S. geologists]	1972 Historic monuments proposed for protection under Antarctic Treaty
		1972–73 Summer, Solo yacht voyage to Antarctic regions [David Lewis, *Ice Bird*, to Antarctic Peninsula]	1972 Jul, Highest confirmed natural Earth wind speed, 202 mph [at Dumont d'Urville base]	1974 Women winter at government Antarctic base [Mary Alice McWhinnie and Sister Mary Odile Cahoon, at U.S. McMurdo Station]
		1977 Feb 13, Tourist overflight of mainland beyond Peninsula Region [Qantas charter over East Antarctica]		**1978 Jan 7, Baby born on Antarctic continent [Emilio de Palma, at Argentine Esperanza Base, Hope Bay]**
		1978–79 Mar–Feb, Private yacht wintering [Jerome and Sally Poncet, *Damien II*, in Marguerite Bay]		1979 Woman winters at South Pole [Michele Eileen Raney, as station doctor]
		1979 Nov 28, Commercial plane crash [Air New Zealand Flight 901, into Mt. Erebus, with 257 deaths]		
		1979 Jan 28–Feb 3, Live television link between Antarctic and home [between Japanese Syowa base and Tokyo]		
1980–1989		1980–81, Private adventure overland Antarctic crossing [Ranulph Fiennes, two others, Transglobe Expedition]	1983 Jul 21, World record low temperature [-129.3° F (-89.2° C) at U.S.S.R. Vostok base]	**1982 Mar–Jun, War in Antarctic regions [Argentina vs. Great Britain at South Georgia and South Sandwich**
			1983–84 Feb, Flowering plant	

APPENDIX B: ANTARCTIC FIRSTS

	1981 Jun 22, Winter mail/supply drop at Amundsen-Scott Station, South Pole 1983–84 Nov, Seven Summits effort climb of Vinson Massif [Richard Bass and Frank Wells] 1985 Commercial air operation to Antarctica founded [Adventure Network International] 1985–86 Summer, Adventure ski traverse to South Pole [Roger Mear, Robert Swan, Gareth Wood, Footsteps of Scott Expedition] 1988 Jan 11, Tourists flown to South Pole [fifteen people, by Adventure Network Intn'l] 1988–89 Summer, Guided commercial ski expedition to South Pole [Mountain Travel Sobek Expedition. Six paying participants, five guides, from Ronne Ice Shelf to South Pole] 1989 Jan 28, Ship with tourists aboard sinks [*Bahia Paraiso*, off Anvers Island] 1989–90 Summer, Ski crossing of Antarctic continent [Reinhold Messner and Arved Fuchs]	found south of Antarctic Circle [by Sally Poncet, Terra Firma Islands, Marguerite Bay] 1984–85 Ozone hole confirmed [by scientists at BAS Halley base]	Islands, Falklands/Malvinas Islands war] **Oct 6–9, International conference in Antarctic regions [organized by Chile, King George Island]** 1985 Jan 7–13, International conference in Antarctic interior [in Transantarctic Mountains; organized by U.S. National Academy of Sciences] 1985–86 Summer, Private environmental monitoring [Greenpeace to Ross Sea] 1987 Antarctic Heritage Trust established [New Zealand] 1988–89 Women reach South Pole on skis [Shirley Metz and Victoria Murden, paying participants with Mountain Travel Sobek expedition] 1989–90 All-woman wintering party [nine led by Monika Puskeppeleit, at West German base, Georg von Neumayer]
1990 and after	**1990 Private solo winterings** [Hugues Delignières and Amyr Klink, aboard yachts off west coast of Antarctic Peninsula] 1991 Aug, Organization formed to control Antarctic tourist industry [IAATO] **1992 Mar–Apr, Ham radio expedition [Eight men on**	**1994–95 Ice shelf disintegrations** [Wordie and northern Larsen Ice Shelves, off west and east coast of Peninsula, respectively] 1995–96 Discovery of vast lake beneath the polar icecap [Lake Vostok, found through deep core ice drilling near Russian Vostok	1998 International Antarctic Gazetteer published [SCAR] 2000–01 Women cross Antarctic continent [Liv Arnesen and Anne Bancroft]

YEARS	DISCOVERY, EXPLORATION, AND OCCUPATION	COMMERCE AND TOURISM; TECHNOLOGY AND TRANSPORT	SCIENCE AND NATURE	WOMEN AND CHILDREN; POLITICS; MISCELLANEOUS
		Thule, South Sandwich Islands]	base]	
		Museum opened in Antarctic regions [South Georgia Whaling Museum]	2000 Mar, record iceberg calves ["B-15," off Ross Ice Shelf, ~4,500 sq. miles area]	
		1992–93 Summer, Icebreaker used for tourist voyages [Quark Expeditions, *Kapitan Khlebnikov*]		
		1996–97 Nov–Jan, Tourist circumnavigation [*Kapitan Khlebnikov*]		
		2000 Jan 27–29, Large passenger cruise ship [*Rotterdam*, to Antarctic Peninsula, where it spent two days]		
		2001 Apr 24, Winter flight to/from South Pole [Physician Ronald Shemenski evacuated, flight took off from and returned to BAS Rothera base on Antarctic Peninsula]		

Appendix C

Glossary of Terms, Abbreviations, and Acronyms

Antarctic Circle: The parallel of latitude (approximately 66° 33' S), where, at the winter solstice, the sun fails to rise for one day and where, at the summer solstice, it fails to set for one day. The length of continuous day or night increases from 24 hours at the Circle to six months at the South Pole. This effect results from the fact that the Earth is tilted on its axis at approximately 23° 27' to the plane of its orbit around the sun. The Antarctic Circle has no ecological or political significance.

Antarctic Convergence: The popular name for what Antarctic scientists now call the Antarctic Polar Front. This is an ecological boundary in the Southern Ocean where cool Antarctic surface waters, flowing north from the coast of the continent, meet and drop below the warmer, less dense, and saltier waters coming south from the subtropics. Circling the Antarctic Continent on an irregular (but semi-fixed) track between roughly 48° and 62° S, the region of the Convergence is typically about 20 to 30 miles wide, a narrow band over which there are abrupt changes in water and air temperatures. North of the Convergence, climates have at least some temperate character. South of it, one is into the Antarctic regions. (Many Antarctic scientific and reference works provide a detailed description and scientific definition of this phenomenon. An excellent one for the layperson can be found in Riffenburgh, ed., *Encyclopedia of the Antarctic*, Vol. 2, pp. 741–43.)

austral: Southern, commonly used to mean in the Southern Hemisphere. Thus, for example, "austral summer" means summer in the Southern Hemisphere.

BAS: British Antarctic Survey (discussed in Chapter 19).

beset: As used here, to surround a ship with ice, so that it is trapped and unable to move freely; also the period when a ship is thus trapped.

BGLE: British Graham Land Expedition (discussed in Chapter 16).

CRAMRA: Convention on the Regulation of Antarctic Mineral Resource Activities (discussed in Chapter 19).

crevasse: A near vertical rift in an ice sheet or glacier, formed where the ice is under sufficient tension to fracture. Open ones are easily seen, but many are thinly bridged by drifted snow and are a serious hazard to travel. Crevasse depths, which depend on ice temperature and pressure, may exceed 100 feet.

fast ice: Sea or lake ice formed in close attachment to a coast, thus "held fast." It is sometimes held in place by islands, rocks, or grounded icebergs.

FID: Falkland Islands Dependencies (discussed in Chapter 10 and after).

FIDS: Falkland Islands Dependencies Survey (discussed in Chapters 17 and 18).

floe: Floating ice on the surface of the ocean, often formed at sea, but may also be a broken piece of fast ice.

IAATO: International Association of Antarctica Tour Operators (discussed in Chapter 19).

iceberg: A large floating mass of freshwater ice that has broken off ("calved") from a glacier or floating ice shelf (see ice shelf, below). Icebergs calving directly from glaciers tend to be under 200 feet long and irregularly shaped. Icebergs born when a portion of a floating ice shelf breaks off are typically far larger and much more regularly shaped (see tabular iceberg, below). Depending on their density, icebergs show one-fifth to one-seventh of their volume above the surface of the water, and may rise as high as 200 feet above that surface. All icebergs break up and decay over time as they drift from their point of origin, driven primarily by water currents. Although most disappear within two to three years, some of the largest tabular bergs have retained their identity much longer, traveling far beyond the Southern Ocean before finally disintegrating.

ice pack (or pack ice): A concentration of floating sea ice made up of ice floes. In contrast to icebergs, which originate from land ice, pack ice forms at sea when surface water freezes during the winter. Broken-off pieces of fast ice can also become part of the pack. Pack ice that fails to disperse during the summer remains and grows thicker the following winter, becoming more difficult for ships to break through. "Open pack ice" consists of separate floes with extensive leads (see lead, below) between them. "Close pack ice" is composed of floes mainly in contact with one another. In "consolidated pack ice," the floes are frozen together, completely covering the underlying ocean.

ice shelf: A thick floating sheet of freshwater ice that extends over the sea, originating from glaciers on land. Small areas within an ice shelf may be grounded on islands or submarine banks. Ice shelves typically end in a cliff edge at the sea front that rises as much as 150 to 200 feet above the water and extends hundreds of feet down below the surface. Ice shelves fringe much of the Antarctic coast. The largest Antarctic ice shelves are the Ross Ice Shelf, at the southern edge of the Ross Sea; the Ronne and Filchner Ice Shelves, at the south of the Weddell Sea; and the Amery Ice Shelf, in East Antarctica.

IGY: International Geophysical Year. An international cooperative effort that took place from July 1, 1957, to December 31, 1958 (discussed in Chapter 18).

IPY: International Polar Year. The first took place in 1882–83, the second in 1932–33 (discussed in Chapters 5 and 18, respectively). The IGY was, in effect, the third IPY. A fourth (not discussed in *The Storied Ice*) ran from March 2007 to March 2009.

lead: An open lane of water through pack ice. The constant movement of the sea ice resulting from winds and currents causes leads to open and close over time.

pack ice: See ice pack.

RARE: Ronne Antarctic Research Expedition (discussed in Chapter 17).

rookery: A breeding place for seals or nesting place for penguins or other birds.

sastrugi: Hard ridges of compacted snow formed by wind and erosion on a snow surface. Strong winds may produce sastrugi over 3 feet high. (Singular form, seldom used, is sastrugus.)

SCAR: Scientific Committee on Antarctic Research. An international organization formed in 1958 in the wake of the IGY to coordinate and promote scientific research in the Antarctic regions. (For purposes of SCAR, the Antarctic is defined as every part of the Earth south of the Antarctic Convergence plus the sub-Antarctic islands on which IGY observations were made, including Iles Kerguelen and Amsterdam, Marion, Gough, Campbell, and Macquarie Islands.)

scurvy: A vitamin-deficiency disease resulting from going too long without ingesting vitamin C. Unlike most mammals, humans cannot metabolize this vitamin. Thus, it must come from a diet containing it, usually fresh fruit or vegetables, but also raw or lightly cooked meat, in particular

GLOSSARY OF TERMS, ABBREVIATIONS AND ACRONYMS

kidneys or livers (or today, vitamin pills). Scurvy is fatal unless the victim receives timely and sufficient vitamin C, but it is also easily cured with no lasting effects by gaining that vitamin C.

south geographic pole: The southern tip of the axis of the Earth's spin, where the sun never sets for six months a year and never rises for the other six months. All the Earth's lines of longitude meet here, at latitude 90° S. (The place that is popularly known as the South Pole)

south magnetic pole: The place in the Southern Hemisphere where a dip needle (a bar magnet suspended freely on a horizontal axis) points toward the Earth's center. This is also the southern location that a compass points to. The position of this pole varies from year to year. Its estimated location in 2009 was roughly 65° S, 138° E, more than 100 miles north of the coast of Adélie Land, East Antarctica.

tabular iceberg: An iceberg born when a portion of a floating ice shelf breaks off, or "calves." Newly calved tabular bergs are table-topped—i.e., flat on top with sheer vertical cliffs for sides that may rise as much as 200 feet above the water, with up to six times that depth below the surface. The larger ones are found only in the Antarctic. Tabular bergs up to 10 miles long are relatively common, but they can be much larger. As of 2011, the longest recorded one, 208 miles long, was seen in the Ross Sea in 1956. One nearly 200 miles long, with a total surface area of about 4,500 square miles, broke off the Ross Ice Shelf in early 2000.

transit of Venus: The rare, periodic event when the planet Venus moves in its orbit between the Earth and the sun, such that Earth dwellers see it as passing across the face of the sun. By measuring how long it takes Venus to cross the sun, the "transit," it is possible to calculate the distance from the Earth to the sun (roughly 93 million miles). There have been more than 50 transits of Venus since 2000 B.C., including those specifically mentioned in the text (1769, 1874, and 1882).

try-pot: A large pot used by whalers and/or sealers to cook blubber taken from whales or elephant seals in order to extract the oil.

USAS: United States Antarctic Service Expedition (discussed in Chapter 16).

Sources and Notes

My accounts of significant expeditions are drawn principally from primary sources (those written by expedition participants). Other material comes from secondary sources (by historians or others who did not participate), and several of these have been used extensively throughout this book. The most important are the first and second editions of Robert Headland's Antarctic chronology, *Chronological List of Antarctic Expeditions . . .*, 1989, and *A Chronology of Antarctic Exploration*, 2009. Together, these works provided critical input or context for every chapter. Other works used for many chapters include Alberts, *Geographic Names of the Antarctic*, 1995; Bertrand, *Americans in Antarctica . . .*, 1971; Christie, *The Antarctic Problem*, 1951; Fogg, *A History of Antarctic Science*, 1992; Fox, *Antarctica and the South Atlantic*, 1985; and Sullivan, *Quest for a Continent*, 1957. Four recent scholarly reference works were also useful: Mills et al., *Exploring Polar Frontiers*, 2003; Riffenburgh, ed., *Encyclopedia of the Antarctic*, 2007; Rosove, *Antarctica, 1772–1922*, 2001; and Stonehouse, ed., *Encyclopedia of Antarctica*, 2002.

Several classic secondary works provided important material for the Introduction and Chapters 1 through 5. In publication order, these include Fricker, trans. Sonnenschein, *The Antarctic Regions*, 1900; Balch, *Antarctica*, 1902; Mill, *The Siege of the South Pole*, 1905; and Hayes, *Antarctica*, 1928. The last also contributed material for Chapters 6 through 13. Hayes, *The Conquest of the South Pole*, 1932, was useful for Chapters 8 through 14.

None of the above mentioned works appear in the chapter-by-chapter listings below, except for cases where they are unusually important.

The Literature Cited section following this Sources and Notes provides full titles and publication information for all works listed above and in what follows.

Introduction

[1] Cook, *Through the First Antarctic Night, 1898–1899*, pp. 340–41.
[2] Scott, *The Voyage of the* Discovery, p. 14.

Chapter 1. In Search of a Southern Continent

The most important primary sources include Cook, *A Voyage Towards the South Pole . . .*, 1784; [Fletcher], *The World Encompassed . . .*, 1966; Schouten, *The Relation of a Wonderfull Voiage . . .*, 1966; Halley, ed. Thrower, *The Three Voyages of Edmond Halley . . .*, 1981; Cook, ed. Beaglehole, *The Journals of Captain Cook . . .*, 1961; Elliott and Pickersgill, ed. Holmes, *Captain Cook's Second Voyage*, 1984; and Fanning, *Voyages Round the World*, 1833.

Many secondary books and articles cover the major events described in this chapter, especially the voyages of Magellan, Drake, and Cook. Among the multiple works consulted, the most important secondary sources used are Beaglehole, *The Exploration of the Pacific, third edition*, 1966; Gurney, *Below the Convergence*, 1997; Headland, *The Island of South Georgia*, 1992; Hough, *The Blind Horn's Hate*, 1971; Jones, *Antarctica Observed*, 1982, and "Voyages to South Georgia 1795–1820," 1973; Knox-Johnston, *Cape Horn*, 1995; Riesenberg, *Cape Horn*, 1939; Silverberg, *The Longest Voyage*, 1972; and Stackpole, *The Sea Hunters*, 1953, and *Whales and Destiny*, 1972. Sobel, *Longitude*, 1995, provides information on the development of the chronometer.

[1] See Menzies, *1421*, 2003, pp. 145–91, for the most developed argument for these claims for the Chinese. See, inter alia, McIntosh, *The Piri Reis Map of 1513*, 2000, for counter arguments.
[2] Silverberg, *The Longest Voyage*, pp. 94–95.
[3] Vasco Nuñez de Balboa, who had crossed the Isthmus of Panama in 1513, had been the first European to see the western shores of the Pacific Ocean. He had simply called his discovery the South Sea, a name that continued in general use for several centuries.

4 [Fletcher], *The World Encompassed* . . . , p. 35. Spellings for this and other historical quotes have been modernized for ease of reading.
5 Carder's story was originally published in 1625 in Samuel Purchas, ed, *Hakluytus Posthumus, or Purchas His Pilgrimes*. . . . James MacLehose and Sons, Glasgow, Scotland, published a modern edition of Purchas's 20-volume opus in 1906. Neider, ed., *Great Shipwrecks and Castaways*, pp. 6–14, reprints the 1906 version of Carder's tale.
6 [Fletcher], op. cit., p. 44.
7 Schouten, *The Relation of a Wonderfull Voiage* . . . , p. 23.
8 Burney, *Chronological History of the Voyages* . . . , vol. 5, p. 38 (quoted in).
9 Wafer, *A New Voyage* . . . , p. 18.
10 Halley, ed. Thrower, *The Three Voyages of Edmond Halley* . . . , vol. 1, pp. 162–63.
11 Modern research has determined that it is primarily the offal (liver and kidney) in animal food, rather than the muscle, that contains ascorbic acid. (Williams, *With Scott in the Antarctic*, p. 136)
12 Cook, ed. Beaglehole, *The Journals of Captain Cook* . . . , p. 619.
13 Ibid., p. 625.
14 Ibid., pp. 621–22.
15 Ibid., p. 632.
16 Ibid., p. 646.
17 King, ed., "An Early Proposal . . . ," p. 315 (reprinted in).
18 Fanning, *Voyages Round the World*, pp. 25–28.
19 Weddell, *A Voyage Towards the South Pole* . . . , p. 54. James Weddell, the source of this oft-quoted number, was a sealer who worked the beaches of the South Shetlands in the 1820s and made a historic voyage into the Weddell Sea in 1822–23 (described in Chapter 3).

Chapter 2. The Continent Found

The most important primary sources used are Bellingshausen, ed. Debenham, *The Voyage of Captain Bellingshausen* . . . , 1945; Fanning, *Voyages Round the World*, 1833; Weddell, *A Voyage Towards the South Pole* . . . , 1970; and reprints of documents, diaries, and articles in Campbell, ed., *The Discovery of the South Shetland Islands*, 2000.

Significant secondary sources include Bertrand, *Americans in Antarctica* . . . , 1971; Campbell, interpretive text in *The Discovery of the South Shetlands*, 2000; Gurney, *Below the Convergence*, 1997; Hobbs, *The Discoveries of Antarctica* . . . , 1939; Jones, *Antarctica Observed*, 1982, "British Sealing on New South Shetland . . . ," 1985, and "Captain William Smith . . . ," 1975; Mitterling, *America in the Antarctic to 1840*, 1959; and Stackpole, *The Sea Hunters*, 1953, *The Voyage of the* Huron *and* Huntress, 1955, and *Whales and Destiny*, 1972.

1 Miers, communicated to Mr. Hodgskin, "Account of the Discovery . . . ," p. 367.
2 Brue, *Mappe Monde en deux Hemispheres* . . . , 1820.
3 Campbell, *The Discovery of the South Shetland Islands*, p. 71 (quoted in).
4 Brown, *A Naturalist at the Poles*, p. 289 (quoted in).
5 [Young], "Notice of the Voyage of Edward Barnsfield [sic] . . . ," p. 45.
6 Miers, op. cit., p. 365.
7 Bellingshausen's name has been transliterated from the Cyrillic alphabet in a number of ways. Other renderings include Faddey Faddeyevich and Thaddeus Thaddevich. The version used here matches the name recorded in the Latin alphabet on Bellingshausen's baptismal records: "Fabian Gottlieb Benjamin." (Headland, "Bellingshausen," p. 328)
8 In 1819, Russia was still using the Julian calendar, which by then was approximately 12 days behind the modern, Gregorian, calendar. All dates quoted in *The Storied Ice* have been converted to their Gregorian equivalents, thus putting them on the basis used by Bellingshausen's English and American contemporaries.

9. Bellingshausen, ed. Debenham, *The Voyage of Captain Bellingshausen* . . . , p. 92.
10. Ibid., p. 110.
11. These sketches would continue to appear in editions of *The Antarctic Pilot* through the fourth, 1974, edition.
12. Bellingshausen, op. cit., p. 420.
13. Jones, "British Sealing on New South Shetland . . . ," p. 298 (quoted in).
14. Bond et al., *Antarctica* . . . , p. 11 (quoted in).
15. Weddell, *A Voyage Towards the South Pole* . . . , p. 145.
16. Ibid., p. 141.
17. Bellingshausen, op. cit., p. 425.
18. Bertrand, *Americans in Antarctica* . . . , p. 80 (quoted in).
19. The only primary source for this voyage is log of the *Hero*. Unfortunately it is sometimes ambiguous and thus subject to a variety of interpretations. The text presents today's generally accepted one.
20. Fanning, *Voyages Round the World* , pp. 436–37 (quoted in).
21. Bellingshausen, op. cit., pp. 425–26.
22. Ibid., p. 438.
23. Nordenskjöld and Andersson, trans. Adams-Ray, *Antarctica* . . . , pp. 70–71.

Chapter 3. The Sealers' Age of Discovery

Important primary sources, in order used, are Weddell, *A Voyage Towards the South Pole* . . . , 1970; Webster, *Narrative of a Voyage* . . . , 1834; Eights, "A Description of the New South Shetland Isles . . . ," 1970; Fanning, *Voyages to the South Seas* . . . , 1970; and Biscoe, "From the 'Journal of a Voyage . . .'," 1901.

Significant secondary sources include Bertrand, *Americans in Antarctica* . . . , 1971; Gurney, *Below the Convergence*, 1997; Hobbs, *The Discoveries of Antarctica* . . . , 1939; Jones, "John Biscoe's Voyage . . . ," 1971, "British Sealing on New South Shetland . . . ," 1985, "Captain George Powell . . . ," 1983, and "New Light on James Weddell," 1965; McKinley, *James Eights* . . . , 2005; Marr, "The South Orkney Islands," 1935; and Mitterling, *America in the Antarctic to 1840*, 1959.

1. Weddell, *A Voyage Towards the South Pole* . . . , p. 28.
2. Ibid., p. 36.
3. Ibid., p. 44.
4. Ibid., p. 50.
5. Ibid., pp. 55–56.
6. Murphy, *Logbook for Grace*, p. 212. Murphy, a renowned early twentieth-century ornithologist whose visit to South Georgia is described briefly in Chapter 10, had Weddell's book with him when he was making his own penguin studies at the island.
7. Weddell, op. cit., p. 314.
8. Pp. 45–70 cover Morrell's 1822–23 voyage. Pp. 66–69 cover the Weddell Sea portion of the voyage.
9. Webster, *Narrative of a Voyage* . . . , vol. I, p. 140. Webster's is the most complete narrative about the *Chanticleer* expedition. Captain Henry Foster never wrote an account because he died before the voyage ended, drowned in January 1831 while exploring the Isthmus of Panama.
10. Ibid., vol. I, pp. 145–52. The brief passages quoted here are from a lengthy and marvelous description of Deception Island, one that reflects weeks of careful observation.
11. Eights, "A Description of the New South Shetland Isles . . . ," p. 198.
12. Ibid., p. 202.
13. In 1971, Eights's scientific papers were re-published in Quam, ed., *Research in the Antarctic*, pp. 5–40.
14. Biscoe, "From the 'Journal of a Voyage . . .'," p. 331.

Chapter 4. Three Great National Expeditions

Key primary sources are d'Urville, trans. and ed. Rosenman, *Two Voyages to the South Seas . . . , vol. II . . .* , 1987; Wilkes, *Narrative of the United States Exploring Expedition*, 1845; McCormick, *Voyages of Discovery . . .* , 1884; and Ross, J. C., *A Voyage of Discovery and Research . . .* , 1847. Although abridged, d'Urville, trans. and ed. Rosenman, vol. II, is the most complete English translation of d'Urville's expedition narrative available as of 2011. In addition, the editor has provided historical background, biographical material for d'Urville and key expedition participants, and selections from several officers' notes. (The only other significant English translation from d'Urville's narrative appeared as a chapter in Murray, ed., *The Antarctic Manual*, 1901, but the included material is entirely about the 1839–40 foray south.)

Many secondary sources cover these three national expeditions, but most have limited use for this chapter, because even book-long accounts of these expeditions typically pay little attention to the time spent in the Peninsula Region. The most important used are Dunmore, *From Venus to Antarctica: The Life of Dumont d'Urville*, 2007; Gurney, *The Race to the White Continent*, 2000 (about all three expeditions); Rosenman, editorial content in d'Urville, *Two Voyages . . .* ; Palmer, *Thulia*, 1843; Philbrick, *Sea of Glory*, 2003; and Ross, M. J., *Ross in the Antarctic*, 1982. The first part of *Thulia* is a 25-page-long poem about the voyage of the Wilkes Expedition's *Flying Fish* (called *Thulia* in the poem). A substantial appendix describes the *Flying Fish*'s voyage in prose. The author, the surgeon on one of the Wilkes Expedition vessels that did not participate in the 1838–39 southern cruises, based his account on crew journals from the *Flying Fish*'s epic voyage, all of which were aboard the *Peacock* and were lost when that vessel sank off the Oregon coast in July 1841. The author of *Ross in the Antarctic* is James Clark Ross's great-grandson, himself a naval officer who rose to the rank of rear admiral. He used unpublished family documents as well as manuscript material from other expedition participants to write this book, which contains an excellent account of Ross's season in the Peninsula Region.

Other important secondary titles about the Wilkes expeditions, but with limited information for this chapter, include Stanton, *The Great United States Exploring Expedition . . .* , 1975; Tyler, *The Wilkes Expedition*, 1968; and Viola and Margolis, eds., *Magnificent Voyagers*, 1985.

1. Dunmore, *From Venus to Antarctica*, p. 148 (quoted in).
2. D'Urville, trans. and ed. Rosenman, *Two Voyages . . . , vol. II*, p. 334.
3. Ibid., p. 339.
4. Ibid., p. 340.
5. Wilkes, *Narrative of the United States Exploring Expedition*, vol. I, pp. 394.
6. Hudson, in Wilkes, ibid., vol. I, p. 406. Wilkes described the February–March 1838 Antarctic voyages of the *Peacock*, *Flying Fish*, and *Sea Gull* using reports from their captains. He also included copies of these reports as appendixes to his narrative.
7. Walker in Wilkes, ibid., vol. I, p. 411.
8. Palmer, *Thulia*, p. 70.
9. On January 30, 1840, Wilkes could clearly see a substantial stretch of snow-covered land along the coast of East Antarctica. And so, he wrote, "now that all were convinced of [the land's] existence, I gave the land the name of the Antarctic Continent. . . ." (Wilkes, op. cit., vol. II, p. 316).
10. Bertrand, *Americans in Antarctica . . .* , p. 166.
11. Wilkes, op. cit., vol. I, p. 145. Smyley wrote to Wilkes after finding the *Chanticleer*'s thermometers, and the words quoted here are from Wilkes's paraphrase from the sealer's letter. Unfortunately, Wilkes misspelled Smyley's name "Smiley" in his paragraph about the letter, and for years, secondary accounts that relied on Wilkes perpetuated this error.
12. Ross, J. C., *A Voyage of Discovery and Research . . .* , vol. I, p. 219.
13. Ibid., vol. II, p. 357.
14. Ross, M. J., *Ross in the Antarctic*, p. 206 (quoted in).
15. Ross, J. C., op. cit., vol. II, p. 327.

SOURCES AND NOTES

Chapter 5. Quiet Decades in the South; the New Hunters

Primary sources are von den Steinen, trans. Barr, "Zoological Observations . . . ," 1984; Murdoch, *From Edinburgh to the Antarctic*, 1894; Brown, *A Naturalist at the Poles*, 1923 (chapters by Murdoch); Donald, "The Late Expedition to the Antarctic," 1894; and Bruce, "Antarctic Exploration," 1894.

Key secondary sources include Barr, *The Expeditions of the First International Polar Year, 1882–83*, 1985; Barr, Krause, and Pawlik, "Chukchi Sea, Southern Ocean, Kara Sea," 2004; Christensen, trans. Jayne, *Such Is the Antarctic*, 1935; Headland, *The Island of South Georgia*, 1992, and "The German Station of the First International Polar Year . . . ," 1982; Murray, "Notes on an Important Geographical Discovery . . . ," 1894; and Southwell, "Antarctic Exploration," 1895.

1. Von den Steinen, trans. Barr, "Zoological Observations . . . , Part 1," p. 68.
2. Ibid., Part 2, p. 153.
3. Ibid., Part 2, p. 156.
4. Murdoch, *From Edinburgh to the Antarctic*, p. 244.
5. Ibid., p. 233.
6. Ibid., p. 244.
7. Donald, "The Late Expedition to the Antarctic," p. 67.
8. The *Jason* had already ensured her place in the history books in 1888 when she carried Norway's Fridtjof Nansen to Greenland for what became the first ever crossing of this massive ice-capped island, a daring trip that many regard as the beginning of modern polar exploration.
9. Christensen, trans. Jayne, *Such Is the Antarctic*, p. 83. Lars Christensen, the author of this book, was Christen Christensen's son. In the 1920s, he would himself become a major figure in the Antarctic whaling industry, sending not only commercial whaling voyages south, but also sponsoring five scientific/whaling reconnaissance voyages, including four aboard the *Norvegia*, a ship he purchased specifically for this purpose.
10. Barrett-Hamilton, "Seals," p. 213 (quoted in).
11. Charcot, trans. Walsh, *The Voyage of the "Why Not?"* . . . , p. 108.
12. Mill, *The Siege of the South Pole*, pp. 384–85 (quoted in).

Chapter 6. De Gerlache and the First Antarctic Night

Until 1998, the most important primary accounts of this expedition available in English were Cook, *Through the First Antarctic Night* . . . , 1900 (which includes appendixes by other expedition members); Amundsen, *My Life as an Explorer*, 1927; Arctowski, "Exploration of Antarctic Lands," 1901; and several short, mostly scientific, articles. English-language secondary accounts typically relied on these sources, especially on Cook. There are, however, substantial differences between these works and the two other major primary accounts: De Gerlache, *Quinze Mois dans l'Antarctique*, 1902, and Lecointe, *Au Pays des Manchots*, 1904. (The latter book, by the expedition second-in-command, did appear in a severely abridged English translation in 1904, but it had very limited distribution.) De Gerlache's account did not appear in English until published as de Gerlache, trans. Raraty, *Fifteen Months in the Antarctic*, 1998. A year later, an edited, scholarly translation of Amundsen's expedition diary came out: Amundsen, trans. Dupont and LePiez, ed. Decleir, *Roald Amundsen's Belgica Diary*, 1999. Chapter 6 employs all these English-language sources, reconciling where necessary, and making judgments of what version to accept where reconciliation is impossible.

Important secondary sources for this chapter include Baughman, *Before the Heroes Came*, 1994, and "Hopeless in a Hopeless Sea of Ice," 1998; Bryce, *Cook & Peary*, 1997; articles in Decleir and de Broyer, eds., *The Belgica Expedition Centennial*, 2001; Huntford, *Scott and Amundsen*, 1980; Raraty, translator's introduction to de Gerlache, *Fifteen Months* . . . , 1998; and Yelverton, *Quest for a Phantom Strait*, 2004.

1. Cook, *Through the First Antarctic Night* . . . , p. 131.
2. Ibid., p. 136. Another expedition member also claimed the credit. In "Exploration of Antarctic Lands," Arctowski, the geologist, wrote "I had the pleasure of discovering the first Antarctic insect. . . ." (p. 470). Both de Gerlache and Amundsen, however, support Cook's version that it was

Racovitza.

3. De Gerlache, trans. Raraty, *Fifteen Months in the Antarctic*, p. 68.
4. Furse, *Antarctic Year*, 1986, a beautifully illustrated book by the expedition leader, tells the story of this expedition. François's father had given the expedition a bronze plaque commemorating *his* father's landing on Brabant Island. On July 21, 1984, this British group celebrated Belgium's national day and formally installed the plaque on a boulder above the expedition camp on the island.
5. Cook, op. cit., p. 145.
6. De Gerlache, trans. Raraty, op. cit., p. 104.
7. Ibid., pp. 106–07.
8. Ibid., p. 113.
9. Cook, op. cit., p. 172.
10. Ibid., p. 234.
11. De Gerlache, trans. Raraty, op. cit., pp. 118–19.
12. Amundsen, *My Life as an Explorer*, p. 29.
13. Cook, op. cit., p. 338.
14. Ibid., pp. 340–41.
15. Ibid., p. 356.
16. De Gerlache, trans. Raraty, op. cit., p. 156.
17. Cook, op. cit., p. 402.
18. J. Gordon Hayes, who coined the name "Heroic Age", dates the beginning of the period with Robert Falcon Scott's 1901–04 *Discovery* expedition and the end with Sir Ernest Shackleton's 1914–17 Imperial Trans-Antarctic expedition. (Hayes, *The Conquest of the South Pole*, p. 30) Writers following Hayes, however, have not always used these precise beginning and ending points. For *The Storied Ice*, I have chosen to date the start of the Heroic Age with the 1897–99 *Belgica* expedition, the first major exploring effort following the 1895 International Geographical Congress. I end, as Hayes does, with Shackleton in 1917.

Chapter 7. Nordenskjöld's Saga of Survival

The most important primary source used here is Nordenskjöld and Andersson, trans. Adams-Ray, *Antarctica . . .* , 1905. (Several other primary accounts also exist, including Andersson, *Antarctic*, 1944; Duse, *Bland Pingviner Och Sälar*, 1905; and Sobral, *Dos Años Entre los Hielos 1901–1903*, 1905. Because none of these works are available in English, they have not been used except as noted below.) Hermelo et al., trans. Perales and Perales, ed. Rosove, *When the Corvette* Uruguay *Was Dismasted*, 2004, has primary material re the end stages of the expedition.

Significant secondary sources used include Elzinga et al., eds., *Antarctic Challenge*, 2004; Lewander, "The Representations of the Swedish Antarctic Expedition, 1901–03," 2002, and "The Swedish Relief Expedition," 2003; Liljequist, *High Latitudes*, 1993; and Yelverton, *Quest for a Phantom Strait*, 2004.

1. Sobral, *Dos Años Entre los Hielos 1901–1903*, auto-translated in www.geociies.com/lunesotraves/cache/ancladosenelfindelmundo, p. 2. Sobral learned to speak Swedish fluently while at Snow Hill Island. After the expedition, he resigned his naval commission and went to Uppsala, Sweden, where he eventually received a doctorate in geology.
2. Nordenskjöld and Andersson, trans. Adams-Ray, *Antarctica . . .* , p. 38.
3. Ibid., p. 108.
4. Ibid., pp. 64–66.
5. Ibid., p. 416.
6. Ibid., p. 436.
7. Ibid., p. 54.
8. Ibid., p. 271.

SOURCES AND NOTES

9   Ibid., p. 287.
10 Ibid., p. 449.
11 Ibid., p. 464.
12 Skottsberg, *The Wilds of Patagonia*, p. 320.
13 Nordenskjöld and Andersson, trans. Adams-Ray, op. cit., p. 557.
14 Ibid., p. 479.
15 Ibid., pp. 306–07.
16 Ibid., p. 490.
17 Ibid., pp. 307–08.
18 Ibid., p. 316. As of 2011, there is a modest-sized emperor penguin rookery near the south end of Snow Hill Island on the opposite side from Nordenskjöld's hut. The farthest north known emperor rookery, its existence was only confirmed in the early 1990s.
19 Ibid., p. 504.
20 Ibid., p. 510.
21 Ibid., pp. 582–83.
22 See Vairo et al., trans. Freire, *Antártida . . .* , 2007, for description of the impressive effort Argentina has made to preserve these huts. This book, published in Spanish with accompanying English-language translation, includes many excellent photos of the huts over the years. It also describes the work Argentina has done on Omond House, the *Scotia* expedition hut described in Chapter 8.

Chapter 8. Bruce and the *Scotia*, Bagpipes in the South

Until 1992, the most substantial published primary sources for the story of the *Scotia* expedition were Three of the Staff, *The Voyage of the "Scotia,"* 1906; Bruce, *Polar Exploration*, circa 1911; and Brown, *A Naturalist at the Poles*, 1923 (a biography of Bruce by the *Scotia* expedition's botanist). Bruce also wrote a narrative, intending that it be volume I of the expedition's scientific reports. He later changed his mind and instead used his limited funds to publish the other scientific results. This narrative finally appeared decades later as Bruce, ed. Speak, *The Log of the Scotia Expedition*, 1992. These four books are the key primary sources for this chapter.

The most significant secondary sources used are Speak's substantial editorial content in Bruce, ed. Speak, *The Log of the Scotia Expedition*, 1992, and his biography of Bruce: *William Speirs Bruce*, 2003.

1 Bruce, "Prefatory Note" to Three of the Staff, *The Voyage of the "Scotia,"* p. viii.
2 Brown, *A Naturalist at the Poles*, p. 100 (quoted in).
3 Ibid., p. 103 (quoted in).
4 Three of the Staff, *The Voyage of the "Scotia,"* p. 80.
5 Ibid., p. 101.
6 Brown, op. cit., p. 146.
7 Bruce, ed. Speak, *The Log of the Scotia Expedition*, p. 156. Unfortunately, Bruce's recordings have been lost. (Speak, *Williams Speirs Bruce*, p. 89)
8 Bruce's was the second attempt to record the sounds of Antarctic fauna. The first was by Drygalski's 1901–03 *Gauss* expedition to East Antarctica. In 1902, this German venture successfully recorded the voices of Adélie penguins. (Drygalski, trans. Raraty, *The Southern Ice Continent*, p. 254)
9 Carsten Borchgrevink's 1898–1900 *Southern Cross* expedition, which wintered at Cape Adare in the Ross Sea in 1899, also tried to take movies. Their efforts were a total failure because the film they had brought did not fit their camera. The next after Bruce to try filming in the Antarctic was Ernest Shackleton, on his 1907–08 *Nimrod* expedition to the Ross Sea. He enjoyed significantly better success than Bruce.
10 Three of the Staff, op. cit., pp. 141–42.

11 Ibid., p. 230.
12 Brown, op. cit., p. 173.
13 Bruce, ed. Speak, *The Log of the* Scotia *Expedition*, pp. 216–17.
14 Ibid., p. 224.
15 Brown, op. cit., p. 208 (quoted in).
16 Three of the Staff, op. cit., p. 333 (quoted in).
17 A replica of the *Belgica* may join them. In 1944, the *Belgica* was sunk in a Norwegian harbor during a German bombing raid. The hulk was re-discovered in 1990, and in 2006, interested parties in Belgium and Norway, including Adrien de Gerlache's grandson Jean-Louis de Gerlache, formed the *Belgica* Society. Their goal was to salvage the wreck, or at least significant parts of it, for display in the vicinity of Antwerp. A year later, a related Belgian group formed the "New Belgica" project to build a replica of de Gerlache's vessel. Construction preparations started in early 2011. Once actual construction begins, it is expected that it will take five years to complete the replica. Once done, it is planned that the new ship will "become an ambassador-ship to stimulate public awareness regarding climate change worldwide and offer a platform for renewed expeditions to the polar regions." (Anon, "New 'Belgica' Project," p. 2)

Chapter 9. Charcot and the *Français* Explore the Antarctic Peninsula

The chief primary source used is Charcot, trans. Billinghurst, *Towards the South Pole Aboard the Français*, 2004 (the first English translation of Charcot, *Le Français au Pôle Sud*, 1906). Charcot also wrote a great deal about his first Antarctic expedition in his account of his second: Charcot, trans. Walsh, *The Voyage of the "Why Not?"...*, 1911.

Key secondary sources include Oulié, *Charcot of the Antarctic*, 1938; Lewander, "The Swedish Relief Expedition to Antarctica...," 2003 (an article as much about Charcot as about the *Frithjof*); and Yelverton, *Quest for a Phantom Strait*, 2004.

1 Raraty, translator's introduction to de Gerlache, trans. Raraty, *Fifteen Months in the Antarctic*, p. xx.
2 Charcot, trans. Billinghurst, *Towards the South Pole...*, p. 200 (quoted in).
3 Ibid., p. 5.
4 Ibid., p. 21.
5 Dallmann had named and mapped Booth Island in 1874, along with Krogmann and Petermann Islands, two small landmasses just to the south. De Gerlache had renamed these Wandel, Hovgaard, and Lund respectively. Charcot would later restore Dallmann's names to Booth and Petermann Islands, and these are the names, along with Hovgaard, on today's maps. To minimize confusion with modern maps, I have used the names Booth and Petermann.
6 Charcot, trans. Walsh, *The Voyage of the "Why Not?"...*, p. 144.
7 Charcot, trans. Billinghurst, op. cit., p. 171.
8 Ibid., p. 200.
9 Ibid., p. 222.
10 Ibid., p. 225.
11 That Charcot sent and received telegrams at Puerto Madryn is from his own account, where he writes that he "stayed for twenty-four hours not far from the door of the telegraph office awaiting news." (Charcot, trans. Billinghurst, op. cit., p. 242.) There is, however, another, more romantic and quite different, account of how Charcot first communicated with the outside world. According to Charcot's biographer, Marthe Oulié, there was no telegraph station at Puerto Madryn. Because of this, she writes, Charcot sent Paul Pléneau, the expedition photographer, 45 miles on horseback to the nearest telegraph station. Pléneau made the trip quickly, but then had to wait three days for a reply before he could ride back to Puerto Madryn with the news. (Oulié, *Charcot of the Antarctic*, p. 97)

SOURCES AND NOTES

Chapter 10. Whalers and Politics

Primary sources include Adie and Basberg, "The First Antarctic Whaling Season of *Admiralen...*," 2009 (largely a reprint of Lange's day-by-day account of the voyage); Filchner, trans. Barr, *To the Sixth Continent*, 1994 (Filchner devoted chapter 7 in this *Deutschland* expedition narrative to describing the whaling operations he observed at South Georgia in late 1911); and Murphy, *Logbook for Grace*, 1947.

Key secondary sources include Basberg, *Shore Whaling Stations at South Georgia*, 2004; Elliot, *A Whaling Enterprise*, 1998; Hart, *Antarctic Magistrate*, 2009 (with wonderful contemporary photographs of early South Shetland and South Georgia whaling), *Pesca*, 2001, and *Whaling in the Falkland Islands Dependencies 1904–1931*, 2006; Headland, *The Island of South Georgia*, 1992; Heyburn and Stenersen, "The Wreck and Salvage of S.S. *Telefon*," 1989; Jones, "Three British Naval Antarctic Voyages, 1906–43," 1981; Kohl-Larsen, trans. Barr, *South Georgia*, 2003; Lyon, chairman, *Report of the Interdepartmental Committee . . .* , 1920; Mathews, *Ambassador to the Penguins*, 2003; Matthews, *South Georgia*, 1931; Tønnessen and Johnsen, trans. Christophersen, *The History of Modern Whaling*, 1982.

1. See Anon, "British Letters Patent of 1908 and 1917 . . ." for the verbatim text of these Letters Patent.
2. Matthews, *Penguin*, p. 145 (quoted in).
3. Details re Solveig Jacobsen's birth from Hart, *Pesca*, p. 218. Solveig Jacobsen was not the first child born in the near-Antarctic. The first documented such birth took place around 1818, a daughter born to the wife of a British sealing captain who was working at Iles Kerguelen. (Headland, *A Chronology of Antarctic Exploration*, p. 120)
4. One account of such activity comes from the naturalist Robert Cushman Murphy, who visited the Prince Olav Harbor Station in early March 1913. He wrote of the day he was there, "Because today is Sunday, a good many of the workers at the Prince Olaf [sic] Station [have] gone off across the mountains on a skiing jaunt to the south coast of the island." (Murphy, *Logbook for Grace*, p. 235)
5. Headland, *The Island of South Georgia*, p. 131 (quoted in).
6. Matthews, *South Georgia*, p. 138. Account of the construction of the jail is also from this source.
7. Skottsberg, *The Wilds of Patagonia*, p. 317.
8. Hart, *Whaling in the Falkland Islands Dependencies . . .* , p. 43.
9. The most detailed, published contemporary account of these events comes from Charcot, trans. Walsh, *The Voyage of the 'Why Not?' . . .* , pp. 255–56. Charcot, whose *Pourquoi Pas?* voyage is described in Chapter 11, was at Deception Island in 1909 until one day before the *Telefon* wreck and then again the following summer, when he heard the story from Andresen.
10. Heyburn and Stenersen, "The Wreck and Salvage of S.S. *Telefon*," p. 51.
11. Ibid., p. 51.

Chapter 11. Charcot's Return with the *Pourquoi-Pas?*

The most important primary source used here is Charcot, trans. Walsh, *The Voyage of the 'Why Not?' . . .* , 1911. Charcot, "Charcot Land . . . ," 1930, also provides useful information. The *Pourquoi Pas?*'s third officer, Jules Rouch, also wrote an expedition narrative, *L'Antarctide. Voyage du "Pourquoi-Pas?" (1908–10)*, 1926, but it is only available in French and has not been used other than as a source of maps.

Oulié, *Charcot of the Antarctic*, 1938, is the most important secondary source currently available in English. Naveen, *Waiting to Fly*, 1999, has useful content because of the author's interest in the penguin studies by Louis Gain, Charcot's biologist.

1. Charcot, trans. Walsh, *The Voyage of the "Why Not?" . . .* , p. 32.
2. Ibid., p. 56.
3. Ibid., p. 70.
4. Ibid., p. 80.
5. Ibid., p. 149.

6   Ibid., p. 158.
7   Ibid., p. 269 (quoted in).
8   Ibid., pp. 284–85.
9   Ibid., pp. 291–92.

Chapter 12. Filchner's Battles in the Weddell Sea

Primary sources for Chapter 12 are Filchner, trans. Barr, *To the Sixth Continent*, 1994, and "Exposé," 1994; Kling, trans. Barr, "Report to the Hamburg-Südamerika Line," 1994; and Przybyllok, trans. Barr, "Handwritten Notes . . . ," 1994. Unfortunately, all these accounts are by Filchner or his supporters. No significant narratives from those opposed to Filchner have been published in English. Most important, Vahsel did not live to write his own story of events, and his version thus remains untold.

Important secondary sources used include Barr, Translator's Introduction to Filchner, *To the Sixth Continent*, 1994; and Murphy, *German Exploration of the Polar World*, 2002. (The first significant English-language account of this expedition appeared in Hayes, *The Conquest of the South Pole*, 1932, but this adds nothing to the sources used.)

1   The Kaiser was disappointed more for political reasons than scientific. He had wanted dramatic results like great geographical discoveries and exciting firsts that could compete with those that Robert Falcon Scott's contemporary, 1901–04, *Discovery* expedition brought home to Great Britain. But there had been none of these, only a winter spent locked in the ice off the coast of East Antarctica. Nonetheless, his and the German public's assessment that the expedition was a failure was wide of the mark. Drygalski had in fact returned home with masses of important scientific data, observations, and coastal discoveries and surveys.

2   Filchner was not the first Antarctic explorer to find this ship, originally named the *Bjørn*, appealing. A few years earlier, Ernest Shackleton had wanted to buy her for his 1907–09 expedition. Her price, however, exceeded his budget, and he settled instead for the much smaller *Nimrod*.

3   Filchner's January 13, 1912, sounding was nearly 350 miles from the nearest continental land, the east coast of the Weddell Sea. The existence of what is in fact the deepest part of the Weddell Sea in the vicinity of this sounding has been confirmed by modern oceanographic work. (Riffenburgh, *Encyclopedia of the Antarctic*, Vol. 2, p. 1053)

4   Przybyllok, trans. Barr, "Handwritten Notes . . . , p. 231 (quoted in).

5   Filchner, trans. Barr, *To the Sixth Continent*, p. 114.

6   Ibid., p. 115.

7   Filchner renamed the location Herzog Ernst Bay in recognition of how much it had changed. Most of the men, however, continued to refer it as location as Vahsel Bay, and that is the name that has stuck.

8   Filchner, op. cit., p. 120.

9   Ibid., p. 152.

10 Ibid., p. 172.

11 Filchner, trans. Barr, "Exposé," p. 213.

12 Kohl-Larsen, trans. Barr, *South Georgia*, 2003, tells the story of this small but historic expedition, an effort that began the exploration of South Georgia's interior and included the first female explorer in the Antarctic regions.

13 Barr, Translator's Introduction to Filchner, *To the Sixth Continent*, p. 38 (quoted in).

Chapter 13. *Endurance*, Shackleton's Triumphant Failure

Many, many books and articles have been written in part or in whole about this celebrated expedition, and a large number have been consulted in connection with this chapter. The most important published primary sources used here include Bakewell, ed. Rajala, *The American on the Endurance*, 2004; Hurley, *Argonauts of the South*, 1925; Hussey, *South with Shackleton*, 1949; Shackleton, *South*, 1919; and Worsley, *Endurance*, 1931, and *Shackleton's Boat Journey*, nd (1939 or 1940). Unpublished

SOURCES AND NOTES

sources include expedition diaries of Frank Hurley, Harry McNeish, and Thomas Orde-Lees.

A great many secondary works provided material for this chapter. The most important are: Dunnett, *Shackleton's Boat*, 1996; Fisher and Fisher, *Shackleton*, 1957; Hayes, *Antarctica*, 1928 (includes substantial extracts from the expedition diary of R. W. James, the *Endurance* physicist); Huntford, *Shackleton*, 1986; Jones, "Frankie Wild's Hut," 1982, and "Shackleton's Amazing Rescue 1916," 1982; Mill, *The Life of Sir Ernest Shackleton*, 1924; Mills, *Frank Wild*, 1999; Piggott, ed., *Shackleton*, 2000; Shackleton and Mackenna, *Shackleton*, 2002; Smith, *Sir James Wordie*, 2004 (includes Wordie's *Endurance* expedition diary), and *An Unsung Hero*, 2000; Thomson, *Elephant Island and Beyond*, 2003, and *Shackleton's Captain*, 1999. Fuchs, trans. Sokolinsky, *In Shackleton's Wake*, 2001, an account of a modern expedition described in Chapter 20, provides a valuable, thoughtful analysis and unique perspective.

Among excellent secondary book-length accounts of this expedition, all read in connection with this chapter, the interested reader may want to consult Alexander, *The Endurance*, 1998; Heacox, *Shackleton*, 1999; and Lansing, *Endurance*, 1959. Murphy et al., *South with* Endurance, 2001, contains a spectacular collection of Frank Hurley's expedition photographs.

1. Shackleton, *South*, p. vii.
2. Ibid., p. xv (quoted in).
3. Ibid., p. vii.
4. Huntford, *Shackleton*, p. 384 (quoted in).
5. Shackleton, op. cit., p. 58.
6. Ibid., p. 74.
7. Worsley, *Endurance*, pp. 19–20.
8. Shackleton also kept a few of the prints that Hurley had made on board the ship. He took one of these, a photograph of the *Endurance* pushed on her side by the force of the ice, with him on the voyage of the *James Caird* to serve as proof of his men's desperate situation. (Ennis, *Man With a Camera*, p. 15)
9. Shackleton, op. cit., pp. 122–23.
10. Hurley, *Argonauts of the South*, p. 245.
11. Ibid., p. 245.
12. Shackleton, op. cit., p. 145.
13. Ibid., p. 164.
14. Ibid., pp. 174–75.
15. Worsley, op. cit., p. 146, is the source of these details. Shackleton, *South*, p. 193, says they had 50 feet of rope.
16. Ibid., p. 156.
17. Burley, "Was Shackleton Valley the Passageway to Stromness?" pp. 234–35.
18. Shackleton, op. cit., p. 205. The words Shackleton placed in quotes *plus* the following sentence are, with slightly altered punctuation, from "The Call of the Wild," a poem by Robert Service. Service was a Scottish-Canadian poet, very popular in Shackleton's time. He is probably best known for his verses about the Yukon, including the one that Shackleton quoted here.
19. Shackleton's account in *South* says they were boys. In fact, they were the daughters of the station manager Sørrle. (Gilkes, "It Ain't Necessarily So," p. 65)
20. Shackleton, op. cit., p. 206. Another version of this meeting is reported in Robertson, *Of Whales and Men*, p. 90. According to Robertson, a doctor on a Christian Salvesen Company whaling ship that called at South Georgia in 1950, a whaler aboard told him that he had been there that day. Robertson quotes the whaler as saying "Everybody at Stromness knew Jack [sic] Shackleton well, and we very sorry he is lost in ice with all hands. But we not know three terrible looking bearded men who walk into the office off the mountainside that morning. Manager say: 'Who the hell are you?, and terrible bearded man in the center of the three say very quietly: 'My name is Shackleton.' Me I turn away and weep. I think manager weep, too." Sadly, wonderful as this story is, there is a serious question about its reliability. Gerald Elliot, a Salvesen company officer, writes

that there was, in fact, no whaler with them that season who had been at Stromness in 1916 and that "much of what [Robertson] wrote [in his book] is fiction." (Elliot, *A Whaling Enterprise*, p. 98)

21   The home into which Sørlle invited them was the official manager's villa at the station, built as was the station in the summer of 1912–13. The station manager would continue to live in this building until the late 1920s. In the early 1920s, however, the manager's villa from the abandoned station at Ocean Harbor was moved to Stromness to become the station hospital. When all the medical facilities for the whaling stations at Stromness Bay were centralized at Leith, the Stromness manager moved into the Ocean Harbor Villa, turning his original home, a much more humble structure, over to lower-ranked personnel. Unfortunately, awareness of this change faded over the years, and by the 1990s, when interest in the Shackleton story was exploding, the Ocean Harbor building was incorrectly identified as the manager's villa that Shackleton had come to. (Basberg and Burton, "New Evidence . . . ," passim) This went so far as having the incorrectly identified building designated a World Heritage Site, with a World Heritage sign put up describing the significance of the site and the building. (Gilkes, op. cit., p. 62) That it was not was conclusively determined at the beginning of the twenty-first century by Bjørn Basberg in his archeological studies of South Georgia whaling stations. The original building does still stand, although now in poor condition. (Basberg and Burton, op. cit., passim)

22   The *James Caird* was eventually taken home to England. Several years later, Shackleton gave her to John Quiller Rowett, an old school friend from Dulwich College who financed Shackleton's last expedition (described in Chapter 14). In 1922, after Shackleton's death, Rowett gave the *James Caird* to his and Shackleton's joint alma mater. His doing so greatly upset Lady Shackleton, because she had not understood the ownership *and* Rowett did not tell her of his gift in advance. Lady Shackleton wrote to H. R. Mill of this, ". . . I saw that J. R. had presented the "James Caird" to Dulwich College, and it cut me to the quick. I did not know it was his. . . . The boat is a living sacred thing to me, and I think it is nice for it to go to the old school, failing a museum, if they will take *care* [Lady Shackleton's emphasis] of it? [sic]" (Shackleton and Mill, ed. Rosove, *Rejoice My Heart*, p. 37) The *James Caird* remains in the possession of Dulwich College, though for many years it was lent to the National Maritime Museum in Greenwich and put on display there.

23   Shackleton, op. cit., p. 209.

24   Jones, "Frankie Wild's Hut," p. 385 (quoted in).

25   Jones, "Shackleton's Amazing Rescue," p. 326 (quoted in).

26   Shackleton tells the story of his Ross Sea Party in *South* (pp. 241–337). The following offer modern accounts: Bickel, *Shackleton's Forgotten Men*, 2000; McElrea and Harrowfield, *Polar Castaways*, 2004; and Tyler-Lewis, *The Lost Men*, 2006.

Chapter 14. The Decade Following World War I

The key primary source used for the Cope expedition is Bagshawe, *Two Men in the Antarctic*, 1939. The most important secondary sources are Naveen, *Waiting to Fly*, 1999, and three biographies of Hubert Wilkins: Grierson, *Sir Hubert Wilkins*, 1960; Nasht, *No More Beyond*, 2006; and Thomas, *Sir Hubert Wilkins*, 1961. Naveen has more to say about this expedition than any other secondary source found, a result of the author's admiration for Bagshawe and Lester's penguin observations.

Four primary sources were used for the *Quest* expedition: Hussey, *South with Shackleton*, 1949; Marr, ed. Shaw, *Into the Frozen South*, 1923; Wild, *Shackleton's Last Voyage*, 1923; and Worsley, *Endurance*, 1931. The most significant secondary sources for the *Quest* include the biographies of Shackleton, Wild, and Worsley listed for Chapter 13; the biographies of Wilkins listed above; Beeby, *In a Crystal Land*, 1994; and Erskine and Kjaer, "The Polar Ship *Quest*," 1998.

For the Discovery Investigations and whaling, the chief primary sources used include Hardy, *Great Waters*, 1967; Lyon, chairman, *Report of the Interdepartmental Committee* . . . , 1920; Matthews, *South Georgia*, 1931; and Ommanney, *South Latitude*, 1938. Important secondary works include Bernacchi, *The Saga of the "Discovery,"* 1938; Hart, *Pesca*, 2001, and *Whaling in the Falkland Islands Dependencies*, 2006; Savours, *The Voyages of the Discovery*, 1992; and Tønnessen and Johnsen, trans. Christophersen, *The History of Modern Whaling*, 1982.

SOURCES AND NOTES

1. The Antarctic aerial age had been inaugurated on February 4, 1902, when Scott's *Discovery* expedition had used a tethered balloon to obtain a view of the Ross Ice Shelf from 800 feet up. Almost two months later, Drygalski's *Gauss* expedition sent its own balloon up to about 1,600 feet over the coast of East Antarctica. Both expeditions took advantage of their balloon flights to take aerial photographs.
2. Bagshawe, *Two Men in the Antarctic*, p. 42.
3. Thomas, *Sir Hubert Wilkins*, p. 139 (quoted in).
4. Bagshawe, op. cit., p. 114.
5. Ibid., p. 128.
6. Ibid., on plate facing p. 164.
7. Ibid., p. 123.
8. Parfit, *South Light*, p. 278. Pp. 278–87 of this book by an American who was aboard the rescue ship describe this incident in detail.
9. The Waterboat Point rookery is far smaller today, probably a result of disruption from fact that Chile established its Gabriel González Videla base there in the summer of 1950–51.
10. Bagshawe, op. cit., pp. 168–69.
11. Thomas, *Sir Hubert Wilkins*, p. 144 (quoted in).
12. Wild, *Shackleton's Last Voyage*, p. 59 (quoted in).
13. Ibid., p. 65 (quoted in).
14. Ibid., p. 88.
15. Ibid., p. 153 (quoted in).
16. Ibid., pp. 194–95.
17. Ibid., p. 193 (quoted in).
18. Ibid., p. 193.

Chapter 15. The First Aviators Arrive

The most important primary sources used for Wilkins are Wilkins, "The Wilkins-Hearst Antarctic Expedition, 1928–1929," 1929, and "Further Antarctic Explorations," 1930. Significant secondary sources include the biographies of Wilkins listed for Chapter 14 plus Page, *Polar Pilot*, 1992 (a biography of Carl Ben Eielson); and Tordoff, *Mercy Pilot*, 2002 (a biography of Joe Crosson). Important secondary sources for both the Wilkins and Ellsworth sections of this chapter include Beeby, *In a Crystal Land*, 1994; Burke, *Moments of Terror*, 1994; and Grierson, *Challenge to the Poles*, 1965.

Coleman-Cooke, Discovery II *in the Antarctic*, 1963; Hardy, *Great Waters*, 1967; and Ommanney, *South Latitude*, 1938, all primary accounts, are the chief sources from the Discovery Investigations section.

Primary sources for Ellsworth are Ellsworth, *Beyond Horizons*, 1938, "The First Crossing of Antarctica," 1937, "My Flight Across Antarctica," 1936, and "My Four Antarctic Expeditions," 1939. Balchen, *Come North with Me*, 1958, has information about the 1934–35 attempt. Ommanney, *South Latitude*, 1938, covers the relief voyage of *Discovery II*. (Olsen, *Saga of the White Horizon*, 1972, has not been used because this account by a man who was with Ellsworth's party on all three of his attempts is so exaggerated that it must be deemed unreliable.) Secondary sources include Pool, *Polar Extremes*, 2002 (a biography of Ellsworth); the listed biographies of Wilkins; and biographies of Bernt Balchen, including Bess Balchen, *Poles Apart*, 2004; Glines, *Bernt Balchen*, 1999; and Knight and Durham, *Hitch Your Wagon*, 1950.

1. After he was knighted, Wilkins, whose full name was George Hubert Wilkins, chose to be called "Sir Hubert" rather than "Sir George," a choice he made, he said, because his first name was the same as the then-reigning king of England.
2. The honor of the first significant flight in the polar regions belongs to Russian naval lieutenant I. I. Nagurskiy, who flew a seaplane from Nova Zemlya over the Barents Sea in August 1914 in search of a Russian who had disappeared in 1912 while attempting to walk to the North Pole. (Grierson, *Challenge to the Poles*, p. 48)

[3] Wilkins, "The Wilkins-Hearst Antarctic Expedition . . . ," p. 233.
[4] Page, *Polar Pilot*, p. 316 (quoted in).
[5] Wilkins, "Further Antarctic Explorations," p. 371.
[6] Ibid., p. 371.
[7] At the time, any South Georgia sighting of a fur seal, the animal that nineteenth-century sealers had nearly wiped out, was worth noting. Harrison Matthews, a Discovery Committee scientist working at Grytviken in the late 1920s, wrote that he saw them "once or twice" on the nearby Willis Islands (Matthews, *Sea Elephant*, p. 88). And Ludwig Kohl-Larsen reported seeing a fur seal at Coal Harbor on the east coast of South Georgia in late 1928. (Kohl-Larsen, trans. Barr, *South Georgia*, p. 57)
[8] Ellsworth, *Beyond Horizons*, p. 255.
[9] Ellsworth had personally met both Earp and his wife before Earp died in 1929.
[10] Ellsworth, *Beyond Horizons*, p. 292.
[11] Ibid., p. 294 (quoted in).
[12] Ibid., pp. 298–99.
[13] Ibid., p. 303 (quoted in).
[14] Ibid., p. 315.
[15] Ellsworth, "My Flight Across Antarctica," p. 13.
[16] Ellsworth, *Beyond Horizons*, p. 345.
[17] Ibid., p. 317.
[18] This effort has been attributed in most accounts to British and Australian concerns over supporting their claim to Antarctic territory. Although this was probably the primary motive, there was at least one other thing at work. A year before Ellsworth's flight, Charles Ulm, an Australian pilot, had been lost over the Pacific while attempting to fly from Australia to California, and the United States navy had mounted a major search for him. Even though the effort failed, Australians remembered it. F. D. Ommanney, a Discovery Investigations scientist who participated in the rescue voyage, wrote, "Everyone in Melbourne said, 'It's because of Ulm you see. . . .' " (Ommanney, *South Latitude*, p. 172)
[19] Ommanney, *South Latitude*, p. 203.
[20] Ellsworth, *Beyond Horizons*, p. 362.

Chapter 16. The Wintering Explorers Return

The most important primary sources for the BGLE section are Bertram, *Antarctica, Cambridge . . .* , 1987; Bertram and Stephenson, "Archipelago to Peninsula," 1985; and Rymill, with Stephenson, *Southern Lights*, 1938. Useful secondary sources include Béchervaise, *Arctic and Antarctica*, 1995 (a biography of Rymill); Hunt, *Launcelot Flemming: A Portrait*, 2003; MacDonald and Rymill, *Penola Commemorative Biographies*, 1996; and Riley, *From Pole to Pole*, 1989.

Two secondary sources, Burke, *Moments of Terror*, 1994, and Grierson, *Challenge to the Poles*, 1964, provided important material for the entire chapter.

No member of the USAS expedition wrote a book-length narrative, although an edited edition of one West Base member's diary was published in 1984. Several USAS expeditioners did include long sections in autobiographies. Unfortunately, only one of these covers East Base. The primary sources used here reflect this situation. The most important used are Black, "East Base Operations . . . ," 1970; articles about East Base in Byrd, ed., *Reports on the Scientific Results . . .* , 1945; and Ronne, F., *Antarctica. My Destiny*, 1979. The most important secondary sources used are Gurling, "Some Notes on a Sledge Journey . . . ," 1979, and the two historical surveys with significant accounts of this expedition: Bertrand, *Americans in Antarctica . . .* , 1971, and Sullivan, *Quest for a Continent*, 1957.

[1] Rymill, with Stephenson, *Southern Lights*, p. 91.
[2] Rymill also named the other five islands in the group after Debenham's children, but not all of the children were entirely satisfied with "their" island. Debenham wrote that Rymill was "scorned when he returned by the youngest, Ann, then five years old, for having given her the smallest one, for as

she fiercely contested, she would not always be the smallest." (Debenham, *Antarctica*, p. 108)

3   This is Rymill's version for publication, written at a time when explorers seldom discussed interpersonal difficulties. Reports from British and other Antarctic bases in later years, when wintering personnel also had radio contact with the outside, often paint a different picture, one in which people frequently had difficulty with isolation.

4   Rymill, with Stephenson, op. cit., p. 187.

5   Ibid., p. 196.

6   Walton and Atkinson, *Of Dogs and Men*, p. 37 (quoted in).

7   Rymill, with Stephenson, op. cit., p. 193.

8   Ibid., p. 200.

9   Ibid., p. 201.

10 The first such finding was made in 1961–62 by an American geological team working in the southernmost part of the Peninsula. They detected a sub-glacial strait—thousands of feet below the surface of the ice—several hundred miles south of where Wilkins had reported seeing his Stefansson Strait. (Behrendt, *The Ninth Circle*, p. 194)

11 Hughes's statement was made in response to a question about whether the United States had a valid claim to Wilkes Land by right of discovery, whether the claim had ever been proclaimed, and what possible objections in law or policy might exist to claiming the area. (Gould, *The Polar Regions in their Relation to Human Affairs*, p. 54)

12 Black, "Geographical Operations from East Base . . . ," p. 7.

13 Ronne, F., "The Main Southern Sledge Journey from East Base . . . ," p. 15.

14 Ibid., p. 19.

15 Sullivan, *Quest for a Continent*, p. 168 (quoted in).

16 Black, "East Base Operations . . . ," p. 95.

17 Sullivan, *Quest for a Continent*, p. 169.

18 Dater, "United States Exploration and Research . . . ," footnote 16, p. 50.

Chapter 17. World War II, New Bases, and Political Conflict

Primary sources used include Darlington, as told to McIlvaine, *My Antarctic Honeymoon*, 1956; Dodson, R. H. T., personal communication; James, *That Frozen Land*, 1949; Ronne, E., *Antarctica's First Lady*, 2004; Ronne, F., *Antarctic Conquest*, 1949, *Antarctica, My Destiny*, 1979, "Ronne Antarctic Research Expedition, 1946–1948," 1948, and "Ronne Antarctic Research Expedition, 1946–1948," 1970; Squires, *S.S. Eagle*, 1992; and Walton, *Two Years in the Antarctic*, 1955.

The most important secondary sources used are Anon., "Argentine and Chilean Territorial Claims . . . ," 1946; Armstrong, "The Role of the Falkland Islands and Dependencies . . . ," 1998; Beck, "A Cold War," 1989; Beeby, *In a Crystal Land*, 1994; Christie, *The Antarctic Problem*, 1951; Fox, *Antarctica and the South Atlantic*, 1985; Dodds, *Pink Ice*, 2002; Fuchs, *Of Ice and Men*, 1982 (A history of the Falkland Islands Dependencies Survey, used principally for secondary material, but also contains primary material); Howkins, "Icy Relations. . . . ," 2006; Jones, "Protecting the Whaling Fleet . . . ," 1974; Mericq, *Antarctica: Chile's Claim*, 2004; Roberts and Thomas, "Argentine Antarctic Expeditions . . . ," 1953, and "Chilean Antarctic Expeditions . . . ," 1953; Rose, *Assault on Eternity*, 1980 (about Operation Highjump); Sullivan, *Quest for a Continent*, 1957; and Wordie, "The Falkland Islands Dependencies Survey 1943–46," 1946.

1   In January 1941 the German raider *Pinguin* captured two Norwegian factory ships and most of their catchers off the coast of Queen Maud Land. The most complete account of this dramatic event I have found in English is in Brennecke, trans. Fitzgerald, *Cruise of the Raider HK-33*, pp. 184–202.

2   Fraser, compiled by, *Shetland's Whalers Remember . . .* , p. 16.

3   Jones, "Protecting the Whaling Fleet . . . ," p. 422.

4   In 1906, a prospectus for a Chilean Antarctic expedition that was to create a base on Elephant Island explicitly mentioned the Antarctic Peninsula, South Orkneys, and South Shetlands as part

of Chilean territory. The plan was dropped after a disastrous earthquake hit Chile in August that year, soaking up all the government's attention and resources. (Mericq, *Antarctica. Chile's Claim*, p. 93) Much earlier, in 1831, Chile's first president, Bernardo O'Higgins, had sent a letter to Great Britain indicating that Chile regarded the Antarctic regions to the south of Chile as an integral part of the country. (Pinochet, *Chilean Sovereignty in Antarctica*, pp. 28–29)

5 Anon., "Argentine and Chilean Territorial Claims . . . ," pp. 416–17.
6 Quigg, *A Pole Apart*, p. 117 (quoted in).
7 Dodds, *Pink Ice*, p. 15 (quoted in).
8 Beeby, *In a Crystal Land*, p. 180.
9 Ibid., p. 185 (quoted in).
10 Squires, *S.S. Eagle*, pp. 67.
11 James, *That Frozen Land*, p. 90 (quoted in).
12 Ibid., p. 90.
13 Squires, op. cit., p. 87.
14 James, op. cit., pp 142–43.
15 British reaction from Walton, *Two Years in the Antarctic*, p. 108.
16 Ronne, F., *Antarctic Conquest*, p. 60.
17 Darlington, as told to McIlvaine, *My Antarctic Honeymoon*, p. 120.
18 The one genuine landmass that Ronne could have seen on this flight is what today is known as Berkner Island. This 85-mile wide, 200-mile long, ice-covered island rises to about 3,200 feet as it separates the Ronne and Filchner Ice Shelves.
19 Unfortunately, the positions Ronne determined for his discoveries on this and other flights were often seriously in error, sometimes as much as 100 miles from their true positions. The U.S. Antarctic Peninsula traverse team determined several of these positions accurately in 1961–62 using a ground survey. (Behrendt, *The Ninth Circle*, p. 52)
20 Date of birth from PublicRecords.com. Name from American Geographical Society newsletter, *Ubique*, vol. XXVI, no. 1 (Feb 2006), p. 5.
21 Anon., "Territorial Claims in the Antarctic," p. 361 (quoted in).
22 Beck, "A Cold War," p. 36 (quoted in).
23 Headland, "The Origin & Development of the Antarctic Treaty System," p. 27 (quoted in).

Chapter 18. The International Geophysical Year and the Antarctic Treaty

Key primary sources used for the IGY portion of this chapter include Behrendt, *Innocents on the Ice*, 1998; Dufek, *Operation Deepfreeze*, 1957; Fuchs, "The Crossing of Antarctica," 1959; Fuchs and Hillary, *The Crossing of Antarctica*, 1958; MacDowall, *On Floating Ice*, 1999; Stephenson, *Crevasse Roulette*, 2009; and Ronne, F., *Antarctic Command*, 1961. Significant secondary works used include Belanger, *Deep Freeze*, 2006; and Kemp, *The Conquest of the Antarctic*, 1956.

With regard to the last part of the chapter, many works describe the origins of the Antarctic Treaty. See, for example, Quigg, *A Pole Apart*, 1983, for an excellent account. This work also discusses the evolution of the Treaty in the first two decades after its ratification. An appendix (pp. 219–25) contains the full Treaty text, which may also be found in many other sources.

1 Date from Behrendt, *Innocents on the Ice*, p. 274. The United States never followed up on this or any of the claim documents left by Americans in the Antarctic over the years with a formal government claim.
2 See Stephenson, *Crevasse Roulette*, p. 173, for an English-language account of the activities at Belgrano. Sadly, as Stephenson notes, ". . . these achievements seem to be largely unknown outside Argentina."
3 Arnold, *Eight Men in a Crate*, 2007, which is based on the diary of a member of the Shackleton advance party, vividly tells the story of this brutal 1956 winter.
4 Fuchs, "The Crossing of Antarctica," p. 25.

SOURCES AND NOTES

5   For the reader interested in the Ross Sea side of Fuchs's Trans-Antarctic Expedition, the most important among several accounts are Fuchs's and Hillary's joint account of the entire expedition, *The Crossing of Antarctica*, and Hillary's *No Latitude for Error*, 1961. McKenzie, *Opposite Poles*, 1963, provides a fascinating discussion of the personality issues that arose between Fuchs and Hillary.

6   Fuchs, op. cit., p. 40.

7   Speak, *William Speirs Bruce*, p. 125.

8   Fuchs, op. cit., pp. 46–47.

9   Quigg, pp. 220–21 (quoted in).

Chapter 19. The Antarctic Treaty Era

Because of its discursive content, Chapter 19 draws on a multitude of sources. Among the many primary works used, the most important include Barker, *Beyond Endurance*, 1997; Burley, "Was Shackleton Valley the Passageway to Stromness?" 1992, and *Joint Services Expedition Elephant Island* . . . , 1971; Headland, "Hostilities in the Falkland Islands Dependencies. . . ," 1983; Lindblad, with Fuller, *Passport to Anywhere*, 1983; Sheridan, *Taxi to the Snowline*, 2006; and Vairo et al., trans. Freire, *Antártida: Patrimoni Cultural. . .* , 2007.

Significant secondary works among many consulted include Anon., "Volcanic Eruption Compels Evacuation . . . ," 1968, "Combined Services Expedition to South Georgia, 1964–65," 1966, "Shackleton's Bible Returns Home After Many Years," 1971, "Antarctica's First Baby Warmly Welcomed," 1978, and "Argentina's *Bahía Paraíso* Sinks off Anvers Island," 1989; Antarctic Treaty Consultative Parties, "Protocol on Environmental Protection . . . ," 1993; Beck, "Convention on the Regulation of Antarctic Mineral Resource Activities," 1989; Fuchs, *Of Ice and Men*, 1982; Headland, *The Island of South Georgia*, 1992; Keys and BAS Press Office, "One Small Ice Shelf Dies . . . ," 1995; Perkins, *Operation Paraquat*, 1986; and Walton and Atkinson, *Of Dogs and Men*, 1996. See also specific sources listed in the citations below.

1   Anon., "Agreement on Disputed Antarctic Place-Names." The names "Palmer Land" and "Graham Land" survive, however, now used to designate specific parts of the Peninsula—Palmer Land for the southern portion, Graham Land for the northern.

2   Fuchs, *Of Ice and Men*, pp. 289–91 (quoted in).

3   In the early 1990s, personnel from the Argentine summer base at Deception Island found a cross from the cemetery that had washed up on the western shore of Port Foster. Realizing that this was not just a bit of driftwood, they carefully erected it on a hillside behind their base. In February 2002, Susan Barr, a Norwegian polar historian, identified the cross as having come from the Whalers Bay cemetery. On February 18, following the identification, the Argentines carefully re-erected the cross at Whalers Bay near where it had once stood. (Barr, Downie, and Sánchez, "Whaler's Cross on Deception Island," p. 70)

4   Fuchs, op. cit., p. 294 (quoted in).

5   Argentina et al., *Deception Island Management Package*, pp. 67–70.

6   Data from IAATO: www.IAATO.org.

7   Scott, *The Voyage of the Discovery*, Vol. I, pp. 398–99.

8   Burley, "Was Shackleton Valley the Passageway to Stromness?" p. 235.

9   Burley, *Joint Services Expedition Elephant Island* . . . , p. 16.

10 See Fuchs, trans. Sokolinsky, *In Shackleton's Wake*, for the story of this adventure. This fascinating book also includes an excellent and thoughtful analysis of Shackleton's *Endurance* expedition.

11 Anon., "Antarctica's First Baby Warmly Welcomed," p. 169 (quoted in).

12 Many books have been written about the central action of the Falklands/Malvinas War. Among others, the interested reader may want to consult Sunday Times Insight Team, *War in the Falklands*, 1982; Freedman, *The Official History of the Falklands Campaign*, 2005; or Hastings and Jenkins, *The Battle for the Falklands*, 1983. Unfortunately, no work that fully presents the Argentine perspective has been published in English, but see Moro, trans. Michael Valeur, *The History of the South Atlantic Conflict*, 1989, for some of this. Perkins, *Operation Paraquat*, is the only book

to focus specifically on the action on South Georgia.

13. In February 1996, Great Britain turned Faraday over to the Ukraine. After a short period of joint operations, the Ukranians took full charge and the base was renamed Vernadsky. It has operated continuously since under this name.

14. Peter N. Spotts, "Antarctica's Wilkins Ice Shelf Eroding at an Unforeseen Pace," ¶ 5 (quoted in).

15. Two recent books—Montaigne, *Fraser's Penguins*, 2010, and Hooper, *The Ferocious Summer*, 2007, provide excellent and lucid discussions of the impact and significance of warming along the Antarctic Peninsula. Both center their texts around studies of the effects on the penguin populations in the vicinity of the U. S. Palmer Station on Anvers Island, but each goes far beyond that, to provide excellent discussions of climate change issues in Antarctica generally.

16. This expedition, led jointly by an American, Will Steger, and a Frenchman, Jean-Louis Etienne, began its crossing near the far northern tip of the Antarctic Peninsula on July 27, 1989. The men and their 36 dogs initially traveled down the east coast of the Peninsula at sea level. A month and nearly 300 miles later, they turned west and climbed to the Peninsula plateau. They then sledged south to the main body of the Antarctic Continent. From there they headed for the South Pole and then on across the rest of the continent, at times through areas that no one had crossed overland. They completed their 3,700-mile trip on March 3, 1990, at the Russian Mirnyy base. (See Steger and Bowermaster, *Crossing Antarctica*, 1992, the official expedition narrative, for a full account.)

17. See Vairo et al., trans. Freire, *Antártida*, for description of Argentina's preservation work.

18. Charcot, trans. Billinghurst, *Towards the South Pole . . .* , pp. 235–36.

Literature Cited

The following lists all works cited in Sources and Notes or mentioned in the text. Many more works were consulted in preparing this book. A complete list is available from the author.

Adie, Susan, and Bjørn Basberg, trans. James Adie, "The First Antarctic Whaling Season of *Admiralen* (1905–1906): The Diary of Alexander Lange," *Polar Record*, 45(235), pp. 243–63, 2009.

Alberts, Fred G., comp. and ed., *Geographic Names of the Antarctic, second edition*, Washington: National Science Foundation 1995.

Alexander, Caroline, *The* Endurance: *Shackleton's Legendary Antarctic Expedition*, New York: Knopf 1998.

Amundsen, Roald, *My Life as an Explorer*, London: William Heinemann 1927.

———, trans. Erik Dupont and Christine LePiez, ed. Hugo Decleir, *Roald Amundsen's* Belgica Diary: *The First Scientific Expedition to the Antarctic*, Huntingdon, England: Bluntisham 1999.

Andersson, J. Gunnar, *Antarctic: "Stolt har hon levat, Stolt skall hon dö,"* Stockholm: Saxon & Lindstroms 1944.

Anon., "Agreement on Disputed Antarctic Place-Names," *Polar Record*, 12(79), pp. 470–71, 1965.

———, "Antarctica's First Baby Warmly Welcomed," *Antarctic*, 8(5), pp. 169–70, March 1978.

———, "Argentina's *Bahía Paraíso* Sinks off Anvers Island," *Antarctic*, 11(9&10), pp. 391–93, 1989.

———, "Argentine and Chilean Territorial Claims in the Antarctic," *Polar Record*, 4(32), pp. 412–17, 1946.

———, "British Letters Patent of 1908 and 1917 Constituting the Falkland Islands Dependencies," *Polar Record*, 5(35-36), pp. 241–43, 1948.

———, "Combined Services Expedition to South Georgia, 1964–65," *Polar Record*, 13(82), pp. 70–71, 1966.

———, " 'New Belgica' Project: English Summary," www.steenschuit.be/belgica_project_english_sum.html.

———, "Shackleton's Bible Returns Home After Many Years," *Antarctic*, 6(3), 105, September 1971.

———, "Volcanic Eruption Compels Evacuation of Three Bases on Deception Island," *Antarctic* 5(1), pp. 23–26, March 1968.

Antarctic Treaty Consultative Parties, "Protocol on Environmental Protection to the Antarctic Treaty," SCAR Bulletin, no. 110, July 1993, in *Polar Record*, 29(170), pp. 256–75, 1993.

Arctowski, Henryk, "Exploration of Antarctic Lands," 465–96 in Murray, ed., *The Antarctic Manual*, 1901.

Argentina, Chile, Norway, Spain, the UK and the USA, submitted by, *Deception Island Management Package*, 2005. Available online at www.deceptionisland.aq/package.php.

Armstrong, Patrick, "The Role of the Falkland Islands and Dependencies in Anglo-Argentine Relations in the Early 1950s," *Polar Record*, 34(188), 53–55, 1998.

Arnold, Anthea, *Eight Men in a Crate: The Ordeal of the Advance Party of the Trans-Antarctic Expedition, 1955–1957. Based on the Diaries of Rainer Goldsmith*, Huntingdon, England: Bluntisham 2007.

Bagshawe, Thomas Wyatt, "Notes on the Habits of the Gentoo and Ringed or Antarctic Penguins," *Transactions of the Zoological Society of London*, 24(3), 185–306, 1938.

———, *Two Men in the Antarctic: An Expedition to Graham Land 1920–22*, New York: Macmillan 1939.

———, "A Year Amongst Whales and Penguins," *Journal of the Society for the Preservation of the Fauna of the Empire*, 36, pp. 30–36, 1939.

Bakewell, William Lincoln, ed. Elizabeth Anna Bakewell Rajala, *The American on the* Endurance: *Ice, Sea, and Terra Firma. Adventures of William Lincoln Bakewell*, Munising, MI: Dukes Hall Publishing 2004.

Balch, Edwin Swift, *Antarctica*, Philadelphia: Allen, Lane & Scott 1902.

Balchen, Bernt, *Come North with Me*, New York: Dutton 1958.

Balchen, Bess, *Poles Apart: The Admiral Richard E. Byrd and Colonel Bernt Balchen Odyssey*, Oakland, OR: Red Anvil Press 2004.

Barker, Nick, *Beyond Endurance: An Epic of Whitehall and the South Atlantic Conflict*, Barnsley, England: Leo Cooper, Pen & Sword Books Ltd. 1997.

Barr, Susan, Rod Downie, and Rodolfo Sánchez, "Whaler's Cross on Deception Island," *Polar Record*, 40(212), 69–70, 2004.

Barr, William, *The Expeditions of the First International Polar Year, 1882–83*, Calgary: Univ. of Calgary, Arctic Institute of North America Technical Paper #29, 1985.

———, translator's introduction, pp. 11–38 in Filchner, *To the Sixth Continent*, 1994.

Barr, William, Reinhard Krause, and Peter-Michael Pawlik, "Chukchi Sea, Southern Ocean, Kara Sea: The Polar Voyages of Captain Eduard Dallmann, Whaler, Trader, Explorer, 1830–96," *Polar Record*, 40(212), 1–18, 2004.

Barrett-Hamilton, Gerald E. H., "Seals," pp. 209–24 in Murray, ed., *The Antarctic Manual*, 1901.

Basberg, Bjørn, *The Shore Whaling Stations at South Georgia: A Study in Antarctic Industrial Archaeology*, Oslo: Novus Forlag 2004. Publication No. 30 from Commander Chr. Christensen's Whaling Museum, Sandefjord, Norway.

———, and Robert Burton, "New Evidence on the Manager's Villa in Stromness Harbour, South Georgia, *Polar Record*, 42(221), pp. 147–51, 2006.

Baughman, Tim H., *Before the Heroes Came: Antarctica in the 1890s*, Lincoln: Univ. of Nebraska Press 1994.

———, "Hopeless in a Hopeless Sea of Ice: The Course of the *Belgica* Expedition and Its Impact on the Heroic Era," paper delivered at Symposium on the Centennial of the *Belgica*, Ohio State Univ. 1997 (Reprinted as pp. 437–452 in Cook, *Through the First Antarctic Night*, centennial edition, Pittsburgh, PA: Polar Publishing Co., 1998).

Beaglehole, John C., *The Exploration of the Pacific, third edition*, Stanford, CA: Stanford Univ. Press 1966.

Béchervaise, John, *Arctic and Antarctic: The Will and the Way of John Riddoch Rymill*, Huntingdon, England: Bluntisham 1995.

Beck, Peter J., "A Cold War," *The Falkland Islands Journal*, pp. 36–43, 1989.

———, "Convention on the Regulation of Antarctic Mineral Resource Activities: A Major Addition to the Antarctic Treaty System," *Polar Record*, 25(152), pp. 19–32, 1989.

Beeby, Dean, *In a Crystal Land: Canadian Explorers in Antarctica*, Toronto: Univ. of Toronto Press 1994.

Behrendt, John C., *Innocents on the Ice: A Memoir of Antarctic Exploration, 1957*, Niwot, CO: Univ. Press of Colorado 1998.

———, *The Ninth Circle: A Memoir of Life and Death in Antarctica, 1960–1962*, Albuquerque: Univ. of New Mexico Press 2005.

Belanger, Dian Olson, *Deep Freeze: The United States, the International Geophysical Year, and the Origins of Antarctica's Age of Science*, Boulder, CO: Univ. Press of Colorado 2006.

Bellingshausen, Thaddeus, ed. Frank Debenham, trans. Edward Bullough, N. Volkov, and others, *The Voyage of Captain Bellingshausen to the Antarctic Seas, 1819–1821*, 2 Vols., London: Hakluyt Society 1945 (Originally published in Russian, St. Petersburg 1830).

———, ed. and trans. Harry Gravelius, *Forschungsfahrten im Südlichen Eismeer 1819–1821*, Leipzig: S. Herzel 1902.

Bernacchi, L. C., *The Saga of the "Discovery*,*"* London/Glasgow: Blackie & Sons 1938.

Bertram, George Colin Lauder, *Antarctica, Cambridge, Conservation and Population: A Biologist's Story*, Cambridge: G. C. L. Bertram 1987.

Bertram, Colin, and Alfred Stephenson, "Archipelago to Peninsula," *Geographical Journal*, 151(2), pp. 155–67, 1985.

Bertrand, Kenneth J., *Americans in Antarctica 1775–1948*, New York: American Geographical Society,

Special Publication 39, 1971.

Bickel, Lennard, *Shackleton's Forgotten Men: The Untold Story of the Endurance Epic*, New York: Thunder's Mouth Press 2000 (Originally published as *Shackleton's Forgotten Argonauts*, South Melbourne: Macmillan 1982).

Biscoe, John, "From the 'Journal of a Voyage Towards the South Pole on Board the Brig 'Tula', Under the Command of John Biscoe, with the Cutter 'Lively' in Company'," pp. 305–36 in Murray, ed., *The Antarctic Manual*, 1901.

Black, Richard B., "East Base Operations, United States Antarctic Service, 1939–41," pp. 89–99 in Friis and Bale, eds., *United States Polar Exploration*, 1970.

———, "Geographical Operations from East Base, United States Antarctic Service Expedition, 1939–41," pp. 4–12 in Byrd, ed., *Reports on the Scientific Results . . .* , 1945.

Bond, Creina, Roy Siegfried, and Peter Johnson, *Antarctica: No Single Country, No Single Sea*, New York: Mayflower Books 1979.

Brennecke, Hans Joachim, trans. Edward Fitzgerald, *Cruise of the Raider HK-33*, New York: Crowell 1954.

Brown, R. N. Rudmose, with five chapters by W. G. Burn Murdoch, *A Naturalist at the Poles: The Life, Work & Voyages of Dr. W. S. Bruce, the Polar Explorer*, London: Seeley, Service 1923.

Bruce, William S., "Antarctic Exploration: The Story of the Antarctic," *Scottish Geographical Magazine*, 10(10), pp. 57–62, February 1894.

———, *Polar Exploration*, New York: Henry Holt, Home University Library of Modern Knowledge, circa 1911.

———, ed. Peter Speak, *The Log of the* Scotia *Expedition, 1902–4*, Edinburgh: Edinburgh Univ. Press 1992.

Brue, Adrien, *Mappe Monde en deux Hemispher presentant L'Etat Acuel de la Geographie Par A. H. Brue, Geographe de S. H. R. Monsieur a Paris*, Paris: Charles Simonneua, June 1820.

Bryce, Robert M., *Cook & Peary: The Polar Controversy, Resolved*, Mechanicsburg, PA: Stackpole Books 1997.

Burke, David, *Moments of Terror: The Story of Antarctic Aviation*, Kensington: New South Wales Univ. Press 1994.

Burley, Malcolm K., "Was Shackleton Valley the Passageway to Stromness?" *Polar Record*, 28(166), pp. 234–36, 1992.

———, *Joint Services Expedition Elephant Island, 1970–71*, London: The Author, Printed in the FONAC Printing Office by the Royal Marines Staff, 1971.

Burney, James, *Chronological History of the Voyages and Discoveries in the South Seas or Pacific Ocean*, 5 Vols., Amsterdam: N. Israel and New York: Da Capo Press 1967 (Facsimile edition of work originally published London: G & W Nicol, 1803–17).

Byrd, Richard E., ed., *Reports on the Scientific Results of the United States Antarctic Service Expedition 1939–41*, Philadelphia: Proceedings of the American Philosophical Society, 89(1), April 30, 1945.

Campbell, R. J., ed., *The Discovery of the South Shetland Islands: The Voyages of the Brig* Williams *1819–1820 as Recorded in Contemporary Documents and the Journal of Midshipman C. W. Poynter*, London: Hakluyt Society 2000.

Chapman, Walker, ed., *Antarctic Conquest: The Great Explorers in their Own Words*, Indianapolis: Bobb-Merrill 1965.

Charcot, Jean-B., "Charcot Land, 1910 and 1930," *Geographical Review*, xx(3), July, pp. 389–96, 1930.

———, trans. A. W. Billinghurst, *Towards the South Pole Aboard the* Français: *The First French Expedition to the Antarctic 1903–1905*, Huntingdon, England: Bluntisham 2004 (Originally published in French as *Le 'Français' au Pôle Sud*, Paris: Flammarion 1906).

———, trans. Philip Walsh, *The Voyage of the 'Why Not?' in the Antarctic: The Journal of the Second French South Polar Expedition, 1908–1910*, New York/London: Hodder and Stoughton, 1911 (Originally published in French as *Le Pourquoi-Pas? dans l'Antarctique*, Paris: Flammarion 1910).

Christensen, Lars, trans. E. M. G. Jayne, *Such Is the Antarctic*, London: Hodder and Stoughton 1935.

Christie, E. W. Hunter, *The Antarctic Problem: An Historical and Political Study*, London: Allen & Unwin 1951.

Coleman-Cooke, J., Discovery II *in the Antarctic: The Story of British Research in the Southern Seas*, London: Odhams Press 1963.

Cook, Frederick A., *Through the First Antarctic Night, 1898–1899,* New York: Doubleday & McClure 1900.

Cook, James, *A Voyage Towards the South Pole and Round the World in 1772–5*, 4th ed., 2 Vols., London: W. Strahan & T. Cadell 1784 (Originally published 1777).

Cook, James, ed. J. C. Beaglehole, *The Journals of Captain James Cook on His Voyages of Discovery: The Voyage of the* Resolution *and* Adventure, *1772–1775*, Cambridge: Cambridge Univ. Press for Hakluyt Society 1961.

Darlington, Jennie, as told to Jane McIlvaine, *My Antarctic Honeymoon: A Year at the Bottom of the World*, Garden City, NY: Doubleday 1956.

Dater, Henry M., "United States Exploration and Research in Antarctica through 1954," pp. 43–55 in Friis and Bale, eds., *United States Polar Exploration*, 1970.

De Gerlache de Gomery, Adrien, trans. Maurice Raraty, *Fifteen Months in the Antarctic: Voyage of the Belgica*, Huntingdon, England: Bluntisham 1998 (Originally published in French as *Quinze Mois dans l'Antarctique: Voyage de la Belgica*, Bruxelles: Ch. Bulens 1902).

Debenham, Frank, *Antarctica: The Story of a Continent*, London: Herbert Jenkins 1959.

Decleir, Hugo, and Claude De Broyer, eds., *The* Belgica *Expedition Centennial: Perspectives on Antarctic Science and History. Proceedings of the* Belgica *Centennial Symposium, 14–16 May 1998, Brussels*, Brussels: Brussels Univ. Press 2001.

Dodds, Klaus J., *Pink Ice: Britain and the South Atlantic Empire*, London: I. B. Tauris 2002.

Donald, Charles W., "The Late Expedition to the Antarctic," *Scottish Geographical Magazine*, 10(2), pp. 62–68, February 1894.

Drygalski, Erich von, trans. M. M. Raraty, *The Southern Ice-Continent: The German South Polar Expedition Aboard the 'Gauss' 1901–1903*, Huntingdon, England: Bluntisham 1989 (Originally published in German as *Zum Continent Des Eisigen Südens*, Berlin: Georg Reimer 1904).

Dufek, George J., *Operation Deepfreeze*, New York: Harcourt, Brace 1957.

Dunmore, John, *From Venus to Antarctica: The Life of Dumont d'Urville*, Auckland: Exisle 2007.

Dunnett, Harding McGregor, *Shackleton's Boat: The Story of the* James Caird, Cranbrook, Kent: Neville & Harding 1996.

D'Urville, Jules S. C. Dumont, trans. and ed. Helen Rosenman, *Two Voyages to the South Seas by Captain (later Rear-Admiral) Jules S. C. Dumont d'Urville, of the French Navy. Vol. I:* Astrolabe *1826–1829; Vol. II:* Astrolabe *and* Zélée *1837–40*, Carlton: Melbourne Univ. Press 1987 (Vol. II abridged from the French original: d'Urville, Jules S. C. Dumont et al., *Voyage au Pôle Sud et dans L'Océanie sur les Corvettes l'Astrolabe et la Zélée*, Paris: Gide, 1841–46).

Duse, Samuel A., *Bland Pingviner och Sälar: Minnen från Svenska Sydpolarexpeditionen, 1901–1903*, Stockholm: Beijer Bokförlagsaktiebolag 1905.

Eights, James, "Description of a New Animal Belonging to the Arachinides of Latrelle . . . ," *Boston Journal of Natural History*, 1, pp. 203-06, 1837.

———, "A Description of the New South Shetland Isles . . . ," pp. 195–216 in Fanning, *Voyages to the South Seas . . .* , 1970.

Elliot, Gerald, *A Whaling Enterprise: Salvesen in the Antarctic*, Wilby Hall, Wilby, Norwich: Michael Russell 1998.

Elliott, John, and Richard Pickersgill, ed. Christine Holmes, *Captain Cook's Second Voyage: The Journals of Lieutenants Elliott and Pickersgill*, London: Caliban 1984.

Ellsworth, Lincoln, *Beyond Horizons*, New York: Doubleday Doran 1938.

———, "The First Crossing of Antarctica," *Geographical Journal*, lxxxix (3), pp. 192–213, March 1937.

———, "My Flight Across Antarctica," *National Geographic*, lxx(1), pp. 1–35, July 1936.

———, "My Four Antarctic Expeditions," *National Geographic*, lxxvi(1), pp. 129–38, July 1939.

LITERATURE CITED

Elzinga, Aant, Torgny Nordin, David Turner, and Urban Wråkberg, eds., *Antarctic Challenges: Historical and Current Perspectives on Otto Nordenskjöld's Antarctic Expedition 1901–1903*, Göteborg, Sweden: Royal Society of Arts and Sciences 2004.

Ennis, Helen, *Man With a Camera: Frank Hurley Overseas*, Canberra: National Library of Australia 2002.

Erskine, Angus B., and Kjell-G. Kjaer, "The Polar Ship *Quest*," *Polar Record*, 34(189), pp. 129–42, 1998.

Fanning, Edmund, *Voyages Round the World; with Selected Sketches of Voyages to the South Seas, North and South Pacific Oceans, China, &etc. &etc. . . .* , New York: Collins & Hannay 1833.

———, *Voyages to the South Seas, Indian and Pacific Ocean, China Sea, Northwest Coast, Feejee Islands, South Shetlands, &etc. &etc.*, Fairfield, WA: Ye Galleon Press 1970 (Facsimile of edition originally published New York: William H. Vermilye 1838).

Filchner, Wilhelm, trans. William Barr, *To the Sixth Continent: The Second German South Polar Expedition*, Huntingdon, England: Bluntisham 1994 (Originally published in German as *Zum Sechsten Erdteil: Die Zweite Deutsche Südpolar-Expedition*, Berlin: Ullstein 1922).

———, trans. William Barr, "Exposé," pp. 196–214 in Filchner, *To the Sixth Continent*, 1994 (Originally published in German as "Festellungen," pp. 24–58 in *Dokumentation uber die Antarktisexpedition 1911/12 von Wilhelm Filchner*, Munchen: Bayerischen Akademe der Wissenschaften 1985).

Fisher, Margery, and James Fisher, *Shackleton*, London: James Barrie Books 1957.

[Fletcher, Francis], *The World Encompassed by Sir Francis Drake . . .* , Cleveland: World Publishing Co. 1966 (Facsimile edition of original, first published London: printed for Nicholas Bourne to be sold at his bookshop at the Royall Exchange, 1628).

Fogg, G. E., *A History of Antarctic Science*, Cambridge: Cambridge Univ. Press, 1992.

Fox, Robert, *Antarctica and the South Atlantic: Discovery, Development and Dispute*, London: British Broadcasting Corporation 1985.

Fraser, Gibbie, compiled by, *Shetland's Whalers Remember . . .* , Published by the author, printed by Nevisprint Ltd., Fort William, Scotland 2001.

Freedman, Lawrence, *The Official History of the Falklands Campaign: vol I: The Origins of the Falklands War; vol II: War and Diplomacy*, Abingdon, England: Routledge 2005.

Fricker, Karl, trans. A. Sonnenschein, *The Antarctic Regions*, London: Swan Sonnenschein 1900 (Originally published in German as *Antarktis*, Berlin: Schall & Grund 1898).

Friis, Herman R., and Shelby G. Bale, Jr., eds., *United States Polar Exploration*, Athens: Ohio Univ. Press, 1970.

Fuchs, Arved, trans. Martin Sokolinsky, *In Shackleton's Wake*, Dobbs Ferry, NY: Sheridan House, Inc. 2001 (Originally published in German as *Im Schatten des Pols*, Bielefeld: Verlag Delius, Klasing & Co. KG 2000).

Fuchs, Sir Vivian E., "The Crossing of Antarctica," *National Geographic*, cxv(1), pp. 25–47, January 1959.

———, *Of Ice and Men: The Story of the British Antarctic Survey, 1943–73*, London: Anthony Nelson 1982.

———, and Sir Edmund Hillary, *The Crossing of Antarctica: The Commonwealth Trans-Antarctic Expedition 1955–58*, London: Cassell 1958.

Furse, Chris, *Antarctic Year: Brabant Island Expedition*, London: Croom Helm, 1986.

Gilkes, Dr. Michael, "It Ain't Necessarily So: South Georgia Loose Ends," *James Caird Society Journal*, 5, 62-66, July 2010.

Glines, Carroll V., *Bernt Balchen: Polar Aviator*, Washington: Smithsonian Institution Press 1999.

Gould, Laurence McKinley, *The Polar Regions in Their Relation to Human Affairs*, New York: American Geographical Society (Bowman Memorial Lectures) 1958.

Grierson, John, *Challenge to the Poles: Highlights of Arctic and Antarctic Aviation*, Hamden, CT: Archon 1964.

———, *Sir Hubert Wilkins: Enigma of Exploration*, London: Robert Hale 1960.

Gurling, Paul, "Some Notes on a Sledge Journey from Stonington Island 1940-41," *Polar Record*, 19(123), pp. 613–16, 1979.

Gurney, Alan, *Below the Convergence: Voyages Toward Antarctica 1699–1839*, New York: Norton 1997.

———, *The Race to the White Continent: Voyages to the Antarctic*, New York: Norton 2000.

Halley, Edmond, ed. Norman J. W. Thrower, *The Three Voyages of Edmond Halley in the* Paramore *1698–1701*, 2 Vols., London: Hakluyt Society 1981.

Hardy, Sir Alister, *Great Waters: A Voyage of Natural History to Study Whales, Plankton and the Waters of the Southern Ocean in the Old Royal Research Ship* Discovery *with the Results Brought up to Date by the Findings of the R.R.S.* Discovery II, London: Collins 1967.

Hart, Ian B., *Antarctic Magistrate: A Life Through the Lens of a Camera*, Laurel Cottage, Newton St. Margarets, Herefordshire: Pequena 2009.

———, *Pesca: The History of Compañia Argentina de Pesca Sociedad Anónima of Buenos Aires*, Salcombe, Devonshire: Aidan Ellis Publishing 2001.

———, *Whaling in the Falkland Islands Dependencies 1904–1931*, Laurel Cottage, Newton St. Margarets, Herefordshire: Pequena 2006.

Hastings, Max, and Simon Jenkins, *The Battle for the Falklands*, London: Norton 1983.

Hayes, J. Gordon, *Antarctica: A Treatise on the Southern Continent*, London: The Richards Press 1928.

———, *The Conquest of the South Pole: Antarctic Exploration 1906–1931*, London: Thornton Butterworth 1932.

Heacox, Kim, *Shackleton: The Antarctic Challenge*, Washington: National Geographic Society 1999.

Headland, Robert K., *A Chronology of Antarctic Exploration: A Synopsis of Events and Activities Until the International Polar Years, 2007 to 2009*, London: Bernard Quaritch 2009.

———, "Bellingshausen," *Antarctic*, 12(9), pp. 327–28, October 1992.

———, "Births on South Georgia and Other Antarctic Regions," *The Falkland Islands Journal*, 7(2), pp. 10–13, 1998.

———, *Chronological List of Antarctic Expeditions and Related Historical Events*, Cambridge: Cambridge Univ. Press 1989.

———, "The German Station of the First International Year, 1882–83, at South Georgia, Falkland Islands Dependencies," *Polar Record*, 21(132), pp. 287–301, 1982.

———, "Hostilities in the Falkland Islands Dependencies March–June 1982," *Polar Record*, 21(135), pp. 549–58, 1983.

———, "The Origin & Development of the Antarctic Treaty System," *Nimrod*, 1, pp. 21–39, October 2007.

———, *The Island of South Georgia*, Cambridge: Cambridge Univ. Press 1992 (Originally published in 1984 by Cambridge Univ. Press).

Hermelo, Ricardo S., José M. Sobral, and Felipe Fliess, trans. Gricelda Perales and Lawrence Perales, ed. Michael H. Rosove, *When the Corvette* Uruguay *Was Dismasted: The Return of the* Uruguay *from the Antarctic in 1903*, Santa Monica, CA: Adélie Books 2004.

Heyburn, Henry R., and Gunnar Stenersen, "The Wreck and Salvage of S.S. *Telefon*," *Polar Record*, 25(152), pp. 51–54, 1989.

Hillary, Sir Edmund, *No Latitude for Error*, New York: Dutton 1961.

Hobbs, William Herbert, *The Discoveries of Antarctica Within the American Sector, as Revealed by Maps and Documents*, Philadelphia: Transactions of the American Philosophical Society, 31(1), pp. 1–71, 1939.

Hough, Richard, *The Blind Horn's Hate*, New York: Norton 1971.

Hooper, Meredith, *The Ferocious Summer: Palmer's Penguins and the Warming of Antarctica*, London: Profile Books Ltd. 2007.

Howkins, Adrian, "Icy Relations: The Emergence of South American Antarctica During the Second World War," *Polar Record*, 42(221), pp. 153–65, 2006.

Hunt, Giles, *Launcelot Fleming: A Portrait*, Norwich, Norfolk: Canterbury Press, 2003.

Huntford, Roland, *Scott and Amundsen*, New York: Putnams 1980 (Originally published, London: Hodder & Stoughton 1979).

———, *Shackleton*, New York: Atheneum 1986 (Originally published, London: Hodder & Stoughton

LITERATURE CITED

1985).

Hurley, Captain Frank, *Argonauts of the South: Being a Narrative of Voyagings and Polar Seas and Adventures in the Antarctic with Sir Douglas Mawson and Sir Ernest Shackleton*, London: Putnams 1925.

Hussey, L. D. A., *South with Shackleton*, London: Sampson Low 1949.

Hydrographer of the Navy, *The Antarctic Pilot*, London: Hydrographic Dept., Admiralty 1930.

James, David, *That Frozen Land: The Story of a Year in the Antarctic*, London: Falcon Press 1949.

Jones, A. G. E., *Antarctica Observed: Who Discovered the Antarctic Continent?*, Whitby, England: Caedmon 1982.

———, "British Sealing on New South Shetland 1819–1826," *The Great Circle*, 7(1), pp. 9–22; "Part 2," *The Great Circle*, 7(2), pp. 74–87, 1985 (Reprinted in Jones, *Polar Portraits*, 1992, pp. 294–307; pp. 308–21).

———, "Captain George Powell, Discoverer of the South Orkneys, 1821," *The Falkland Islands Journal*, pp. 4–12, 1983.

———, "Captain William Smith and the Discovery of New South Shetland," *Geographical Journal*, 141(3), pp. 445–61, 1975 (Reprinted in Jones, *Polar Portraits*, 1992, pp. 343–59).

———, "Frankie Wild's Hut," *The Falkland Islands Journal*, pp. 13–20, 1982.

———, "John Biscoe's Voyage Round the World, 1830–33," *Mariner's Mirror*, 57(1), pp. 41–62, 1971 (Reprinted in Jones, *Polar Portraits*, 1992, pp. 61–82).

———, "New Light on James Weddell, Master of the Brig *Jane* of Leith," *Scottish Geographical Magazine*, 81(3), pp. 182–87, 1965 (Reprinted in Jones, *Polar Portraits*, 1992, pp. 375–81).

———, *Polar Portraits. Collected Papers*, Whitby, England: Caedmon 1992.

———, "Protecting the Whaling Fleet During World War II," *BAS Bulletin*, 38, pp. 37–42, 1974 (Reprinted in Jones, *Polar Portraits*, 1992, pp. 419–24).

———, "Shackleton's Amazing Rescue 1916," *The Falkland Islands Journal*, pp. 21–31, 1982.

———, "Three British Naval Antarctic Voyages, 1906–43," *The Falkland Islands Journal*, pp. 29–36, 1981.

———, "Voyages to South Georgia 1795–1820," *BAS Bulletin*, 32, 1973 (Reprinted in Jones, *Polar Portraits*, 1992, pp. 360–65).

Kemp, Norman, *The Conquest of the Antarctic*, London: Allan Wingate 1956.

Keys, Harry, and BAS Press Office, "One Small Ice Shelf Dies, One Giant Iceberg Is Born," *Antarctic*, 13(9), pp. 361–64, March 1995.

King, H. G. R., ed., "An Early Proposal for Conserving the Southern Seal Fishery," *Polar Record*, 12(78), pp. 313–16, 1964.

Kling, Alfred, trans. William Barr, "Report to the Hamburg-Südamerika Line," pp. 233–35 in Filchner, *To the Sixth Continent*, 1994.

Knight, Clayton, and Robert C. Durham, *Hitch Your Wagon: The Story of Bernt Balchen*, Drexel Hill, PA: Bell Publishing 1950.

Knox-Johnston, Robin, *Cape Horn: A Maritime History*, London: Hodder & Stoughton 1995.

Kohl-Larsen, Dr. Ludwig, trans. W. Barr, *South Georgia: Gateway to Antarctica*, Huntingdon, England: Bluntisham 2003 (Originally published in German as *An den Toren der Antarktis*, Stuttgart: Strecker und Schröder 1930).

Lansing, Alfred, *Endurance: Shackleton's Incredible Voyage*, New York: McGraw-Hill 1959.

Lecointe, Georges, *Au Pays des Manchots: Expédition Antarctique Belge,* Bruxelles: Oscar Schepens 1904.

Lester, Maxime Charles, "An Expedition to Graham Land, 1920–22," *Geographical Journal*, 62(3), pp. 174–94, 1923.

Lewander, Lisbeth, "The Representations of the Swedish Antarctic Expedition, 1901–03," *Polar Record*, 38(205), pp. 97–114, 2002.

———, "The Swedish Relief Expedition to Antarctica 1903–04," *Polar Record*, 39(209), pp. 97–110, 2003.

Liljequist, Gösta H., *High Latitudes: A History of Swedish Polar Travels and Research*, Stockholm: The Swedish Polar Research Secretariat 1993.

Lind, James, *A Treatise of the Scurvy . . .* , London: A. Millar 1753.

Lindblad, Lars-Eric, with John G. Fuller, *Passport to Anywhere: The Story of Lars-Eric Lindblad*, New York: Times Books 1983.

Lyon, P. C., chairman, *Report of the Interdepartmental Committee on Research and Development in the Dependencies of the Falkland Islands,* London: His Majesty's Stationery Office, April 1920.

MacDonald, Bruce, and Andrew Rymill, *Penola Commemorative Biographies. The Explorers. Lawrence Allen Wells and John Riddoch Rymill*, Penola: National Trust of South Australia 1996.

MacDowall, Joseph, *On Floating Ice: Two Years on an Antarctic Ice-Shelf South of 75°*, Edinburgh: The Pentland Press 1999.

Marr, James W. S., "The South Orkney Islands," Cambridge, Cambridge Univ. Press, *Discovery Reports*, X, pp. 283–382, 1935.

———, ed. F. H. Shaw, *Into the Frozen South*, London: Cassell 1923.

Maskelyne, Nevil *Nautical Almanac*, 1767.

Mathews, Eleanor, *Ambassador to the Penguins: A Naturalist's Year Aboard a Yankee Whaleship*, Jaffrey, NH: David R. Godine 2003.

Matthews, L. Harrison, *Penguin: Adventures Among the Birds, Beasts and Whalers of the Far South*, London: Peter Owen 1977.

———, *Sea Elephant: The Life and Death of the Elephant Seal,* London: MacGibbon & Kee 1952.

———, *South Georgia: The British Empire's Sub-Antarctic Outpost: A Synopsis of the History of the Island,* London: Simpkin Marshall 1931.

McCormick, Robert, *Voyages of Discovery in the Arctic and Antarctic Seas, and Round the World,* 2 Vols., London: Sampson, Low, Marston, Searle, and Rivington 1884.

McElrea, Richard, and David Harrowfield, *Polar Castaways: The Ross Sea Party (1914–17) of Sir Ernest Shackleton*, Christchurch: Canterbury Univ. Press 2004.

McIntosh, Gregory C., *The Piri Reis Map of 1513*, Athens GA: Univ. of Georgia Press 2000.

McKenzie, Douglas, *Opposite Poles*, London: Robert Hale 1963.

McKinley, Daniel L., *James Eights 1792–1882: Antarctic Explorer, Albany Naturalist, His Life, His Times, His Works*, Albany: New York State Museum Bulletin 505, 2005.

Menzies, Gavin, *1421: The Year China Discovered America*, New York: William Morrow 2003.

Mericq, Luis S., *Antarctica: Chile's Claim*, Honolulu: Univ. Press of the Pacific 2004 (Originally published, Washington: National Defense Univ. Press 1987).

Miers, John, communicated to Mr. Hodgskin, "Account of the Discovery of New South Shetland, with Observations on Its Importance in a Geographical, Commercial, and Political Point of View; with Two Plates," *Edinburgh Philosophical Journal*, 3(6), pp. 367–80, 1820 (Reprinted in *Polar Record*, 5(40), pp. 565–75, 1950).

Mill, Hugh Robert, *The Life of Sir Ernest Shackleton*, London: Heinemann 1924.

———, *The Siege of the South Pole: The Story of Antarctic Exploration*, London: Alston Rivers 1905.

Mills, Leif, *Frank Wild*, Whitby, England: Caedmon 1999.

Mills, William James, et al., *Exploring Polar Frontiers: A Historical Encyclopedia*, 2 Vols., Santa Barbara, CA: ABC-Clio 2003.

Mitterling, Philip I., *America in the Antarctic to 1840*, Urbana: Univ. of Illinois Press 1959.

Montaigne, Fen, *Fraser's Penguins: A Journey to the Future in Antarctica*, New York: Henry Holt, 2010.

Moro, Ruben O., trans. Michael Valeur, *The History of the South Atlantic Conflict: The War for the Malvinas*, London: Praeger 1989.

Morrell, Benjamin, Jr., *A Narrative of Four Voyages . . . . From the Year 1822 to 31*, New York: J. & J. Harper 1832.

Murdoch, W. G. Burn, *From Edinburgh to the Antarctic: An Artist's Notes and Sketches During the Dundee Antarctic Expedition of 1892–93*, London: Longmans Green 1894.

Murphy, David Thomas, *German Exploration of the Polar World: A History, 1870–1940*, Lincoln: Univ. of Nebraska Press 2002.

LITERATURE CITED

Murphy, Robert Cushman, *Logbook for Grace: Whaling Brig* Daisy, *1912–1913*, New York: Macmillan 1947.

Murphy, Shane, Gael Newton, and Michael Gray, *South with* Endurance: *Shackleton's Antarctic Expedition 1914–1917. The Photographs of Frank Hurley*, New York: Simon & Schuster 2001.

Murray, George, ed., *The Antarctic Manual: For the Use of the Expedition of 1901*, London: Royal Geographical Society 1901.

Murray, John, "Notes on an Important Geographical Discovery in the Antarctic Regions," *Scottish Geographical Magazine*, 10(4), pp. 195–99, April 1894.

Nasht, Simon, *No More Beyond: The Life of Hubert Wilkins*, Edinburgh: Birlinn 2006 (Originally published as *The Last Explorer: Hubert Wilkins, Australia's Unknown Hero*, Sydney: Hodder Australia 2005).

Naveen, Ron, *Waiting to Fly: My Escapades with the Penguins of Antarctica*, New York: William Morrow 1999.

Neider, Charles, ed., *Great Shipwrecks and Castaways*, London: Neville Spearman, 1955.

Nordenskjöld, Dr. N. Otto, and Joh. Gunnar Andersson, trans. Edward Adams-Ray, *Antarctica, or Two Years Amongst the Ice of the South Pole*, New York: Macmillan 1905 (Abridged English translation of original Swedish edition, *Antarctic: Två År Bland Sydpolens Isar*, 2 Vols., Stockholm: Albert Bonniers 1904).

Olsen, Magnus L., *Saga of the White Horizon*, Lymington, Hampshire: Nautical Publishing 1972.

Ommanney, Francis D., *South Latitude*, London: Longmans, Green 1938.

Oulié, Marthe, *Charcot of the Antarctic*, London: John Murray 1938 (Originally published in French as *Jean Charcot*, Paris: Gallimard 1937).

Page, Dorothy G., *Polar Pilot: The Carl Ben Eielson Story*, Danville, IL: Interstate Publishers 1992.

Palmer, James C., *Thulia: A Tale of the Antarctic*, New York: Samuel Colman 1843.

Parfit, Michael, "Reclaiming a Lost Antarctic Base," *National Geographic*, 183(3), pp. 110–26, March 1993.

———, *South Light: A Journey to the Last Continent*, New York: Macmillan 1985.

Perkins, Roger, *Operation Paraquat: The Battle for South Georgia*, Beckington, near Bath, Somerset: Picton Publishing 1986.

Philbrick, Nathaniel, *Sea of Glory: America's Voyage of Discovery, the U.S. Exploring Expedition, 1838–1842*, New York: Viking Penguin 2003.

Piggot, Jan, ed., *Shackleton: The Antarctic and* Endurance, London: Dulwich College 2000.

Pinochet de la Barra, Oscar, *Chilean Sovereignty in Antarctica*, Santiago: Editorial Del Pacifico S. A. 1955.

Pool, Beekman H., *Polar Extremes: The World of Lincoln Ellsworth*, Fairbanks: Univ. of Alaska Press 2002.

Powell, George, *Notes on South Shetlands, &c, Printed to Accompany the Chart of Those Newly Discovered Lands . . . ,* London: Printed for R. H. Laurie 1822.

Przybyllok, Erich, trans. William Barr, "Handwritten Notes by Erich Przybyllok on Events During the Filchner Antarctic Expedition," 229–32 in Filchner, *To the Sixth Continent*, 229–32, 1994.

Quam, Louis O., ed., *Research in the Antarctic: A Symposium Presented at the Dallas Meeting of the American Association for the Advancement of Science, December 1968*, Washington DC: American Association for the Advancement of Science, Pub. 93, 1971.

Quigg, Philip W., *A Pole Apart: The Emerging Issue of Antarctica (A Twentieth Century Fund Report)*, New York: McGraw-Hill 1983.

Riesenberg, Felix, *Cape Horn: The Story of the Cape Horn Region . . .* , New York: Dodd, Mead 1939.

Riffenburgh, Beau, ed., *Encyclopedia of the Antarctic*, 2 Vols., Abingdon Oxon: Routledge 2007.

Riley, Jonathan P., *From Pole to Pole: The Life of Quintin Riley 1905–1980*, Huntingdon, England: Bluntisham 1989.

Roberts, Brian B., and Ena Thomas, "Argentine Antarctic Expeditions, 1942, 1943, 1947, and 1947–48," *Polar Record*, 6(45), pp. 656–62, 1953.

———, "Chilean Antarctic Expeditions, 1947 and 1947–48," *Polar Record*, 6(45), pp. 662–67, 1953.

Robertson, R. B., *Of Whales and Men*, New York: Alfred A. Knopf 1954.

Ronne, Edith M. ("Jackie"), *Antarctica's First Lady: Memoirs of the First American Woman to Set Foot on the Antarctic Continent and Winter-Over*, Beaumont, TX: Clifton Steamboat Museum and Three Rivers Council #578, BSA, 2004.

———, "Woman in the Antarctic, or, the Human Side of a Scientific Expedition," *Appalachia*, 28(1), pp. 1–15, 1950.

Ronne, Finn, *Antarctic Command*, Indianapolis: Bobbs-Merrill 1961.

———, *Antarctic Conquest: The Story of the Ronne Expedition, 1946–1948*, New York: Putnams 1949.

———, *Antarctica, My Destiny: A Personal History by the Last of the Great Polar Explorers*, New York: Hastings House 1979.

———, "The Main Southern Sledge Journey from East Base, Palmer Land, Antarctica," pp. 13–20 in Byrd, *Reports on the Scientific Results . . .* , 1945.

———, "Ronne Antarctic Research Expedition, 1946–1948," pp. 159–71 in Friis and Bale, eds., *United States Polar Exploration*, 1970.

———, "Ronne Antarctic Research Expedition, 1946–1948," *Geographical Review*, 38(3), pp. 355–91, 1948.

Rose, Lisle A., *Assault on Eternity: Richard E. Byrd and the Exploration of Antarctica, 1946–47*, Annapolis, MD: Naval Institute Press 1980.

Rosove, Michael H., *Antarctica, 1772–1922: Freestanding Publications through 1999*, Santa Monica, CA: Adélie Books 2001.

Ross, James Clark, *A Voyage of Discovery and Research in the Southern and Antarctic Regions, During the Years 1839–43*, 2 Vols., London: John Murray 1847.

Ross, M. J., *Ross in the Antarctic: The Voyages of James Clark Ross in Her Majesty's Ships* Erebus *and* Terror *1839–1843*, Whitby, England: Caedmon 1982.

Rouch, Jules, *L'Antarctide. Voyage du "Pourquoi-Pas?" (1908–10)*, Paris: Societe d'Editions Geographiques, Maritimes et Coloniales 1926.

Rymill, John, with two chapters by A. Stephenson, *Southern Lights: The Official Account of the British Graham Land Expedition, 1934–1937*, London: Chatto and Windus 1938.

Savours, Ann, *The Voyages of the Discovery: The Illustrated History of Scott's Ship*, London: Virgin Publishing 1992.

Schouten, William C., *The Relation of a Wonderfull Voiage Made by William Cornelison Schouten of Horne. . .* , Cleveland: World Publishing 1966 (Facsimile of original English-language edition, London: T. Dawson for Nathanaell Newbery 1619).

Scott, Robert F., *The Voyage of the* Discovery, 2 Vols., London: Smith Elder 1905.

Sellick, Douglas R. G., ed., *Antarctica: First Impressions, 1773–1930*, Fremantle, Western Australia: Fremantle Arts Centre Press 2001.

Shackleton, Emily, and Hugh Robert Mill, ed. Michael Rosove, *Rejoice My Heart: The Making of H. R. Mill's "The Life of Sir Ernest Shackleton." The Private Correspondence of Dr. Hugh Robert Mill and Lady Shackleton, 1922-33*, Santa Monica, CA: Adélie Books 2007.

Shackleton, Sir Ernest H., *The Imperial Trans-Antarctic Expedition, Prospectus*, London: Office of the Expedition, for private circulation, 1913.

———, *South: The Story of Shackleton's Last Expedition 1914–1917*, London: Heinemann 1919.

Shackleton, Jonathan, and John Mackenna, *Shackleton: An Irishman in Antarctica*, Madison: Univ. of Wisconsin Press 2002.

Sheridan, Guy, *Taxi to the Snow Line. Mountain Adventures on Nordic Skis*, 11340 Carmurac, France: White Peak Publishing, 2006.

Silverberg, Robert, *The Longest Voyage: Circumnavigations in the Age of Discovery*, Indianapolis: Bobbs-Merrill Company 1972.

Skottsberg, Carl, *The Wilds of Patagonia: A Narrative of the Swedish Expedition to Patagonia, Tierra del Fuego and the Falkland Islands in 1907–1909*, London: Edward Arnold 1911.

Smith, Michael, *An Unsung Hero: Tom Crean Antarctic Survivor*, Wilton, Cork: Collins 2000.

———, *Sir James Wordie, Polar Crusader*, Edinburgh: Birlinn 2004.

Sobel, Dava, *Longitude: The True Story of a Lone Genius Who Solved the Greatest Scientific Problem of His*

Time, New York: Walker 1995.

Sobral, José M., *Dos Años Entre los Hielos 1901–1903*, Buenos Aires: J. Tragant 1904.

Southwell, Thomas, "Antarctic Exploration," *Natural Science*, vi(36), pp. 97–107, 1895.

Speak, Peter, *William Speirs Bruce: Polar Explorer and Scottish Nationalist*, Edinburgh: National Museum of Scotland 2003.

Spears, John R., *Captain Nathaniel Brown Palmer: An Old-Time Sailor of the Sea*, New York: Macmillan 1922.

Spotts, Peter N., "Antarctica's Wilkins Ice Shelf Eroding at an Unforeseen Pace," www.csmonitor.com/2008/0328/p25s10-wogi.html.

Squires, Harold, *S.S. Eagle: The Secret Mission 1944–45*, St John's, Newfoundland: Jesperson Press 1992.

Stackpole, Edouard A., *The Sea Hunters: The New England Whalemen During Two Centuries 1635–1835*, Philadelphia: Lippincott 1953.

———, *The Voyage of the Huron and the Huntress: The American Sealers and the Discovery of the Continent of Antarctica*, Mystic, CT: Marine Historical Association, Number 29, November 1955.

———, *Whales and Destiny: The Rivalry Between America, France, and Britain for Control of the Southern Whale Fishery, 1775–1825*, Amherst: Univ. of Massachusetts Press 1972.

Stanton, William, *The Great United States Exploring Expedition of 1838–1842*, Berkeley: Univ. of California Press 1975.

Steger, Will, and Jon Bowermaster, *Crossing Antarctica*, New York: Alfred A. Knopf 1992.

Stephenson, Jon, *Crevasse Roulette: The First Trans-Antarctic Crossing 1957–58*, Dural Delwey Centre, NSW: Rosenberg Publishing Pty. Ltd. 2009.

Stonehouse, B. ed., *Encyclopedia of Antarctica and the Southern Oceans*, Chichester, West Sussex: John Wiley & Sons 2002.

Sullivan, Walter, *Quest for a Continent*, New York: McGraw-Hill 1957.

Sunday Times Insight Team (Paul Eddy, Magnus Linklater, and Peter Gillman), *War in the Falklands: The Full Story*, New York: Harper & Row 1982.

Thomas, Lowell, *Sir Hubert Wilkins: His World of Adventure*, New York: McGraw-Hill 1961.

Thomson, John Bell, *Elephant Island and Beyond: The Life and Diaries of Thomas Orde Lees*, Huntingdon, England: Bluntisham 2003.

———, *Shackleton's Captain: A Biography of Frank Worsley*, Oakville, Ontario: Mosaic Press 1999.

Three of the Staff (R. N. Rudmose Brown, J. H. H. Pirie, and R. C. Mossman), *The Voyage of the "Scotia": Being the Record of a Voyage of Exploration in Antarctic Seas*, Edinburgh and London: Blackwood 1906.

Tønnessen, J., and A. O. Johnsen, trans. R. I. Christophersen, *The History of Modern Whaling*, Berkeley, CA: Univ. of California Press 1982.

Tordoff, Dirk, *Mercy Pilot: The Joe Crosson Story*, Kenmore, WA: Epicenter Press 2002.

Tyler, David, *The Wilkes Expedition: The First United States Exploring Expedition (1838-1842)*, Philadelphia: American Philosophical Society 1968.

Tyler-Lewis, Kelly, *The Lost Men: The Harrowing Saga of Shackleton's Ross Sea Party*, New York: Viking 2006.

Vairo, Carlos, Ricardo Capdevila, Verónica Aldazábal, and Pablo Pereyra, trans. Iraí Rayén Freire, *Antártida: Patrimonio Cultural de la Argentina. Museos, sitios y refugios historicos de la Argentina / Antarctica: Argentina Cultural Heritage. Museums, Sites and Shelters of Argentina*, Ushuaia: Zagier & Urruty 2007.

Viola, Herman J., and Carolyn Margolis, eds., *Magnificent Voyagers: The U.S. Exploring Expedition, 1838–1842*, Washington: Smithsonian Institution Press 1985.

Von den Steinen, Karl, trans. William Barr, "Zoological Observations, Royal Bay, South Georgia. 1882–1883. Part 1," *Polar Record*, 22(136), pp. 57–71, 1984; "Part 2. Penguins," *Polar Record*, 22(137), pp. 145–58, 1984.

Wafer, Lionel, *A New Voyage & Description of the Isthmus of America*, London: James Knapton 1699

(Reprinted in part as p. 18, Chapman, ed., *Antarctic Conquest*, 1965).

Walton, E. W. Kevin, *Two Years in the Antarctic*, London: Lutterworth Press 1955.

——, and Rick Atkinson, *Of Dogs and Men: Fifty Years in the Antarctic: The Illustrated Story of the Dogs of the British Antarctic Survey 1944–1994*, Malvern Wells, England: Images Publishing 1996.

Webster, W. H. B., *Narrative of a Voyage to the Southern Atlantic Ocean, in the Years 1828, 29, 30, Performed in H. M. Sloop* Chanticleer . . . , 2 Vols., London: Richard Bentley 1834.

Weddell, James, *A Voyage Towards the South Pole Performed in the Years 1822–24* . . . , Annapolis: Naval Institute Press 1970 (Facsimile of second edition, originally published, London: Longman, Rees et al. 1827; first edition, London: Longman, Hurst et al. Paternoster-Row 1825).

Wild, Commander Frank, from the Official Journal and Private Diary Kept by Dr. A. H. Macklin, *Shackleton's Last Voyage: The Story of the* Quest, London: Cassell 1923.

Wilkes, Charles, *Narrative of the United States Exploring Expedition: During the Years 1838, 1839, 1840, 1841, 1842* . . . , second edition, second issue, 5 Vols., Philadelphia: Lea and Blanchard 1845 (First edition, Philadelphia: C. Sherman 1844).

Wilkins, George Hubert, "Further Antarctic Explorations," *Geographical Review*, 20(3), pp. 357–88, July 1930.

——, "The Wilkins-Hearst Antarctic Expedition, 1928–1929," *Geographical Review*, 19(3), pp. 353–76, July 1929 (Reprinted in part as pp. 230–39, Sellick, ed., *Antarctica: First Impressions* . . . , 2001).

Williams, Isobel, *With Scott in the Antarctic: Edward Wilson: Explorer, Naturalist, Artist*, Stroud, England: The History Press 2008.

Wordie, James M., "The Falkland Islands Dependencies Survey, 1943–46," *Polar Record*, 4(32), pp. 372–84, 1946.

Worsley, Frank Arthur, Endurance: *An Epic of Polar Adventure*, London: Philip Allan 1931.

——, *Shackleton's Boat Journey*, London: Philip Allen nd (1939 or 1940).

Yelverton, David E., *Quest for a Phantom Strait: The Saga of the Pioneer Antarctic Peninsula Expeditions 1897–1905*, Guildford, England: Polar Publishing 2004.

[Young, Adam], "Notice of the Voyage of Edward Barnsfield [sic], Master of His Majesty's Ship *Andromache*, to New South Shetland," *Edinburgh Philosophical Journal*, 4(8), pp. 345–348, April 1821 (Reprinted in part, pp. 43–46, Chapman, ed., *Antarctic Conquest*, 1965).

Index

Bold page numbers denote pages with maps relevant to the index listing. (When a location appears on multiple maps, only the first or most important maps for it are noticed.) ***Bold italic*** page numbers indicate pages with an illustration of the index listing.

This index does not include entries for the material in Appendixes A and B. A separate index for these Appendixes is available from the author.

Active (ship), 77–79
Adams, Charles, 263
Adams, John Quincy, 51
Adelaide Island, **5**, 54, 137, 155–56, 234–35, 298
Adélie Land, 61, 214, 329
Admiralen (ship), 147–48
Adventure (ship), 19
Adventure expeditions, 284–88
Africa, 9, 19, 21, 89
Agreed Measures for Conservation of Antarctic Flora and Fauna, 282
Alexander Island (Alexander I Land), **5**
    Discovery, early sightings of, 36, 82, 90, 137, 156, 161
    Insularity, determination of, 237-40, 245
Almirante Brown base, **200**, 202
America, Americans, *see* United States
American Museum of Natural History, 147, 222
Americas
    North America, 13, 61, 66
    South America, 2, 9–11, 13–15, 26, 251
Amundsen, Roald, 97, 177, 221, 126
    Belgica expedition, 84, 88, 91, 93–95
Amundsen-Scott Station, 268, 275
Andersson, Axel, 108
Andersson, Johann Gunnar, 140, 254
    Antarctic expedition, 99, 104–05, 107–08, 110–11, ***112***, 114
Andresen, Adolf Amandus, 139–40, 147–49, 152–53, 160
Andresen, Wilhelmine, 147, 153, 160
Animals on expeditions
    Cats, 93, 106, 113, 129, ***134***, 147, 152, 167, 182–83, 223, ***235***
    Dogs, *see*
    Other, 39, 74, 129, ***134***, 136, 147, 167
    Ponies, 165–67, ***170***, 172, ***173***
    Rats, 95, 223
Annawan (ship), 50–53

Annenkov Island, **9**, 190
Antarctic Circle, 3, 72, 327
    Crossing of, firsts, 19, 72
    Land south of, early sightings, 35, 56
    Winters south of, firsts, 1, 91–94, 97
Antarctic Continent, **endpaper**, 21, 31–32, 66, 72, 162, 240
    Claims to, *see* Claims to Antarctic territory
    Circumnavigations of, *see* Bellingshausen, Biscoe, Cook (James), *Discovery II*
    Crossings/attempts/plans to, 164, 176, **178**, 213, 221, 225–27, **226**, 271–76, **273**
    Landings, early, 38, 54, 72, 83, 103
    Sightings, early, 31–32, **34**, 35, 37–38, 40, 54, 61, 66, 68
Antarctic Convergence, **endpaper**, 2–3, 16–17, 50, 327
    Discovery, study of, 210, 220
    Scientific work south of, significant early, *see Chanticleer* and German IPY
Antarctic (ship), 83, 99, **105**
Antarctic expedition, 99–115, **101**, **107**, 222, 254–56
    Expedition narratives, 114, 336
Antarctic Heritage Trusts, New Zealand and United Kingdom, 299
Antarctic Peninsula, vii, 2, **4**, **5**, **6**, 9, 21, 118, 197, 348
    Channels through, 73, 87, 136, 216–17, **226**, 236, 239–40, 345
    Claims to, *see* Argentina, Chile, Great Britain
    Climate change along, 294–95, 348
    Crossings of, early, 218–19, 246
    Exploration of, early, **217**, **233**, **243**, **262**
        Aerial, 216–17, 224–25, 233–34, 247
        Inland, 159, 239, 244–46
    Landings on, early, 38
    Names for, 38–39, 54, 265, 278, 347
    Sightings of, early, 31, 37–38, 40
    Weather stations on, 245, 248, 261

Antarctic Pilot, The 35, 333
Antarctic Polar Front, *see* Antarctic Convergence
Antarctic Sound, 100, **101**, 107
Antarctic Treaty, 276–77, 296
    Annexes, *See* Agreed Measures for Conservation of Antarctic Flora and Fauna, Convention for Conservation of Antarctic Seals, Convention on the Regulation of Antarctic Mineral Resource Activities, Madrid Protocol
    Historic Monuments, 73, 115, 125, 203, 298–99
Antarctic: Två År Bland Sydpolens Isar (Nordenskjöld et al.), 114
Antarctica, East, **endpaper**, 35, 47, 54, 61, 66, 166, 204, 212, 229, 241, 249, 268, 334
Antarctica, West, **endpaper**, 5
Anvers Island, **6**, 54, 89, 279, 297, 348
Arctic, 72, 74, 97, 203, 213, 221, 268
Arctowski base, 97
Arctowski, Henryk, 84, 86, 88, 97, 336
Argentina, 100, 122, 266, 283, 288
    Claims by, 143, 214–15, 241, 250–51, **251**
    Conflict with Great Britain, 250–52, 264–67, 289. *See also* Falklands/Malvinas war
    Explorers/expeditions from
        Government-sponsored, 138, 257, 265–67, 268, 271, **273**, 278–80, 288, 294, 297. *See also* Almirante Brown, Deception Island, and Laurie island bases, *Primero de Mayo*, Uruguay
        Other, *see* Cia. Argentina de Pesca, Sobral
    Preservation efforts, 115, 125, 299, 337
    "Settlers," 288–89
    Support for explorers, 86, 100, 111–14, 125–26, 128–29, 138, 152, 197
Argentine Islands, 232–33, 281, 294. *See also* Winter Island
Artists on expeditions, 19, 39, 60, 77, 100, 102
    Works by, **21**, **35**, **50**, **60**, **78**, **112**, **183**, **194**
Artuso, Felix, 292–93
Astrolabe (ship), 58–61, **60**
Atlantic Ocean, 9, 12–15, 17, 61
Aurora (ship), 177, 193–94, 196
Aurora Islands, 29
Austral (ship, formerly *Français*), 138
Australia, 2, 35, 61, 228, 268, 272, 283, 344
    Claims by, 241, **251**
    Explorers/expeditions from, 97, 166, 298. *See also* Hurley and Wilkins
Avery, George, 53–54
Aviation in Antarctica
    Early flights, 219, 343. *See also* BGLE, Byrd, Ellsworth, Wilkins, USAS
    Early plans, 198–99, 200, 204
    Post World War II, 257–58, **262**, 261–64, 270–75

Bagshawe, Thomas, 198–203, ***202***
Bahía Paraíso (ship), 291, 297
Balaena (ship), 77–79
Balboa, Vasco Nuñez de, 331
Balchen, Bernt, 221–24
Balleny Islands, 56, 219
Balleny, John, 56
Barker, Nick, 290–93
Barry Island, **233**, 236–40, 242
Bay of Whales, 213, 219, 222, **226**, 228, 241
Beagle Channel, **11**, 86
Bear (ship), 241–42, 247–48
Beaufoy (ship), 43–47, **46**
Belgica (ship), **cover**, 84, **85**, **92**, 338
    New *Belgica* project, 338
Belgica expedition, 1, 67, 84–97, **88**
    Beset drift, **88**, 90–96
    Expedition narratives, 96–97, 335
    Winter night, 91–94
Belgium, 97, 268
    Expeditions from, *see Belgica* expedition
Bellingshausen Station, 279
Bellingshausen Sea, **4**, 35, 63, 88, 90, 156
Bellingshausen, Fabian Gottlieb von, 32–33, **33**, 40, 332
    Antarctic circumnavigation, 32–36, **34**, 38–39
    Expedition narrative, 39, 332
Berkner Island, 271, 346
Berthelot Islands, 136, 154
Bertram, Colin, 236
Beset ships, *see Antarctic*, *Belgica*, *Deutschland*, *Endurance*, and *Gauss* expeditions
Betsey (ship), 24–25
BGLE, *see* British Graham Land Expedition
Bible, 183, 286
Bingham, Edward, 231–32, 234, 237–40, **240**, 256
Bird Island, **8**, 220, 287
Births in Antarctic regions, 144, **288**, 339
Biscoe, John
    Antarctic circumnavigation, 53–55, **55**
Bjørvik, Paul, 170
Black, Richard, 242, 244–47
Blackborrow, Perce, 179, 195
Blaiklock, Ken, 272–74
Blériot, Louis, 160
Blijde Boodschap (ship), 14
Bongrain, Maurice, 154–56
Booth Island, 73, **131**, ***133***, 138, 153–54, 157, ***158***, 338
    Français expedition at, 132–37
Borchgrevink, Carsten, 96, 337
Bouvet Island, **endpaper**, 17, 47
Bouvet, Jean-Baptiste-Charles, 17

INDEX

Boy Scouts, 204
Brabant Island, **6**, **87**, 88–89, 336
Bransfield Strait, vii, **6**, 37, 284
Bransfield, H. M. S. (ship), 252–53
Bransfield, Edward, 30–32, 42, 252
Brazil, 61, 296
Brisbane, Mathew, 43
British Antarctic Survey (BAS), 278, 279–81, 290–94, 298
British Antarctic Territory (BAT), **251**, 278
British Graham Land Expedition (BGLE), 231–41, **233**, 344–45
    Expedition narratives, 344,
Brown, R. N. Rudmose, 119, 121, 125
Bruce, William Speirs, 31, 52, 99, 116, ***119***, 126, 164, 167, 177, 275, 337
    Dundee whaling expedition, 77–80, **78**
    Scotia expedition, 116–26
Brue, Adrien, 30
Bull, Henrik, 83, 99
Burley, Malcolm, 191–92, 285–87
Byrd, Richard E., 213–14, 221–22, 224, 241–42, 247

Caird Coast, 180, **226**
Caird, James, 180
Calendars, Julian and Gregorian, 332
Canada, 57, 203
    Explorers from, *see* Cheesman, Hollick-Kenyon, Lymburner
Cape Adare, 83, 97, 337
Cape Disappointment, **9**, 22
Cape Flying Fish, **63**, 66
Cape Horn, **11**, 15–17, 66
Cape of Good Hope, 9–10, 14, 19
Cape Town, 22, 124, 204, 209, 210
Cape Valentine, Elephant Island, 187, 286
Cape Well Met, Vega Island, **107**, 111
Carder, Peter, 13, 332
Carnarvon Castle (ship), 252
Carse, Duncan, 231, 285, 288
Castor (ship), 81–82
Cemeteries, *see* Deception Island, Grytviken
Challenger expedition, 72
Chanticleer expedition, 48–50, **50**, 63, 67
    Expedition narrative, 333
Charcot Island, **5**, 161, 218, 238, 257, 264
Charcot, Jean-Baptiste, 82, 127, ***129***, ***136***, 300–01, 338
    Death, 163, 239
    Français expedition, 125–26, 127–38
    Pourquoi-Pas? expedition, 151–63
Cheesman, Al, 218
Chile, 195–96, 258, 276, 278, 283
    Claims by, 143, 250, **251**, 346
    Conflicts with Great Britain, 264–67
    Government expeditions, 200, 257, 264–67, 268, 278–81, 289, 296, 343
China, 9, 23, 296
Christensen, Christen, 80–81, 83, 147–48
Christensen, Lars, 335
Chronometer, 18–19, 43, 191
Church, Grytviken, ***145***, 208, 291, 299
Churchill, Winston, 179,
Claims to Antarctic territory, **251**
    Antarctic Treaty, treatment under, 277
    Explorers/others, by, 21, 28, 31, 42, 54, 69, 216, 219, 225, 241, 243, 246, 257, 270
    Governments, formal claims by, *see* Argentina, Australia, Chile, France, Great Britain, Norway, New Zealand
Clarence Island, **6**, **42**, 60
Cleveland, Benjamin Dunham, 147
Clifford, Miles, 265–66
Climate change/global warming, 294–95, 348
Coats Land, **117**, 123, 143, 168, 179, 197
Coats, James and Andrew, 116, 122
Cockburn Island, **69**
Commonwealth Trans-Antarctic Expedition, 176, 271–76, **273**, 347
    Expedition narratives, 346–47
Convention for Conservation of Antarctic Seals, 282
Conservation proposals, early, 23–24, 47
Convention on the Conservation of Antarctic Marine Living Resources (CCAMLR), 282
Convention on the Regulation of Antarctic Mineral Resource Activities (CRAMRA), 296
Cook, Frederick, 84–89, 91–97, 160
Cook, James, 19, 45
    Antarctic circumnavigation, 19–22, **20**, 32–33
    Expedition narratives, 331
    Record south latitude, 19, 45, 62, 64, 162
Cooper, Mercator, 72
Cope expedition, 198–203, **200**
    Expedition narratives, 203, 342
Cope, John Lachlan, 198–200, 203
Coronation Island, **7**, 42
Cramer, Parker, 218
Crean, Tom, 178, 188, 191–93
Crevasse, 191, 238–39, 245, 271–72, 274–75, 327
    Falls into, 88, 170, 261
Crosson, Joe, 214–15, 218
Crozier, Francis, 69–70
Cumberland Bay, **8**, 140, 142–43, 208, 290–93

D'Urville Island, **6**, 60, 104–05
D'Urville, Jules Sébastian-César Dumont, 57, 61
    Antarctic expedition, 57–61, **59**
    Expedition narrative, 61, 334

Daisy expedition, 146–47, 339
Dallmann, Eduard, 338
   Grönland expedition, 73–74
Danco, Émile, 84, 88, 93
Darlington, Harry, 248, 258–60, **260**
Darlington, Jennie, 259–60, **260**, 264, 270
David Copperfield, 191
Davidoff, Constantino, 289–90
Davidsen, Nokard, 148–49
Davis, John (American sealer), 38, 86
Davis, John (British explorer), 13
De Gerlache, Adrien, 84, **85**, 97, 127–28, 177, 338
   Belgica expedition, 84–98
De Gerlache, François, 89, 336
De Gerlache, Jean-Louis, 338
Deaths of expedition participants, 11, 12, 54, 64, 148, 205, 219, 292
   Heroic Age, 86, 93, 109, 120, 167, 174, 177, 196
Debenham Islands, 236, 344–45
Debenham, Frank, 236, 344–45
Deception Island, **6**, **7**, 49, 150, 210, 282, 347
   Cemetery, 148, *149*, 281, 347
   Eruptions/earthquakes at, 6, 67, 219, 279–82, **280**
   Expeditions at, visits to, 7, 38, 48–50, 62–63, 66–67, 125, 152–53, 159–61, 199, 213, 215–19, 222, 224, 232, 235, 250–52.
   Government bases, 253–55, 266–67, 279–81
   Whaling at, 147–49, 152–53, 160, 203, 213, 215–16, 220
Deutschland (ship), 164, *173*, 175, 340
Deutschland expedition, 164–76, **169**, 208
   Beset drift, **169**, 171–75
   Expedition narratives, 176, 340
Dickens, Charles, 191
Discovery Committee/Investigations, 209–11, 212, 217, 219–20, 344
Discovery expedition, 98, 112, 116, 121, 177, 336, 340, 343
Discovery II (ship), 210, 219–20, 228–29, 232
Discovery (ship), 126, 209–10
Discovery Reports, 211
Dobrowolski, Anton, 84
Dodson, Robert, 261
Dogs in Antarctica
   Heroic Age expeditions, 74, 100, 102, 113–14, 121, 129, 132, *134*, 152, 165, 171–73, 179–83, *182*, 185
   Post Heroic Age expeditions, 199, 201, 203, 231–32, ***234***, 237–39, 242, 244–48, 253, 255, 259, 263, 273–75, ***274***
   Removal from Antarctica, 298
Donald, Charles, 77–80
Dove (ship), 41

Drake Passage, vii, **endpaper,** 13, 17
Drake, Francis, 12–14
Drygalski, Erich von, 98, 121, 164. *See also Gauss* expedition
Dudley Docker (boat, *Endurance*), 183, 185–87, ***194***
Dufek, George, 257–58
Dundee Island, vii, **6**, 60, 79, 224–25
Dundee whaling expedition, 77–80
   Expedition narrative, 80, 335
Duse, Samuel, 104–05, 107, 110–12, ***112***
Dutch expeditions, 14–16
Dyer, Glenn, 245–46

Eagle, S. S. (ship), 253–55, ***255***
Earp, Wyatt, 221, 224
Eendracht (ship), 14–16
Eielson, Carl Ben, 213–18, ***216***
Eights Station, 53, 278–79
Eights, James, 51–53
Ekelöf, Erik, 102, 109
Eklund, Carl, 245–46
Elephant Island, **6**, 41, 51, 207, 346
   Endurance expedition at, 187–88, 194–96
   Joint Services Expedition (Burley) at, 286–87
Ellsworth Station, 269–71, **273**, 278
Ellsworth, Lincoln, 221, ***229***
   Antarctic expeditions, 221–30, 241
   Trans-Antarctic flight, 225–27, **226**, **227**
   Expedition narrative, 343
Emma (ship), 195
Enderby Brothers, 53, 55–56
Enderby Land, 54, **55**, 56
Endurance (ship), 177, 182, ***182***, 184
Endurance expedition, 177–97, **178**, **181**, 272, 285–86, 341–42,
   Beset drift, 180–82, **181**
   James Caird voyage to South Georgia, **181**, 188–89, ***189***
   Elephant Island, at, 187–88, 194–96, ***194***
   Expedition narratives, 196, 340–41
   Recreations, 285, 287–88, 292
   Ross Sea Party, 177, 193–94, 196, 198, 342
   South Georgia crossing, 191–93, **192**, 341
Endurance, H. M. S., (ship), 286, 290–93
England, *see* Great Britain
Erebus (ship), 67, 70
Erebus and Terror Gulf, **6**, 69, 77, 79–80
Esperanza base, 265–66, 288
Espírito Santo (sealing ship), 28–29
Eternity Range, 225, 235, 245–46
Evensen, Carl Julius, 81–82
Explorer, M.V. (ship), *see Lindblad Explorer*

Fairweather, Alexander, 77–79
Falkland Islands Dependencies (FID), 143, 208, **226**, 243, 250–54, 264

INDEX

Government regulations, licenses, 143, 146, 148, 211
Magistrates, South Georgia and Deception islands, 126, 148, 150, 289–91. *See also* Wilson (James Innes)
Original claim (letters patent), 143
Redefinition, 278
Falkland Islands Dependencies Survey (FIDS), 256, 259, 261–64, 265–67, 269, 278, 287
Falkland Islands/Islas Malvinas, **endpaper**, 9, 13–14, 23–25, 29, **30**, 100, 104, 117, 121–22, 142, 188, 195, 232, 253
War over, 141, 289–94, **290**
Fanning, Edmund, 24–25, 29, 38, 40, 51, 53, 61
Faraday Base, 294, 348
Filchner Ice Shelf, vii, **4**, 168–71, 197, 270–71, 273, 328, 346
Filchner, Wilhelm, 164, **165**, 175–76, 191
Deutschland expedition, 164–76
Fishing industry, 282
Fitzroy, S. S. (ship), 253–54
Flanders Bay, **87**, 130
Flying Fish (ship), 62–66, **64**, 334
Fortuna (ship), 140
Fossils, 52–53, 102, 105, 238–39
Seymour Island, from, 80–81, 101, 103
Foster, Capt. Henry, 48–50, 333
Foyn Coast (Foyn Land), **81**, 82
Foyn, Svend, 82, 83
Fram (ship), 126
Français (ship), 128, **134**, 138
Français expedition, 125, 127–38, **130**, **131**, 151, 153, 155, 285
Expedition narratives, 138, 338
France, 268
Claims by, 214, 241, **251**
Expeditions/explorers from, 17, 74, 97, 348. *See also* d'Urville, *Français*, *Pourquoi-Pas?*
Franklin, Sir John, 72
Fricker, Karl, 46
Frithjof (ship), 112, 114, 128
From Edinburgh to the Antarctic (Murdoch), 80
Fuchs, Arved, 287–88
Fuchs, Vivian, 176, 271
Commonwealth Trans-Antarctic Expedition, 271–76
Furneaux, Tobias, 19

Gain, Louis, 159
Galíndez, Don Ismael, 125
Gauss expedition, 98, 121, 164, 174, 210, 337, 340, 343. *See also* Drygalski
Gauss, Friedrich, 57
General Belgrano Station, 269, 271, **273**, 346
Geology/geologists, 39, 49, 52, 67, 72, 84, 99, 103–05, 129, 168, 209, 238, 243, 263, 270, 281, 345. *See also* fossils
George IV Sea, *see* Weddell Sea
George VI Sound, **5**, 233, 239, 243, 245–46
Gerlache, Adrien de, *see* De Gerlache, Adrien
Gerlache Strait, vii, **6**, 73, **87**, 87–89, 104, 200
German South Georgia IPY expedition, 74–76, 268
Germany, 150, 241, 249–50, 296,
Antarctic expeditions from, 74, 97, 175–76, 241, 287, 345. *See also* Dallmann, *Deutschland*, *Gauss*, German South Georgia IPY, *Meteor*
Scientists/geographers from, 46, 57, 74, 103, 210
Gherritz, Dirk, 14, 16, 29
Glydén, Olof, 112, 114, 128
Gobernador Bories (ship), 147–49, 152, 160
Godfroy, René, 154, 158–59
Goeldel, Wilhelm, 166, 174
Golden Hind (ship), 12–13
Gomez, Estavo, 10–11
Goupil, Ernest, 60
Gourdon, Ernest, **129**, 133, 151–54, 159
Graham Land, 54, **130**, 278, 347
Gray, David and John, 74
Great Britain, 41, 141–44, 146, 150, 249, 258, 268, 278
Claims, **251**
Government, by, 21, 31, 214, 250–52. *See also* British Antarctic Territory, Falkland Islands Dependencies
Others, by, 28–29, 42, 54, 217–18
Conflicts with Argentina and Chile, 141, 249–52, 264–67, 289. *See also* Falklands/Malvinas, war
Explorers/expeditions from
Government/government-sponsored
Early, 74, 142–43. *See also* Bransfield, *Challenger*, *Chanticleer*, Cook (James), Halley, Ross
Modern, 249–52, 285–87, 288. *See also* British Antarctic Survey, Discovery Investigations, Falklands Islands Dependencies Survey, Operation Tabarin
Others, 13, 16, 56, 97–98, 116, 152, 177, 285. *See also* Biscoe, British Graham Land, Commonwealth Trans-Antarctic, Cope expedition, Drake, Dundee whaling, *Endurance*, Powell, *Quest*, Royal Society Expedition, *Scotia*, sealers/sealing, Smith (William), Weddell
Historic preservation by, 299–300
Green, Charles, 184, 187, 286
Greenland, 27, 84, 99, 231, 335
Greenland, New South, 48, 172–73

Greenstreet, Lionel, 286
Greenwich Island, 257, 265
Greenwich, England, 18
Grönland (ship), 73–74
Grunden, Toralf, 104–05, 110
Grytviken, **8**, *140*, 140–46, 148, 167, 175, 179, 205, 208, 210, 249, 282, 291–93, 299–30
   Cemetery, 144, 167, **208**, 292, ***293***

Halley Bay/Station, 269–70, 272, **273**, 294
Halley, Edmond, 17, 269
Hampton, W. E., 231–33, 235, 237
Hawkins, Capt. Geoffrey, 250
Hearst Press, 213, 217
Heim, Fritz, 168–71
Hektoria (ship), 215
Hero (ship), 37, 333
Heroic Age, 97, 196–97, 336
   Expeditions of, *see* individual expeditions, Chapters 6–9, 11–13
Hersilia (sealing ship), 29, 37
Hertha (ship), 81–83
Herzog Ernst Bay, *see* Vahsel Bay
Hillary, Sir Edmund, 272, 275, 347
Historic monuments, *see* Antarctic Treaty
Hitler, Adolf, 238
Hobart, Tasmania, 54, 60–61, 67, 68
Hodges, Michael, 142–43
Hodges, William, 21
Hollick-Kenyon, Herbert, 224–29, ***229***
Hooker, Joseph Dalton, 69–70
Hoorn (ship), 14
Hope Bay, **107**, 199, 288
   Antarctic expedition at, 101, 104–10, ***109***, 113
   Conflict over bases at, 265–66
   Operation Tabarin at, 252–56, ***254***
Hopeful (ship), 56
Hovgaard Island, 73, 135, 338
How, Walter, 286
Hudson, Hubert, 186
Hudson, William, 63–65
Hughes Bay, **6**, 38, 86–87
Hughes, Charles Evans, 241, 345
Hurley, Frank, 178, 184–85, ***184***, 187, 194–96 341
Hussey, Leonard, 183, 204–05, 207–08

IAATO, *see* International Association of Antarctica Tour Operators
Ice shelf, 68, 262, 328. *See also* Filchner, Larsen, Ronne, Ross, Wilkins, Wordie ice shelves
Iceberg, 45, ***46***, 49, 52, ***64***, 137, 328
   Tabular, 17, ***78***, 170–71, 329
Icebreaker, 4, 157, 162, 256–57, 264, 269–70
IGY, *see* International Geophysical Year
Imperial Trans-Antarctic Expedition, *see Endurance* expedition
Indian Ocean, 18, 47, 67, 72, 74
Insects in Antarctica, 3, 87, 336
Instituto de Pesca (ship), 195
International Association of Antarctica Tour Operators (IAATO), 283
International Geographical Congress, Sixth, 83
International Geophysical Year (IGY), 268–76, 283, 294, 328
International Polar Year (IPY), 74–76, 268, 328
Irízar, Julián, 111–13, 129, 182

Jackson, Andrew, 51
Jacobsen, Solveig Gundjörb, 144–45, 339
Jacquinot, Charles, 58
Jail, Grytviken, 145–46
James Caird (boat, *Endurance*), 183, 185–87, 193, 204, 287, 341, 342
   Voyage to South Georgia, **181**, 188–90, ***189***
   Recreations of, 287–88
James Monroe (ship), 41
James, David, 255–56
James, Reginald, 194
Jane (ship), 43–47, ***46***
Japan, 97, 268, 297
Jason (ship), 80, 335
Jason expeditions, 77–78, 80–83, **81**, 139, 147
Jerez, Gregorio, 17, 21,
Johnson, Robert, 63, 67
Joinville Island, **6**, 60, 79, 100, 104–05, 186
Jonassen, Ole, 102–03, 109–10

Kaiser Wilhelm I, 74, 164, 340
Keeler, Cape, 161–63, ***162***
Kendall, Lt. Edward Nichols, 50
Kerguelen, Iles, 47, 74, 328, 339
Kerr, Gilbert, 120, 123, ***124***
Killingbeck, John, 298
King Edward Point, 143, 208, 289–94, ***290***
King George Island, **6**, 28, 31, 37, 73, 148, 160
   Bases on, 6, 97, 279, 289
King Haakon Bay, South Georgia, **8**, 144, 190, 193, 285, 288, 292
   Pegotty camp at, 191
Kinnes, Robert, 77, 79
Kling, Alfred, 166–68, 172–75, ***173***
Knowles, Paul, 246–47
Kohl (Kohl-Larsen), Ludwig, 166–67, 175–76, 340, 344
Kohl-Larsen, Margit, 175–76, 340
König, Felix, 165, 168, 170, 172–73, 176
Korea, South, 296
Kosmos (ship), 219

Lancing (ship), 211
Lange, Alexander, 147
Larsen Ice Shelf, vii, **5**, **81**, 82, 295

INDEX

Larsen B collapse, *295*
Larsen, Carl Anton, 80, 128, *140*, 166–67, 175, 211
    Jason voyages, 77–78, 80–83
    Antarctic expedition, 99–102, 104–06, 108–09, 113–14
    Pesca and South Georgia, 139–42, 144–46
Larsen, Lauritz, 142, 144
Lassiter, James, 263–64
Latady, William, 263
Laurie Island, *7*, 118, 256
    Meteorological base, 7, 122, 125, 138, 150, 250, 299
    Scotia expedition at, 118–22, 125
Laurie, R. H., publisher, 42
Lazarev, Mikhail, 38
Le 'Français' au Pôle Sud (Charcot), 138
Le Maire, Isaac, 14, 16
Le Maire, Jacob, 14–16
Le Maire, Strait of, *11*, 15–16
Le Pourquoi-Pas? dans L'Antarctique (Charcot), 163
Leard, John, 23–24
Lecointe, Georges, 84, 86–87, 89–91, 93–94
Leith Harbor whaling station, 249, 282, 289–93
Lemaire Channel, *6*, 89, 130, 132–35, *133*, *158*
Lemaire, Charles, 89
León (ship), 17, 21
Leskov Island, *8*, 33–34, *35*, 54
Lester, Maxime, 199–203, *202*
Letters Patent, British, 143
Lewis, John, 272–73
Lier, Leif, 219
Lind, James, 18
Lindbergh, Charles, 214
Lindblad Explorer (ship), *284*
Lindblad, Lars-Eric, 283–84, *284*
Liouville, Dr. Jacques, 152–53
Little America, 213, 221–22, 224–25, *226*, 227–28
Lively (ship), 53–55
Livingston Island, *6*, 287
Logbook for Grace (Murphy), 147
Løken, Kristen, 145
Longitude, methods for finding, 18–19
Lord Melville (ship), 237
Lorenz, Wilhelm, 174–75
Los Angeles (airplane), 214–15, 218, 221
Loubet Coast (Land), 137, 155
Luitpold Coast (Land), 168, **169**, 179, 197, *226*
Lummo (BGLE cat), *235*
Lymburner, James Harold, 224–25, 229

Macklin, Alexander, 185, 194–95, 204–05
Madrid Protocol, Antarctic Treaty, 297–98
Magellan, Ferdinand, 10–12
Magellan, Strait of, *11*, 9–16, *15*, 58, 139

Malvinas, Islas, *see* Falkland Islands/Islas Malvinas
Maps of Antarctic regions, early, 9, 16, *30*, 36, 39, 74, 148
    Explorers maps, early, *15*, 31, *42*, *87*, 104, *130*, *178*, 190, *192*, *226*, *262*
Marathon, Antarctic, 6
Marguerite Bay, *5*, 156–57, 162
    BGLE at, *233*, 235–40
    RARE/FIDS at, 256, 259–64
    USAS at, 242–48, *233*
Marr, James, 204, 252–54
Marsten, George, 183, 194
Maskelyne, Nevil, 18
Matha Bay, 155, 157
Matha, André, 128, *129*, 130, 134–36
Matthews, Harrison, 244
Maury, Matthew Fontaine, 74
Mawson, Douglas, 166, 178
McCarthy, Timothy, 188
McCormick, Robert, 69
McIlroy, James, 194–95, 204–05
McLeod, Michael, 42–43
McLeod, Thomas, 286
McMurdo Sound, 98, 152, 164, 177, 193, 196, 272, 297
McMurdo Station, 276
McNeish, Harry, 182, 188
Meinardus Line, 210
Meinardus, Wilhelm, 210
Melchior Islands, 251–52, 257
Meteor expedition, 210
Meteorology, 61, 75, 77, 80, 96, 117, 131, 141, 159, 204, 243, 253, 269, 271, 291
    Meteorologists, 84, 119, 121, 128, 165, 183, 210, 231
    Stations in Antarctic, 119, 121–22, 138, 245, 248, 261. *See also* Laurie Island
Michotte, Louis, 85, 92–93, 95
Mid-winter (solstice) party, 119, *244*
Miers, John, 27–28, 31–32
Mikkelsen Island, 247–48
Mirnyy (ship), 33, 38
Moltke (ship), 74–75
Moltke Harbor, 74, 83
Moneta House, 125, 299
Montevideo, 27–28, 85, 162, 200, 205, 207
Mooney, Norman, 204
Morrell, Benjamin, 47–48, 172
Mossman, Robert, 121–22, 124–25
Mountaineering, 285–86
Murdoch, W. G. Burn, 77–78, 80
Murphy, Robert Cushman, 47, 147, 333, 339
My Antarctic Honeymoon (Darlington), 270

Nansen, Fridtjof, 335
Narrative of Four Voyages . . . (Morrell), 48, 333
Nautical Almanac (Maskelyne), 18

Neko Harbor, 148
Neptunes Bellows, **7**, 49, 250
Neumayer Channel, **6**, **87**, 89, 131
Neumayer, Georg von, 74
New York Lyceum for Natural History, 51
New Zealand, 19–20, 61, 268, 272, 299
    Claims by, 214, 241, **251**
    Explorers from, 272. *See also* Worsley
Nordenskiöld, Baron Adolf Erik, 99
Nordenskjöld, Nils Otto, 40, 99, 114, 167
    Antarctic expedition, 99–115, *112*
North Star (ship), 241, 247
Northeast Passage, 99
Northrop Aircraft Corporation, 221
Northrup, Jack, 214, 221
Northwest Passage, 72
Norway, 84, 108, 139, 145, 164–65, 213, 252, 268, 338
    Claims by, 241, **251**
    Explorations/explorers from, 83, 85–86, 99, 165, 170, 335. *See also* Amundsen, *Jason* expeditions, Kohl-Larsen (Margit), Larsen (Carl Anton)
    Whalers/whaling, 139–50, 152–53, 211, 213, 249–50, 335, 345. *See also* Larsen (Carl Anton)

O'Higgins, Bernardo, 265, 346
Ocean Camp (*Endurance* expedition), 184–85
Ocean Harbor, South Georgia, 144, 342
Oceanography/hydrography, 47, 67, 72, 100, 117, 166, 168, 209, 220, 340
Omond House, 119, 125, 299, 337
Omond, Robert Traill, 119
Operation Highjump, 257–58
Operation Tabarin, 252–56, 299
Operation Windmill, 264
Orange Harbor, 62–63, 65
Oscar II Coast (King Oscar II's Land), **81**, 82, 103
Ozone hole, 294

Pacific Ocean, 10–16, 18–20, 23, 50, 57–58, 61, 66, 87, 89, 331
Paget, Mount, South Georgia, **8**, 286
Palma, Jorge, Emilio, and Silvia de, ***288***
Palmer Archipelago, **6**, 14, 54, 73, 89, 128, 130
Palmer Land/Palmer's Land/Palmer Peninsula, **39**, **42**, 42, 64, **262**, 278, 347
Palmer Station, 279, 297, 348
Palmer, Alexander, 50–53
Palmer, Nathaniel, 29, 37–40, 41–42, 50–53, 61, 242
    Sighting of Antarctic Peninsula, 37–38, 40
Palmer-Pendleton expedition, 50–53
Paradise Harbor (Bay), 148, **200**, 201–02
    Bagshawe/Lester at, 199–03

Paramore (ship), 17
Pardo, Luis, 196
Patagonia, 10, 12–14, 23–24, 99, 137
Patience Camp (*Endurance* expedition), **184**, 185–86
Paulet Island, vii, 69, **101**, 182, 186
    Antarctic expedition at, 100–01, 106–09, **107**, *109*, 113–15
Paulet, Captain Lord George, 69
Paulsen, Olava, 148–49
Peacock (ship), 62–66, 161, 334
Peary, Robert, 160
Pedersen, Morten, 81
Pegotty Camp, South Georgia, 191
Pendleton, Benjamin, 37, 50–51, 53
Pendulum Cove, Deception Island, **7**, 49–50, **50**, 63
Penguin (ship), 50–53
Penguins, 40, 103, 132, 135, 287, 337, 348
    Descriptions of, early, 12, 46
    Food and fuel, source of, 37, 49, 92, 94, 106–08, 111, 118–19, 133, 135, 172, 185, 194
    Music and, 123, **124**, 136
    Species
        Adélie, 61, 101, **114**, **136**, 159, **173**, 295, 337
        Chinstrap, 8, 33, 202
        Emperor, 103, 111, 123, **124**, **173**, 182, 337
        Gentoo, 8, 202, 295, 343
        King, 8, 46, 75–76, **76**
        Macaroni, 8
    Study of, 80, 120–21, 159, 202–03, 348
Penola (ship), 231–33, 235–36, 240
Penola Strait, 235
Perón, Isabel, 288
Perón, Juan Domingo, 266
Pesca, Cia. Argentina de, 139–43, 146, 291
Peter I Island, **endpaper**, 35, 64, 161
Petermann Island, 73, **156**, 338
    Pourquoi-Pas? expedition at, 154–59, *158*
Peterson, Harris-Clichy, 261
Photography/photographers
    Photography
        Movie, 121, 184, 196, 298, 337
        Still, 76, 79–80, 89, 92, 184, 196, 279, 286, 341, 343
    Photographers, 80, 89, 128, 178, 263, 265, 291. *See also* Hurley
Pierce-Butler, Kenelm, 259, 263
Piloto Pardo (ship), 280–81
Pinguin (ship), 345
Pinochet, President Augusto, 288
Piri Reis map, 9
Pirie, John, **119**, 120–22, 125
Pléneau, Paul, 128–29, ***129***, 338

INDEX

Point Wild, 187, 194, 207, 286
Poland, 84, 97, 241, 296
Polar Star (airplane), 221–30, **227**
Polar Star (ship), 77–78
Pole, geographic
   North, 97, 160, 213, 221
   South, 13, 97, 160, 177, 197, 213, 268, 275, 284, 329
Pole, magnetic, 57, 61, 66–67, 85, 329
Porpoise (ship), 62–63
Port Charcot, **131**, 132, 153–54, 157
   Français expedition at, 133–37
Port Circumcision, 154
   Pourquoi-Pas? expedition at, 154, 157–59
Port Foster, Deception Island, **7**, 48–50, 63, 66–67, 250, 279, 281, 347. *See also* Whalers Bay
Port Lockroy, **6**, 131–32, 137, 148, 153, 218, 232–33, 252–56, 299
   Tourists at, **299**
Port of Beaumont (ship), 258–59, 264
Port Stanley, Falklands, *see* Stanley
Possession Bay, South Georgia, **8**, **21**, 191
Potter Cove, King George Island, 73
Pourquoi-Pas? (ship), 151, **153**, **155**, **158**, 163
Pourquoi-Pas? expedition, 151–63, **156**
   Expedition narratives, 162, 339
Powell, George, 41, 119
   South Orkneys, discovery of, 41–42
Presidential visits, 265, 288–89
Primero de Mayo (ship), 250–53
Prince Olav Harbor whaling station, 144, 190, 339
Przybyllok, Erich, 165, 168, 174
Puerto Madryn, Argentina, 138, 338
Punta Arenas, Chile, **11**, 85, 95–96, 139–40, 162, 195–96, 286

Queen Elizabeth II (ship), 293
Queen Maud Land, **226**, 241, 345
Queen of Bermuda (ship), **250**
Quest (ship), 203–04, **205**
Quest expedition, 203–09, **206**
   Expedition narratives, 342
   Shackleton, death of and burial, 205, 207–08
Quinze Mois dans l'Antarctique (de Gerlache) 97, 335

Racovitza, Émile, 84, 86–87, 91, 336
Ramsay, Allan, 120
Ramsay, Mount, 120, 125
RARE, *see* Ronne Antarctic Research Expedition
Record south latitude, 14, 19, 45, 68, 98, 160, 177
Reindeer, South Georgia, 144–45, **145**, 290
Reports on the Scientific Results of the United States Antarctic Service Expedition 1939–41, 248
Resolution (ship), 19–21
Rey, J.-J., 128–29, **129**
Reynolds, Jeremiah, 51, 53, 61
Riley, Quintin, 231, 237
Rio de Janeiro, 62, 85, 204
Roberts, Brian, 234, 236
Robertson, R. B., 341–42
Robertson, Thomas, 77, 79–80, 116, 121, 123
Roché, Anthony de la, 16–17, 20–21, 197
Romania, 84
Ronne Antarctic Research Expedition (RARE), 258–64, **262**, 346
   Expedition narratives, 270, 345
   Women on expedition, 258–60, **260**, 264
Ronne Ice Shelf, vii, **4**, 264, 270, 328, 346
Ronne, Edith "Jackie," 259–60, **260**, 264
Ronne, Finn, 242, 245–46, **260**
   RARE, 258–64, 346
   Ellsworth Station, 270–71,
Roosevelt, President Franklin, 241
Rose (ship), 56
Ross Deep, 70, 123–24
Ross Ice Shelf, 68, 98, **226**, 227, 328, 343
Ross Island, 121, 152, 178, **273**
Ross, James Clark, 57, 67, 110
   Antarctic expedition, 67–71, **68**, 123–24
      Expedition narratives, 334
   Whales, report of, 70, 74, 77–78
Ross Sea, **endpaper**, 44, 53, 68, 162, 211–13, 227–28, 268, 272, 328
   Ross Sea Region, 2, 72, 83, 85, 97–98, 152, 164, 177–78, 213–14, 221, 241, 327
Ross Sea Party, *see Endurance* expedition
Rothera Base, 298
Rowett, John Quiller, 204, 209, 342
Royal Geographical Society, British, 56, 77, 83, 286
Royal Bay, **9**, 74
Royal Society, Britain, 48, 67
   Expedition (Halley Bay), 269–70, 272, **273**
Rozo (*Français* cook), 129, 133
Russia/Soviet Union, 32, 40, 164, 282, 332, 343–44
   Expeditions from, 268, 276, 278–79, 296. *See also* Bellingshausen
Ryder, R. E. D., 231, 234–35
Rymill, John Riddoch, 231, **240**, 345
   BGLE, 231–41

San Francisco (airplane), 214–17
Sappho, voyage of, 142–43
Schouten, Willem, voyage of, 14–16, **15**
Schrader, Dr. Karl, 74, 76
Schreiner, Ingvold, 219
Scotia (ship), 116
Scotia Bay, Laurie Island, **7**, 118, 121, 125

Scotia expedition, 52, 116–26, **117**, 167, 299, 337
    Expedition narratives, 125, 337
    Weddell Sea voyage, 122–24
Scotia Ridge (Scotia Arc), 117–18, 125
Scotia Sea, vii, **4**, 125
Scotland, 74, 77
    Expeditions/explorers/whalers from, 250.
        See also Dundee whaling and *Scotia* expeditions
Scott, Robert Falcon, 2, 98, 116, 126, 152, 177, 285. *See also Discovery* and *Terra Nova* expeditions
Scurvy, 2, 18–19, 54, 93, 134, 328–29
Sea Gull (ship), 62–63, 66, 334
Sea spider, **52**
Seal Nunataks (Islands), **81**, 82
Sealing/sealers, 23–27 **25**, 36–38, 41–48, 50–56, 66, 72–73, 77–79, **78**, 82–83, 282, 329
    Elephant sealing, 23, 27–28, 72–73, 104, 140, 146–47
    Exploration/discovery resulting from, 50–53, 73–74, 81–82. *See also* Biscoe, Palmer (Nathaniel), Powell, Weddell
    Fur sealing, 23–26, 27–30, 36–38, 72–73
Seals, **25**, 39, 75, 101, 133
    Elephant seal, 8, 23, **24**, 75, 295
    Food/fuel from
        Heroic Age, during, 92–94, 106–08, 118, 133, 158, 172, 185, 187, 194, 196
        Other expeditions, during, 37, 51, 75, 201, 206–07
    Fur seal, 8, **24**, 26, 37, 48, **287**, 295, 344
        Population recovery, 220, 287
    Leopard seal, 44, 85
    Weddell seal, 43, **44**, 75, 120, **134**
Seelheim, Heinrich, 165–66
Seraph (ship), 50–53
Service, Robert, 341
Seymour Island, **6**, 69, 99, 101, 111–12, 256, 288
    Fossils found at, 80–81, 101, 103
Shackleton Base, 176, **273**, 273–74, 346–47
Shackleton Mountain Range, 271
Shackleton, Emily (Lady Shackleton), 205, 207, 342
Shackleton, Lord E. E. A., 289
Shackleton, R.R.S. (ship), 280–81, 289
Shackleton, Sir Ernest, 98, 112, 152, 156, 160, 165, **184**, 203, 337, 340, 341, 342
    Endurance expedition, 177–96
    Quest expedition, 203–05
        Death and burial, 205, 207–08
Sheffield, James, 29, 40
Shirreff, William, 27, 30
Skis/skiing, 78, 82, 88, **89**, 119, 130, 144, 284, 287
Skottsberg, Carl, 108, 113–14, 146

Slossarczyk, Walter, 167, 208
Smith, Bill, 122, 125
Smith, William, 30–31, 40, 296
    South Shetlands, discovery of, 27–28, 32, 35
Smithsonian Institution, 66, 230
Smyley, William Horton, 66–67, 334
Sno-Cat, 270, 273–75, **274**
Snow Hill Island, **6**, 69, 182, 185, 199, 222–23, 255–56, 299, 337
    Antarctic expedition at, 101–07, **107**, **109**, 109–11, 113–15
Snow, Ashley, 247
Sobral, José, 100, 102–03, 336
Sörling, Erik, 141
Sørlle, Petter, 211
Sørlle, Thoralf, 193, 341
South (Shackleton), 196, 341
South Africa, 268, 272
South Georgia Island, 2, **4**, **8**, **21**, 118, 126, 150, 220, 249, 282–83, 287, 299–300, 342, 344
    Claims to, 21, 143, 215, 251, 278, 291
    Discovery of, 16–17
    Explorers/expeditions at, 9, 21–22, 33, 46–47, 104, 166–67, 175–76, 179, 190–93, 196, 205–10, 285–86, 287–88, 340. *See also* German IPY expedition
    Falklands/Malvinas war at, 289–94, **290**
    Sealing at, 23, 26–27, 146–47
    Whaling/whalers at, 82–83, 139–46, 193, 208, 220, 249, 282
South Georgia Joint Services expedition (Burley), 191–92, 285–86
South Georgia Survey expedition (Carse), 285
South Ice, **273**, 274–75
South Orkney Islands (Orcadas del Sud), 2, **4**, **7**, **42**, 52, 118, 211, 220. *See also* Coronation and Laurie islands
    Claims to, 42, 118, 122, 143, 214, 346.
    Discovery of, 41–43
    Expeditions/explorers at, 43, 58, 60, 73, 118–21, 122, 125
South Pole, *see* Pole, Geographic
South Sandwich Islands, 2, **4**, **8**, **30**, **35**, 47, 53–54, 118, 167, 197, 220. *See also* Leskov, Thule, Visokoi, and Zavodovski islands
    Claims to, 143, 215, 251, 278
    Conflict over/war at, 289, 293–94
    Discovery of, 22, 33–35
South Shetland Islands, vii, 2, **4**, **6**, 7, 28, **30**, **42**, 118, 197, 250. *See also* Deception, Elephant, Greenwich, and King George islands
    Claims to, 28–29, 31, 143, 346
    Discovery of, 14, 27–28
    Exploration of/expeditions to, 30–31, 38–39, 48–53, 60, 62–63, 89, 210

INDEX

Sealing/sealers at, 28–29, 36–38, 41–42, 47–48, 51–53, 54–56, 72–73
Whaling/whalers at, 73, 143, 147–50. *See also* Deception Island, whaling
Southern Continent, theory of, 1, 9, 16–17, 19, 22, 32
Southern Cross expedition, 97, 337
Southern Sky (ship), 195
Soviet Union, *see* Russia/Soviet Union
Spain, 10–12, 14, 17, 281
Spitsbergen, 116, 165, 213–14, 221
Squires, Harold, 254–55
Stancomb Wills (boat, *Endurance*), 183, 185–87, **194**
Stanley, Falkland Islands, 142, 188, 232, 255, 293
Staten Island, **11**, 15–16, 51, 86, 114, 129, 138
Staten Island (ship), 269, **270**
Station Iceberg (Stationeisberg), 170–71
Stefansson Strait, 216, 236–37, 239–40, 345
Stephenson, Alfred, 231–32, 235, 237–39
Stockholm Museum of Natural History, 141
Stokes, Frank W., 100, 102
Stonington Island, 242, **243**, 295
Expeditions at, 242–48, **243**, 256–57, 258–64, **262**, 271
Stromness Bay/whaling station, **8**, 145, 190, **192**, 192–93, 285–86, 342
Sweden, expeditions/explorers from, 97, 336. *See also Antarctic* expedition
Sweeney, John, 298
Swinhoe, Ernest, 141–43
Sydney, Australia, 35, 66, 68
Symmes, John Cleve, 50–51

Tasmania, *see* Hobart, Tasmania
Taylor, Andrew, 253–56
Telefon (ship), 148–49, 160, 339
Telefon Bay, **7**, 149, 279
Temperature records, 67, 95
Terra Nova expedition, 177, 236. *See also* Scott
Terror (ship), 67, **69**, 70
Theron (ship), 271–73
Through the First Antarctic Night (Cook), 97, 335
Thule Island, **8**, 289, 293–94
Thulia (narrative poem, Palmer), 334
Thurston Island, **63**, 66
Tierra del Fuego, **11**, **15**, 15–16, 20, 57, 74, 104
Exploration of, 10–13, 56, 61–62, 104, 109
Tierra O'Higgins, 265, 278
Tierra San Martín, 265, 278
Toby (*Français* pig), 120, 129, **134**, 136
Tourism/tourists, 283–84, 297, **299**
Transit of Venus, 19, 74, 75, 329
Treatise of the Scurvy, A (Lind), 18
Tula (ship), 53–55
Turquet, J., **129**

Two Men in the Antarctic (Bagshawe), 203
United States (ship), 23, 25
United States, 72, 74, 200, 214, 252, 258, 268, 276, 278, 283, 294
Claims/claim polity, 225, 241, 243, 246, 257, 270, 346
Explorers/expeditions from, 97
Government
Early, 74. *See also* USAS and Wilkes expeditions
Modern, 53, 202, 257–58, 264, 268, 275–76. *See also* Amundsen-Scott, Eights, Ellsworth, McMurdo, and Palmer Stations
Other, 66–67, 72, 100, 146–47, 348. *See also* Byrd, Cook (Frederick), Ellsworth, Palmer (Nathaniel), Palmer-Pendleton, RARE, sealers
United States Antarctic Peninsula Traverse Team, 345, 346
United States Antarctic Services expedition (USAS), 241–48
East Base, 241–48, **243**, 252, 257–58
Expedition narrative, 248, 344
United States South Seas Exploring expedition, *see* Wilkes expedition
Uruguay (ship), 112–14, 122, 125–26, 128–29
Uruguay, 195, 207, 296
USAS, *see* United States Antarctic Services
Ushuaia, **11**, 86, 103, 112, 128–29

Vahsel Bay, **169**, 257, 264, 270, 340
Explorers/expeditions at, 168–71, 176–78, 180, 272–73
Vahsel, Richard, **174**, 340
Deutschland expedition, 164–72, 174
Valparaíso, Chile, 27–28, 30–32, 66, 259
Vega Island, **107**, 110
Venus de Milo, 57
Vernadsky Station, 348
Vespucci, Amerigo, 9
Victoria Land, 72, 83, 85, 94
Videla Station, 200, 343
Videla, President Jorge, 288–89
Videla, President Gabriel González, 265
Vincennes (ship), 62
Visokoi Island, **8**, 33–34, **35**, 54
Von den Steinen, Karl, 75–76, 83
Vostok (ship), 33, 38–39
Voyage of the 'Scotia', The (Three of the Staff), 125
Voyage Towards the South Pole, (Weddell), 47

Wafer, Lionel, 17
Walker, William, 63–65
Wandel Island, *see* Booth Island
Waterboat Point, 199–203, **200**, 343

Webster, William H. B., 48–49, 333
Weddell Sea, vii, 2, **4**, 44, 73, 132, 164, 176, 181, 216, 221, 237, 328, 340
    Exploration of, expeditions/voyages into, 47–48, 58–60, 69–70, 175, 250, 257–58, 262–64, 269–74, **273**. *See also Deutschland, Endurance, Jason, Scotia,* Weddell expeditions
Weddell, James, 37, 43, 46–47, 332
    Weddell Sea voyage, 43–46, **45**, 60, 69–70
        Expedition narrative, 47
        Record south latitude, 45, 69, 169
Wegener, Alfred, 103
Wennersgaard, Ole Christian, 109, 113, 120
Whalers Bay, Deception Island, **7**, 148, 153, 219, 279
Whales, 22, 65, 70, 73, 77–78, **141**, 209, 282,
Whaling/whalers, 23, 50, 126, 139–50, **141**, 219, 220, 240, 335, 342
    Early Antarctic attempts, 73, 74, 83. *See also* Dundee whaling and *Jason* expeditions
    End of, 282
    Explorers, help to
        Heroic Age, 150, 152–53, 160, 166–67, 175, 179, 193, 195, 341–42
        Post–Heroic Age, 199–200, 203, 205–06, 208, 213–16, 212
    International whaling convention, 220
    Pelagic, 211–12, 249
    South Georgia, at, *see*
    Whaling stations, 139–46, 148, 190, 219–20, 249, 299–300. *See also* Grytviken, Deception Island, Leith Harbor, Ocean Harbor, Prince Olav Harbor, Stromness
    World War I and II, 150, 198, 249–50, 345
Wiencke Island, vii, **6**, 89, 125, 129–30, 253
Wiencke, Auguste-Karl, 86
Wild, Frank, ***207***
    Endurance expedition, 178–79, 186–88, 194–96
    Quest expedition, 204–09
Wilkes, Charles, 61–62, 66
    Wilkes expedition, 61–66, **63**, 161
        Expedition narratives, 334
Wilkes Land, 66, 345
Wilkins Ice Shelf, **5**, 218, 295
Wilkins, Sir George Hubert, 198–201, 204, 209 **216**, 343
    Ellsworth expeditions, 220–25, 228–29
    1928–29, 1929–30 expeditions, 213–19, **217**
        Channels through Antarctic Peninsula, 216–17
        First flight in Antarctic, 215
William Scoresby, R. R. S. (ship), 210, 217–19, 253–54
Williams (ship), 27–28, 30–31

Willis Islands, **8**, 21, 344
Wilson, Edward, 98
Wilson, James Innes, 143–44
Wilton, David, 116, ***119***
Winter Island, **233**, 233–35, 251, 256
Winter, John, 13
Wireless/radio transmissions, 149, 165–67, 237–38
Women in the Antarctic, 144, 147, 148–49, 258–60, 264, 268, 339
Wordie Ice Shelf, **233**, 237, 243, 295
Wordie, James, 233
World War I, 114, 150, 175, 178–79, 193, 195–96, 198
World War II, 66, 208, 241, 247–48
Worsley, Frank, 178, 182, 186, 188–93, 195–96, 204
Wyandot (ship), 269–70, ***270***
Wyatt Earp (ship), 221–24, 228–29, 241

Yelcho (ship), 195–96
Yelcho (ship, modern), 279
Young, Dr. Adam, 30–32

Zavodovski Island, **8**, 33–34, ***35***, 54, 167
Zavodovski, Ivan, 33
Zélée (ship), 58–61, ***60***
Zum Sechsten Erdteil (Filchner), 174, 176

ABOUT THE AUTHOR

Joan N. Boothe has been fascinated with stories of Antarctic adventure and exploration since childhood. In 1995, after many years working in the worlds of economics, finance, and teaching business administration to graduate business students, she at last made her first trip to Antarctica and saw where so many things she had read about had taken place. Ms. Boothe has returned to the Antarctic regions many times since, including making a 67-day circumnavigation of the entire Antarctic continent aboard an icebreaker. In 2010, she taught a course on Antarctica's Heroic Age for Stanford University's continuing studies program. Ms. Boothe has two children, both raised in San Francisco, California, where she and her husband have lived since 1970.

The author welcomes any comments or questions about *The Storied Ice* or other matters Antarctic. She may be contacted at joannboothe@joannboothe.com.

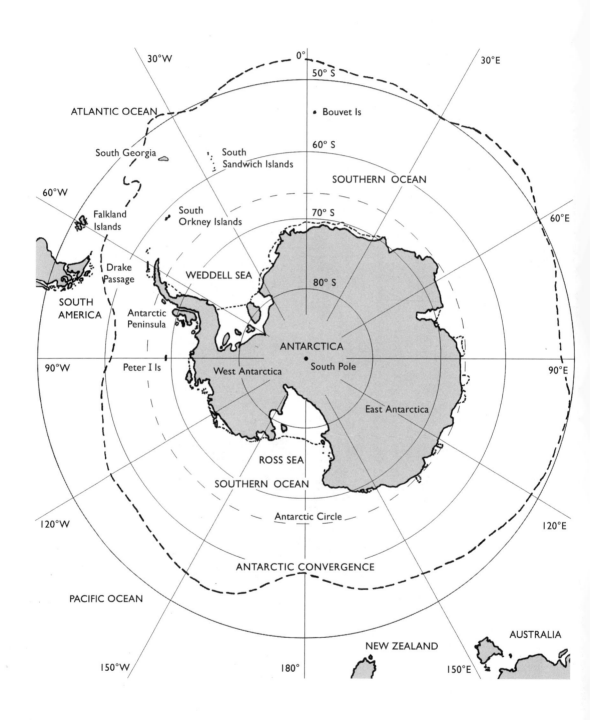

THE ANTARCTIC REGIONS